# Quantum Mechanics

# Quantum Mechanics

**HENDRIK F. HAMEKA**
***University of Pennsylvania***

A WILEY-INTERSCIENCE PUBLICATION

**JOHN WILEY & SONS** New York · Chichester · Brisbane · Toronto

*To Charlotte*

Copyright © 1981 by John Wiley & Sons, Inc.

All rights reserved. Published simultaneously in Canada.

Reproduction or translation of any part of this work beyond that permitted by Sections 107 or 108 of the 1976 United States Copyright Act without the permission of the copyright owner is unlawful. Requests for permission or further information should be addressed to the Permissions Department, John Wiley & Sons, Inc.

***Library of Congress Cataloging in Publication Data:***

Hameka, Hendrik F.
Quantum mechanics.

"A Wiley-Interscience publication."
"Based in part on an earlier book, Introduction to quantum theory, which was published in 1967"--Pref.
Includes bibliographies and index.
1. Quantum theory. I. Title.

QC174.12.H35 530.1'2 81-3430
ISBN 0-471-09223-1 AACR2

Printed in the United States of America

10 9 8 7 6 5 4 3 2 1

# Preface

The present book is based in part on an earlier book, *Introduction to Quantum Theory*, which was published in 1967. At that time I intended that the book be used as a text for junior or senior undergraduates in physics and for senior undergraduates and first-year graduate students in chemistry. In the preface I outlined my approach to the teaching of quantum mechanics and the goals that I hoped to achieve in a one-semester course on the subject. Since then the book has been used for 12 years, and I have modified some of my original ideas, either because of my own observations or due to suggestions made by colleagues. In order to describe the present book I quote the relevant parts of the preface to the old book and then discuss the changes in my philosophy and in the new book.

"The approach is semihistorical. First, it is shown how classical mechanics became inadequate for the explanation of certain experimental findings. This is followed by a discussion of the wave nature of free particles from which the Schrödinger equation is more or less derived. In the historical discussion I felt free to omit certain developments that had no pedagogic value, although they might be important from a historical point of view. I tried to keep the discussion closely linked to physical ideas. Whenever there was a conflict between physical understanding and mathematical rigor, I always decided in favor of the former.

"An important consideration in teaching quantum theory at the elementary level is the inadequate mathematical background of the students. In order to understand quantum theory and to apply it, the student must have some knowledge of many branches of mathematics: differential and integral calculus, Fourier analysis, differential equations, vector analysis, complex numbers, matrices and determinants, linear equations and eigenvalue problems, and the theory of special functions. I expect students to be acquainted with elementary differential and integral calculus, but the other mathematical topics listed above are discussed here. Naturally the teacher is free to omit any of them from his discussion if he feels the students are already familiar with them.

"I hope the present work can be used for a variety of courses, particularly junior and senior physics courses and senior and first-year graduate chemistry courses. Its length and contents make it suitable for a one-semester course designed as a formal introduction to quantum mechanics. It is also suitable for the first half of a two-semester course in quantum chemistry. In this case, it needs to be supplemented by another text for the second semester."

Today, students in physical chemistry need to know more about quantum mechanics than they did in the past. Specifically, time-dependent quantum mechanics and the interaction between radiation and matter are essential for the understanding of newly developed areas in spectroscopy and kinetics. I added these topics to the new book. I also rounded off the chapter on atomic structure. Consequently the new book is about 25% larger than the old book, and it is not possible to cover all its material in one semester. The new version is suitable for a two-semester course rather than a one-semester course in quantum mechanics.

I found that most students appreciate the detailed mathematical derivations in the book. The better qualified students should be familiar with the more elementary derivations, but they still like to be able to review them. Some of the readers criticized the organization of the material; they felt that the relevance of some mathematical discussions to the quantum theory was not immediately clear because they were separated into different chapters. I felt that this criticism was justified, and I rearranged the material so that each mathematical derivation was immediately followed by its quantum mechanical application. I left the chapter on matrices as a separate entity. Usually I do not discuss the matrix algebra while teaching the course, but most students like to have the material available for review purposes.

I revised and expanded the problem sets at the end of each chapter; most of the new problems are taken from exams that I gave. I also added a list of recommended books at the end of each chapter. Naturally these lists are far from complete; they are mostly books that I found useful myself and the selection reflects my personal taste.

Finally, I wish to thank Dr. O. Zamani-Khamiri for her help in correcting the manuscript.

HENDRIK F. HAMEKA

*Philadelphia, Pennsylvania*
*May 1981*

# Contents

CHAPTER ONE

# Preliminaries

## 1-1 Introduction

A beginning student usually has more difficulty in learning quantum mechanics than classical mechanics, although the complexities of the two theoretical approaches are not widely different. Certain simple systems can be treated exactly in either quantum mechanics or classical mechanics; examples are one-dimensional motion, a particle in a central force field, or some two-particle systems. More complex systems cannot be treated exactly in either classical or quantum mechanics. They may at best be described by means of approximate mathematical methods, which are just as complicated and laborious in classical as in quantum mechanics.

The difficulty in learning quantum mechanics is caused mostly by the fact that everyone is much more familiar with the concepts and everyday applications of classical mechanics than with those of quantum mechanics. For example, in driving a car we must be able to predict the positions of the other cars on the road at future times and we adjust the future positions of our car accordingly by using the steering wheel, the gas pedal, and the brake. All this involves applications of the laws of classical mechanics.

Another example of applied classical mechanics is a baseball game. Every aspect of a baseball game is related to classical mechanics because it involves predictions about the orbit of the ball. A major league baseball player must have an intuitive understanding of classical mechanics because he can judge and anticipate the flight of the ball much quicker and much more effectively than any theoretician can calculate it.

As we go through life, we use the results and the concepts of classical mechanics all the time; we use them when we walk, when we drive a car, when we play, and even when we eat.

The fundamental laws of classical mechanics were first proposed by Newton during the seventeenth century. These same laws were transformed into more sophisticated mathematical form during the eighteenth and nineteenth centuries by Lagrange and Hamilton. They were supplemented by Maxwell during the late nineteenth century in order to describe the behavior of electrically charged particles.

It is not surprising that, when the structure of the atom was first discovered, scientists expected the particles within the atom to obey the same laws of classical mechanics as did all other systems that they had been able to observe. In particular, Lorentz published extensive calculations on the behavior of the electrons within the atom, using a combination of classical mechanics and the Maxwell equations. This work had a certain degree of success; for instance, Lorentz explained the Zeeman effect in this way. However, as more experimental information on atomic structure became available, it showed conclusively that classical mechanics was not valid within the atom. In order to explain all these new experimental observations in a logical and consistent manner, it became necessary to derive a new mechanics.

Since the evidence of experiments and the authority of leading scientists support the necessity of using quantum mechanics for the description of atomic motion, the beginning student has no choice but to accept this situation. Yet emotionally he has difficulty in believing that a baseball game and a hydrogen molecule are governed by different laws of motion, and he clings to the classical concept of electrons orbiting around the nuclei as long as he can. Therefore, before discussing quantum mechanics, we think it is useful to discuss briefly the arguments and experiments that led to the abandonment of classical mechanics for atomic motion. This means that we will use the historical approach in teaching quantum mechanics.

We feel that the historical approach constitutes the best method for teaching quantum mechanics because it provides a smooth transition from the old classical mechanics that we are familiar with to the new quantum mechanics that we wish to learn. We will only discuss those theoretical advances that were important in the development of quantum mechanics because our purpose is to teach the subject and not to give its complete history. Even so, we will mention some old theories that are now obsolete but that were important at the time as long as these theories have pedagogical value.

In these early chapters we also review some of the main features of classical mechanics so that we will be able to recognize where it differs and where it agrees with quantum mechanics.

Throughout the book we explain the various topics in mathematics that are necessary for expressing the physical concepts in classical and quantum mechanics. For example, in Sections 2 and 5 of this chapter we describe some aspects of vector analysis; in Chapter 3 we discuss Fourier analysis; in Chapter 5 we give a brief review of the theory of differential equations, and so on. We feel that this mixing of mathematics and physics makes the book more readable than the other alternative of segregating all mathematics into separate chapters.

## 1-2 Classical Mechanics

The basic equation of classical Newtonian mechanics for a particle in three-dimensional space is most conveniently expressed in terms of vectors. We

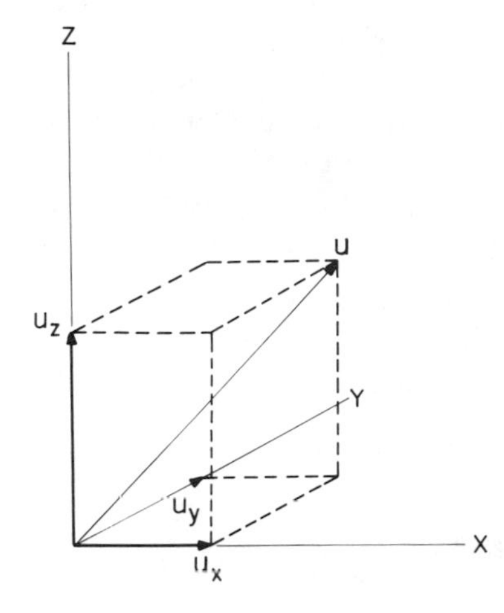


**Fig. 1-1** Graphic representation of a vector **u** and its three components $u_x$, $u_y$, and $u_z$.

briefly discuss a few vector properties in this section; some additional vector properties are discussed in Section 5 of this chapter.

A vector **u** can be defined as a directed line segment. It is determined by its three components $u_x$, $u_y$, and $u_z$ along the $x$, $y$, and $z$ axes (see Fig. 1-1). Therefore, the boldface vector symbol **u** actually represents three quantities, which can be denoted by $\mathbf{u}=(u_x, u_y, u_z)$. The direction of **u** is determined by the three direction cosines, and its magnitude, which is the length of the line segment and which is denoted by $|\mathbf{u}|$ or $u$, is given by

$$u=\left(u_x^2+u_y^2+u_z^2\right)^{1/2} \tag{1-1}$$

The sum **w** of two vectors **u** and **v**,

$$\mathbf{w}=\mathbf{u}+\mathbf{v} \tag{1-2}$$

is defined such that each component of **w** is the sum of the two corresponding components of **u** and **v**,

$$\begin{aligned} w_x &= u_x + v_x \\ w_y &= u_y + v_y \\ w_z &= u_z + v_z \end{aligned} \tag{1-3}$$

When a vector **u** is a function of a parameter $t$, that is, each component of **u** is a function of $t$, we can define the derivative of **u** with respect to $t$:

$$\frac{d\mathbf{u}(t)}{dt}=\lim_{\Delta t\to 0}\frac{\mathbf{u}(t+\Delta t)-\mathbf{u}(t)}{\Delta t} \tag{1-4}$$

This definition is again equivalent to the three equations:

$$\left(\frac{d\mathbf{u}}{dt}\right)_x=\lim_{\Delta t\to 0}\frac{u_x(t+\Delta t)-u_x(t)}{\Delta t} \quad \text{and so on} \tag{1-5}$$

In classical mechanics we can represent the position of a point particle in three-dimensional space by means of a vector. In Fig. 1-1 we show how the position of the point $P$ is determined by the vector $\mathbf{r}$, which is identical with the line segment $OP$. The Cartesian coordinates $(x, y, z)$ of the point $P$ are the three components of the vector $\mathbf{r}$. The motion of the point $P$ is described by its orbit, which is the time dependence $\mathbf{r}(t)$ of its position vector $\mathbf{r}$.

The velocity $\mathbf{v}(t)$ of the particle is defined as

$$\mathbf{v}(t)=\frac{d\mathbf{r}}{dt} \tag{1-6}$$

and the acceleration $\mathbf{a}(t)$ of the particle is defined as

$$\mathbf{a}(t)=\frac{d\mathbf{v}(t)}{dt}=\frac{d^2\mathbf{r}(t)}{dt^2} \tag{1-7}$$

The fundamental equation of Newtonian mechanics is

$$\mathbf{F}=m\mathbf{a} \tag{1-8}$$

In other words, if a particle (or a body) is subject to a force $\mathbf{F}$, it will experience an acceleration $\mathbf{a}$ that is proportional to $\mathbf{F}$. The proportionality constant $m$ is defined as the mass of the particle. Clearly, the particle will have zero acceleration $\mathbf{a}=\mathbf{0}$ if it is not subjected to any exterior forces. In that case it will move through space with a constant velocity $\mathbf{v}$.

We may rewrite Eq. (1-8) in a different form by introducing the momentum $\mathbf{p}$ of the particle, which is defined as

$$\mathbf{p}=m\mathbf{v} \tag{1-9}$$

It is easily seen that

$$\mathbf{F}=\frac{d\mathbf{p}}{dt} \tag{1-10}$$

The equations of motion (1-8) and (1-10) may be solved for simple systems, such as one-dimensional motion, a particle in a central force field, and so on. For more complex systems it is useful to make use of a more general mathematical formulation of the laws of motion. This formulation can be derived from Hamilton's principle or the principle of least action, and it leads to a set of differential equations that are known as the Lagrangian equations of motion or the Hamiltonian equations of motion. The latter are quite important in the formulation of quantum mechanics and we feel that it is helpful to discuss them here. However, we do not present the rigorous derivation of Hamilton's equations from the principle of least action; instead we just illustrate their validity for a simple one-particle system.

We consider a system of one particle with mass $m$ moving in a conservative force field. Such a force field is defined by the condition that the three components $F_x$, $F_y$, and $F_z$ of the force acting on the particle can all be represented as the derivatives of a single function $V(x, y, z)$ of the position coordinates $x$, $y$, and $z$.

$$F_x = -\frac{\partial}{\partial x} V(x, y, z)$$

$$F_y = -\frac{\partial}{\partial y} V(x, y, z)$$

$$F_z = -\frac{\partial}{\partial z} V(x, y, z) \tag{1-11}$$

Equations (1-11) can be combined into a vector equation,

$$\mathbf{F} = -\nabla V(x, y, z) \tag{1-12}$$

where the three components $(\partial/\partial x)$, $(\partial/\partial y)$, and $(\partial/\partial z)$ are symbolically represented by the vector symbol $\nabla$. Each component of $\mathbf{F}$ is a function of the position coordinates $x$, $y$, and $z$. We write $\mathbf{F}$, therefore, as $\mathbf{F}(x, y, z)$ and we call it a vector field. The quantity $\nabla V$ is called the gradient of the function $V$ and we can write Eq. (1-12) also as

$$\mathbf{F}(x, y, z) = -\text{grad } V(x, y, z) \tag{1-13}$$

A vector field that can be expressed as the gradient of a function of position is called an irrotational field; it should be noted that not all vector fields are irrotational.

We will now reformulate the equations of motion (1-10) into the Hamiltonian form. By substituting the set of equations (1-12) we find

$$\frac{dp_x}{dt} = -\frac{\partial V}{\partial x} \qquad \frac{dp_y}{dt} = -\frac{\partial V}{\partial y} \qquad \frac{dp_z}{dt} = -\frac{\partial V}{\partial z} \tag{1-14}$$

We define the kinetic energy $T$ of the particle either in terms of the velocity $v$ or in terms of the momentum $p$ as

$$T = \frac{m}{2}\left(v_x^2 + v_y^2 + v_z^2\right) = \frac{1}{2m}\left(p_x^2 + p_y^2 + p_z^2\right) \tag{1-15}$$

Obviously,

$$\frac{dx}{dt} = v_x = \frac{p_x}{m} = \frac{\partial T}{\partial p_x} \tag{1-16}$$

and we have the three equations,

$$\frac{dx}{dt}=\frac{\partial T}{\partial p_x} \qquad \frac{dy}{dt}=\frac{\partial T}{\partial p_y} \qquad \frac{dz}{dt}=\frac{\partial T}{\partial p_z} \tag{1-17}$$

We next introduce the function

$$H(x, y, z; p_x, p_y, p_z)=T(p_x, p_y, p_z)+V(x, y, z) \tag{1-18}$$

which is called the Hamiltonian function and which is the energy of the particle written as a function of the three position coordinates $x$, $y$, and $z$ and of the momentum components $p_x$, $p_y$, and $p_z$. Since $T$ does not depend on $x$, $y$, and $z$, we have

$$\frac{\partial H}{\partial x}=\frac{\partial V}{\partial x} \qquad \frac{\partial H}{\partial y}=\frac{\partial V}{\partial y} \qquad \frac{\partial H}{\partial z}=\frac{\partial V}{\partial z} \tag{1-19}$$

and since $V$ does not depend on $\mathbf{p}$, we have

$$\frac{\partial H}{\partial p_x}=\frac{\partial T}{\partial p_x} \qquad \frac{\partial H}{\partial p_y}=\frac{\partial T}{\partial p_y} \qquad \frac{\partial H}{\partial p_z}=\frac{\partial T}{\partial p_z} \tag{1-20}$$

Hence Eqs. (1-14) and (1-17) can be reformulated as

$$\frac{dx}{dt}=\frac{\partial H}{\partial p_x} \qquad \frac{dy}{dt}=\frac{\partial H}{\partial p_y} \qquad \frac{dz}{dt}=\frac{\partial H}{\partial p_z}$$

$$\frac{dp_x}{dt}=-\frac{\partial H}{\partial x} \qquad \frac{dp_y}{dt}=-\frac{\partial H}{\partial y} \qquad \frac{dp_z}{dt}=-\frac{\partial H}{\partial z} \tag{1-21}$$

In this way the motion of the particle can be derived mathematically from a single function, the Hamiltonian function $H$. We note that in Eq. (1-21) the coordinates and momenta have been "paired off": the first pair is $p_x$ and $x$, the second is $p_y$ and $y$, and the third is $p_z$ and $z$. We say that the momentum $p_x$ is conjugate to the coordinate $x$, and so on.

Equations (1-21) are called Hamilton's equations of motion. We use them to show that the Hamiltonian function $H$ is time independent. We have

$$\frac{dH}{dt}=\left(\frac{\partial H}{\partial x}\frac{dx}{dt}+\frac{\partial H}{\partial p_x}\frac{dp_x}{\partial t}\right)+\left(\frac{\partial H}{\partial y}\frac{dy}{dt}+\frac{\partial H}{\partial p_y}\frac{dp_y}{dt}\right)+\left(\frac{\partial H}{\partial z}\frac{dz}{dt}+\frac{\partial H}{\partial p_z}\frac{dp_z}{dt}\right)=0 \tag{1-22}$$

The Hamiltonian function $H$ represents the energy $E$ of the system; since the Hamiltonian function is time independent, the energy of the system remains a constant in time. We call the energy a constant of the motion.

We mentioned already that Hamilton's equations of motion are generally valid; they are valid in different coordinate systems, they are valid for describing $N$-particle systems, and so forth. For example, they can be used to describe the vibrational motion in polyatomic molecules such as methane, benzene, and so on. We will describe these generalized Hamilton's equations of motion, but we do not derive them.

First, we consider again a particle in three-dimensional space, but now we assume that its position is described by three generalized coordinates $(q_1, q_2, q_3)$ that are determined in a certain way by the Cartesian coordinates $(x, y, z)$. In other words, the coordinates $q_i$ are functions of $x$, $y$, and $z$:

$$\begin{aligned} q_1 &= q_1(x, y, z) \\ q_2 &= q_2(x, y, z) \\ q_3 &= q_3(x, y, z) \end{aligned} \tag{1-23}$$

For example, we may wish to use polar, elliptical, or cylindrical coordinates, instead of the Cartesian coordinates. The transformation of Eq. (1-23) also leads to a new set of momenta $(p_1, p_2, p_3)$ and to a new Hamiltonian:

$$H = H(q_1, q_2, q_3; p_1, p_2, p_3) \tag{1-24}$$

The Hamiltonian equations are now given by

$$\frac{dq_i}{dt} = \frac{\partial H}{\partial p_i} \qquad \frac{dp_i}{dt} = -\frac{\partial H}{\partial q_i} \qquad (i = 1, 2, 3) \tag{1-25}$$

We see that the momenta $p_i$ and the coordinate $q_i$ are coupled; we say that $p_1$ is conjugate to $q_1$, $p_2$ is conjugate to $q_2$, and so on.

The above description is also applicable to systems that are determined by $N$ coordinates $(q_1, q_2, q_3, \ldots, q_N)$ with $N$ either smaller or larger than 3. An example of such a system is the vibrational motion of a polyatomic molecule. Again we introduce a set of momenta $(p_1, p_2, p_3, \ldots, p_N)$, so that $p_1$ is conjugate to $q_1$, $p_2$ is conjugate to $q_2$, and so forth, and a Hamiltonian function

$$H = H(q_1, q_2, q_3, \ldots, q_N; p_1, p_2, \ldots, p_N) \tag{1-26}$$

The Hamiltonian equations of motion are now given by

$$\frac{dq_i}{dt} = \frac{\partial H}{\partial p_i} \qquad \frac{dp_i}{dt} = -\frac{\partial H}{\partial q_i} \qquad (i = 1, 2, 3, \ldots, N) \tag{1-27}$$

If we can solve the Hamiltonian equations, we obtain the solution as a set of expression $q_i(t)$ and $p_i(t)$ as functions of time. However, each of the Eqs. (1-27) contributes an arbitrary integration constant, so that the solution for an $N$-coordinate system contains $2N$ undetermined parameters. This result agrees with our experience that we cannot know the exact behavior of a particle in three-dimensional space unless we know its position **r** and its momentum **p** at a certain time $t_o$. The six quantities $x(t_o)$, $y(t_o)$, $z(t_o)$, $p_x(t_o)$, $p_y(t_o)$, and $p_z(t_o)$ can then be used to determine the six unknown integration constants.

We have described the Hamiltonian equations of motion here because they are important for the mathematical formulation of quantum mechanics. In some ways we can look upon them as a bridge between the two types of mechanics. The derivation of the generalized Hamiltonian equations (1-27) is rather involved and we do not describe it here. Instead we refer the reader to one of the books on classical mechanics that are listed at the end of this chapter.

## 1-3 The Classical Harmonic Oscillator

The harmonic oscillator played a very important role in the theoretical developments at the beginning of this century, since it was an essential part of many of the theories that led to quantum mechanics. Once it became known that an atom consisted of a nucleus and a certain number of electrons that moved about in the vicinity of the nucleus, the harmonic oscillator represented a simple model reproducing the essential features of the electronic motion as they were known at that time. In fact, the harmonic oscillator was often used as a starting point for theories involving electronic motion in atoms, molecules, or crystals.

In one dimension a harmonic oscillator is a particle of mass $m$, oscillating back and forth around an origin 0. At any time it is subject to a force $F$ which tends to move it back toward the origin and which is proportional to the displacement $x$ from the origin:

$$F = -kx \tag{1-28}$$

Since

$$F = -\frac{\partial V}{\partial x} \tag{1-29}$$

it is easily found that the potential energy of the particle is

$$V = \tfrac{1}{2}kx^2 \tag{1-30}$$

if we require that $V=0$ at the point $x=0$. The Hamiltonian of the one-dimensional harmonic oscillator is, therefore,

$$H(x, p) = \frac{p^2}{2m} + \frac{kx^2}{2} \tag{1-31}$$

It follows from Hamilton's equations of motion (1-25) that $x$ and $p$ are determined by the equations

$$\frac{\partial x}{\partial t}=\frac{p}{m} \qquad \frac{\partial p}{\partial t}=-kx \tag{1-32}$$

If we differentiate the first Eq. (1-32) with respect to $t$ and eliminate $\partial p/\partial t$ with the second Eq. (1-32), we obtain

$$\frac{\partial^2 x}{\partial t^2}=-\frac{k}{m}x \tag{1-33}$$

It is convenient to introduce the angular frequency

$$\omega=\left(\frac{k}{m}\right)^{1/2} \tag{1-34}$$

so that Eq. (1-33) is written as

$$\ddot{x}+\omega^2 x=0 \tag{1-35}$$

It is easily verified that the solutions of this differential equation are

$$x=e^{\pm i\omega t} \tag{1-36}$$

so that the general solution for the harmonic oscillator is

$$x(t)=\lambda e^{i\omega t}+\mu e^{-i\omega t} \tag{1-37}$$

The solution contains two undetermined parameters $\lambda$ and $\mu$. If we impose the condition that $x(0)=0$, then we can eliminate one of the parameters, and we obtain the solution

$$x(t)=A\sin\omega t \tag{1-38}$$

The motion of the oscillator is determined by two quantities: $\omega$, which is called the angular frequency, and $A$, which is the amplitude of the oscillator. The frequency $\nu$ of the oscillations, that is, the number of complete cycles of motion that are covered per unit time, is given by

$$\omega=2\pi\nu \tag{1-39}$$

It follows from the first Eq. (1-32) that

$$p(t)=Am\omega\cos\omega t \tag{1-40}$$

If this result, together with Eq. (1-38), is substituted into Eq. (1-31), it is found that the Hamiltonian, and consequently the energy, of the system is

$$E=H(x,p)=\tfrac{1}{2}mA^2\omega^2 \tag{1-41}$$

It is obvious that for fixed $m$ and $A$ the energy $E$ is a continuous function of $\omega$ that can assume all positive values.

The three-dimensional harmonic oscillator is treated in a manner similar to the one-dimensional case. The Hamiltonian is now

$$H(\mathbf{r},\mathbf{p})=\frac{1}{2m}\left(p_x^2+p_y^2+p_z^2\right)+\frac{1}{2}\left(k_xx^2+k_yy^2+k_zz^2\right) \tag{1-42}$$

Hamilton's equations of motion are

$$\begin{aligned}
\frac{\partial x}{\partial t}&=\frac{p_x}{m} & \frac{\partial p_x}{\partial t}&=-k_xx\\
\frac{\partial y}{\partial t}&=\frac{p_y}{m} & \frac{\partial p_y}{\partial t}&=-k_yy\\
\frac{\partial z}{\partial t}&=\frac{p_z}{m} & \frac{\partial p_z}{\partial t}&=-k_zz
\end{aligned} \tag{1-43}$$

Each pair of Eqs. (1-43) is identical to Eq. (1-32), and the solutions are easily shown to be

$$\begin{aligned}
x(t)&=\lambda_xe^{i\omega_xt}+\mu_xe^{-i\omega_xt}\\
y(t)&=\lambda_ye^{i\omega_yt}+\mu_ye^{-i\omega_yt}\\
z(t)&=\lambda_ze^{i\omega_zt}+\mu_ze^{-i\omega_zt}
\end{aligned} \tag{1-44}$$

with

$$\omega_\alpha=\left(\frac{k_\alpha}{m}\right)^{1/2} \qquad \alpha=x,y,z \tag{1-45}$$

If, again, we impose the conditions that at $t=0$ the particle is at the origin, we have

$$\begin{aligned}
x(t)&=A_x\sin\omega_xt\\
y(t)&=A_y\sin\omega_yt\\
z(t)&=A_z\sin\omega_zt
\end{aligned} \tag{1-46}$$

and the energy of the particle is

$$E=\tfrac{1}{2}m\left(A_x^2\omega_x^2+A_y^2\omega_y^2+A_z^2\omega_z^2\right) \tag{1-47}$$

It should be noted that we did not really need to use the Hamiltonian formalism in order to solve the harmonic oscillator equations. We can write Eq. (1-28) as

$$F = -kx = ma = m\frac{d^2x}{dt^2} \tag{1-48}$$

which is identical with Eq. (1-33) that we derived from the Hamiltonian. We mentioned already that the Hamiltonian formalism is not really necessary for dealing with simple systems. On the other hand, we felt that it might be useful to give an example of a solution of the Hamiltonian equations of motion and the harmonic oscillator seems to be suitable for this purpose.

## 1-4 The Quantization of the Harmonic Oscillator

At the end of the nineteenth century classical Newtonian mechanics was still the basis of every theoretical description of physical and chemical phenomena, as it had been for almost 200 years. It was generally believed that all fundamental laws of nature were well established and well understood and that it would be only a matter of time before every natural phenomenon could be described mathematically in terms of the classical motion of electrons. This optimistic attitude was supported by the work of the great physicist H. A. Lorentz, who set out to explain all electrodynamic and optical phenomena in terms of electronic motion, described, of course, in terms of classical mechanics. The problems that were foremost in the minds of theoreticians at that time were concerned with the nature and properties of the world aether. It was hard for them to foresee that the next few decades would witness the abandonment of classical mechanics for the description of atoms and molecules and a complete revolution of theoretical physics. First, it was shown from the considerations of Lorentz, Poincaré, Einstein, and others that high-speed phenomena could not be adequately described in terms of classical mechanics. Their description was based on entirely new mechanical principles contained in the theory of relativity. Different lines of thought, initiated by Planck, Bohr, Heisenberg, Schrödinger, Dirac, and others, led to the realization that the Newtonian laws of motion also fail when they are applied to atomic phenomena. Here, it was necessary to introduce quantum mechanics.

The "birth" of quantum mechanics, if we may use this term, is usually taken as the fall of 1900. At that time the German physicist Max Planck presented two papers to the German Physical Society, the first one at the meeting of the Society on October 19, 1900, and the second paper at a subsequent meeting on December 14, 1900. Science historians cannot agree as to which of these two dates should be considered the true "birthday" of quantum mechanics, and we prefer not to take sides in this controversy.

The two papers by Planck dealt with some inconsistencies in the description of black-body radiation. It is generally agreed that both the experimental description and the theoretical interpretation of black-body radiation are fairly involved, and it is unfortunate that we must first explain such a sophisticated physical phenomenon in order to explore the historical development of quantum mechanics. However, if we want to relate the true course of events, we have no other choice.

The concept of a black body or a perfectly black body was introduced by Gustav R. Kirchhoff in 1859. A year earlier Kirchhoff had presented some important theoretical considerations on the relations between the emission and absorption of light and heat radiation by material bodies. A black body is defined as a body that absorbs all incident radiation. This became an important theoretical concept because the radiation emission of a black body should be a universal function, in other words, the radiation emission of different black bodies should be the same, independent of the material of the black body. Subsequently, Kirchhoff showed that the emissive power of a black body is proportional to the energy density of cavity radiation. Consequently, the energy density of radiation in a cavity should be independent of the nature of its walls. The function $\rho(\nu, T)$, which describes the dependence of this energy density on the temperature $T$ and on the frequency $\nu$ of the radiation, should therefore be a universal function, independent of the material of the cavity walls.

It may be helpful to interrupt our history in order to give a simple experimental description of black-body radiation. The phenomenon has been observed by anyone who has ever watched an electric heater or a kitchen stove. If we watch carefully what happens after an electric heater has been turned on, we first notice some heat radiation. A little later, when the heating element becomes warmer, we notice a red glow that slowly changes to orange, yellow, and white as the temperature of the electric coil increases. This change in color is caused by the change in spectral intensity distribution of the radiation that is produced by the heating coil. The intensity distribution of black-body radiation, that is, the energy density of black-body radiation as a function of its frequency at different temperatures, is shown in Fig. 1-2. We show three curves, one at a temperature $T_2$ and two others at slightly higher and lower temperatures $T_3$ and $T_1$. It may be seen that the energy density is considerably higher at the higher temperature and that the position of the maximum is shifted slightly to the right. The total energy of the emitted light, that is, the total area under one of the curves, is proportional to the fourth power of the temperature. This feature had already been predicted by Josef Stefan in 1879.

Let us now resume our historical description. Toward the end of the nineteenth century various attempts were made to derive theoretical representations of the intensity distribution curves of Fig. 1-2. These attempts were based on different theoretical viewpoints. For example, Wilhelm Wien believed that the radiation emitted by molecules was due to the molecular motion, consequently the wavelength of the radiation should be related to the molecular velocity $v$. If it is then assumed that the molecular velocities are described by the Maxwell-Boltzmann distribution function, the energy density $\rho(\nu, T)$ of

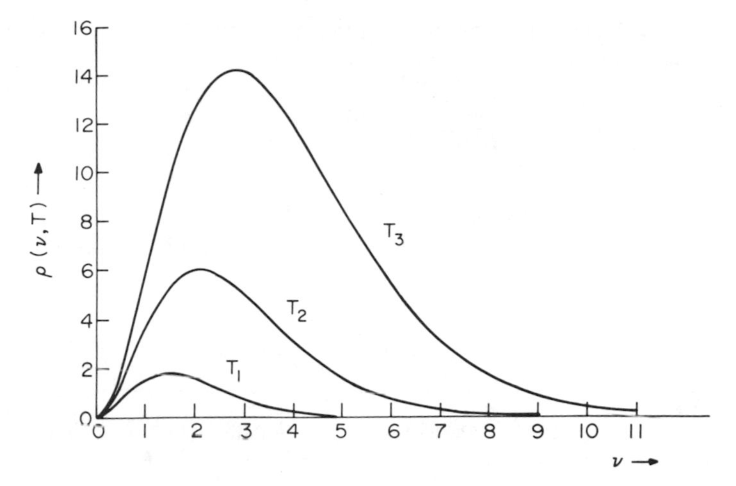


**Fig. 1-2** The spectral intensity distribution of black-body radiation $\rho(\nu, T)$ as a function of the frequency $\nu$ at three different temperatures $T_1$, $T_2$, and $T_3$. The temperature $T_2$ is higher than $T_1$ but lower than $T_3$.

black-body radiation is described by the expression

$$\rho(\nu, T) = a\left(\frac{8\pi\nu^3}{c^3}\right)\exp\left(\frac{-b\nu}{T}\right) \tag{1-49}$$

A few years later, Lord Rayleigh derived the energy density function from the equipartition theorem of statistical mechanics. First, he derived the relation between the energy density $\rho(\nu, T)$ and the energy $U$ of a mode of vibration. This is given by

$$\rho(\nu, T) = \left(\frac{8\pi\nu^2}{c^3}\right)U \tag{1-50}$$

According to the equipartition theorem the energy $U$ is equal to $kT$, and by substituting this result, Rayleigh obtained the theoretical expression

$$\rho(\nu, T) = \left(\frac{8\pi\nu^2}{c^3}\right)kT \tag{1-51}$$

where $k$ is Boltzmann's constant.

It may be interesting to derive the energy $U$ of a harmonic oscillator in order to prove the equipartition theorem for this particular case. According to Maxwell-Boltzmann statistical mechanics $U$ is given by

$$U = \frac{\iint E(x, p)\exp[-E(x, p)/kT]\, dx\, dp}{\iint \exp[-E(x, p)/kT]\, dx\, dp} \tag{1-52}$$

where $E(x, p)$ is the energy expression of the harmonic oscillator. According to Eq. (1-31) it is given by

$$E(x, p) = \frac{p^2}{2m} + \frac{fx^2}{2} \tag{1-53}$$

We denote the force constant now by $f$ in order to distinguish it from Boltzmann's constant $k$. By substituting Eq. (1-53) into Eq. (1-52) we obtain $U$ as a sum of two integrals:

$$U = \frac{1}{2m} \frac{\int p^2 \exp[-p^2/2mkT]\, dp}{\int \exp[-p^2/2mkT]\, dp} + \frac{f}{2} \frac{\int x^2 \exp[-fx^2/2kT]\, dx}{\int \exp[-fx^2/2kT]\, dx} \tag{1-54}$$

We cannot evaluate the individual integrals by simple integration procedures, but we can calculate the ratio without too much difficulty. We define

$$I(\alpha) = \int_{-\infty}^{\infty} \exp(-\alpha x^2)\, dx \qquad J(\alpha) = \int_{-\infty}^{\infty} x^2 \exp(-\alpha x^2)\, dx \tag{1-55}$$

We have

$$J(\alpha) = -\frac{\partial I(\alpha)}{\partial \alpha} \tag{1-56a}$$

and

$$\frac{J(\alpha)}{I(\alpha)} = -\frac{1}{I(\alpha)} \frac{\partial I(\alpha)}{\partial \alpha} = -\frac{\partial \log I(\alpha)}{\partial \alpha} \tag{1-56b}$$

By changing the variable in Eq. (11-55) we obtain

$$I(\alpha) = \left(\frac{1}{\sqrt{\alpha}}\right) \int_{-\infty}^{\infty} \exp(-t^2)\, dt \tag{1-57a}$$

Since the integral in Eq. (1-57a) is independent of $\alpha$ we find that

$$\frac{\partial}{\partial \alpha} \log I(\alpha) = -\frac{1}{2\alpha} \tag{1-57b}$$

or

$$\frac{J(\alpha)}{I(\alpha)} = \frac{1}{2\alpha} \tag{1-58}$$

We use this result to evaluate $U$ from Eq. (1-54),

$$U=\frac{1}{2m}\frac{2mkT}{2}+\frac{f}{2}\frac{2kT}{2f}=kT \tag{1-59}$$

which is in agreement with the equipartition theorem.

Let us return again to the theoretical description of black-body radiation. Clearly, the situation was far from satisfactory. Starting from different viewpoints, Wien and Rayleigh had derived two different theoretical expressions for black-body radiation, namely, Eqs. (1-49) and (1-51). The former, Wien's result, is in agreement with the experimental curve for high frequencies and the latter, Rayleigh's result, is in agreement for small frequencies. Neither one of the two results represents the experimental data on the complete frequency range. Rayleigh's expression is based on a sounder and more logical theoretical foundation than Wien's. On the other hand, Rayleigh predicts infinite energy densities for large frequencies. This is not only in conflict with the experimental data, it is totally impossible.

We see that there were serious conflicts between theory and experiment in black-body radiation when Planck became interested in the problem in 1900. Planck approached the problem from yet another point of view. He had specialized in thermodynamics and he had great confidence in the general validity of thermodynamic theories. Therefore, Planck approached the problem by studying the entropy of an assembly of harmonic oscillators. As a starting point we consider the well-known thermodynamic expression

$$\frac{dS}{dU}=\frac{1}{T} \tag{1-60}$$

If we use Wien's expression (1-49), in combination with Eq. (1-50), we obtain

$$\frac{1}{T}=\frac{dS}{dU}=\frac{-1}{b\nu}\log\frac{U}{\alpha\nu} \tag{1-61a}$$

If, on the other hand, we use Rayleigh's expression (1-51) together with Eq. (1-50) we obtain

$$\frac{1}{T}=\frac{dS}{dU}=\frac{k}{U} \tag{1-61b}$$

By differentiating both expressions (1-61a) and (1-61b) with respect to $U$ we obtain

$$\frac{d^2S}{dU^2}=-\frac{\text{const.}}{U} \tag{1-62a}$$

according to Wien's result and

$$\frac{d^2S}{dU^2} = -\frac{\text{const.}}{U^2} \tag{1-62b}$$

according to Rayleigh's result.

In this way Planck derived two different expression (1-62a) and (1-62b) for the second derivative. It appeared from the experiments that the first expression (1-62a) is asymptotically correct for small values of $U$ and that the second expression (1-62b) agrees with the experiments for large $U$ values. Planck proposed now that the second derivative be represented by the expression

$$\frac{d^2S}{dU^2} = -\frac{A}{U(U+\beta)} = \frac{A}{\beta}\left(\frac{1}{U+\beta} - \frac{1}{U}\right) \tag{1-63}$$

This equation has the correct asymptotic behavior for both large $U$ and small $U$, and it can be used for all values of $U$. If we integrate it we obtain

$$\frac{1}{T} = \frac{dS}{dU} = \frac{A}{\beta}\log\frac{U+\beta}{U} \tag{1-64}$$

or

$$U = \frac{\beta}{\exp(\alpha/T) - 1} \tag{1-65}$$

From additional experimental data it was derived further that $U$ should have the form

$$U = \frac{c_1\nu}{\exp(c_2\nu/T) - 1} \tag{1-66}$$

and that $\rho$ is given by

$$\rho(\nu, T) = \frac{8\pi\nu^2}{c^3} \quad \frac{c_1\nu}{\exp(c_2\nu/T) - 1} \tag{1-67}$$

Here $c_1$ and $c_2$ are unknown parameters. The last result [Eq. (1-67)] is Planck's radiation law.

We might say now that the birth of quantum mechanics coincides with the introduction of Eq. (1-63), but at the time this equation was nothing more than a mathematical interpolation expression without much physical justification. In fact, if we want to be very critical we might say that it was no more than a lucky guess. However, it gave a very good representation of the experimental data and it seemed worth the effort to find further justification for Eq. (1-67).

We now turn to Planck's second paper, which was read on December 14, 1900; here Planck offered a physical explanation of his radiation formula. He now introduced the concept of quantization by assuming that the energy of an oscillator cannot be emitted in infinitely small amounts. Instead, the energy that is emitted must be an integral amount of "energy quanta." The magnitude of such an energy quantum $\varepsilon$, which is the smallest amount of energy that can be emitted, is given by

$$\varepsilon = h\nu \tag{1-68}$$

where $h$ is a newly introduced natural constant.

Starting from the above assumption, Planck now set out to calculate the entropy $S$ of a set of $N$ harmonic oscillators, having a total energy $U$. According to Boltzmann's entropy definition

$$S = k \log W \tag{1-69}$$

the entropy is determined from the number of ways $W$ that the total energy $U$ can be realized from different distributions. In other words, we have $P = U/\varepsilon$ energy elements, and $W$ is the number of ways that these $P$ energy elements can be distributed over $N$ oscillators.

We can calculate $W$ by determining the number of ways that $P$ objects can be distributed over $N$ boxes. It can be seen from Fig. 1-3 that $W$ is given by

$$W = \frac{(N+P-1)!}{P!(N-1)!} \tag{1-70}$$

We represent the boxes by $N-1$ vertical lines and the objects by crosses. The total number of distributions is given by the number of permutations of both lines and crosses, that is $(N+P-1)!$ But obviously we have the same distribution if we permute the lines or the crosses among themselves, and therefore we must divide by $(N-1)!$ and by $P!$

By making use of Sterling's approximation it is easily found that

$$\begin{aligned} S = k \log W &= k \log(N+P-1)! - k \log(N-1)! - k \log P! \\ &= k\left[(N+P)\log(N+P) - N \log N - P \log P\right] \end{aligned} \tag{1-71}$$

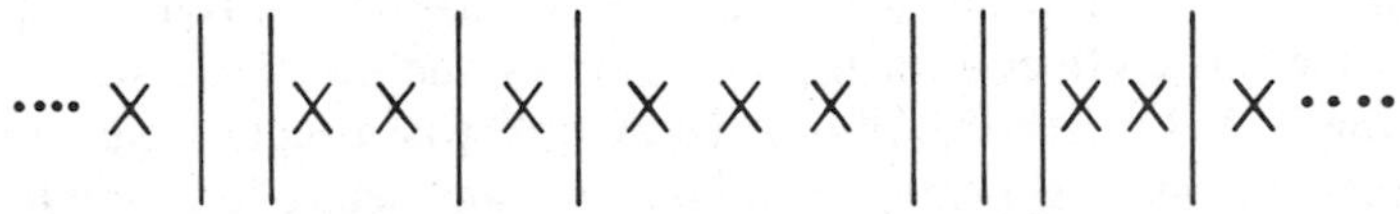

···· X | | X X | X | X X X | | | X X | X ····

**Fig. 1-3** In order to calculate the number of possible distributions of $P$ objects over $N$ boxes, we consider all possible permutations of $N-1$ partitions (the vertical bars) and $P$ objects (the crosses), a total of $N+P-1$ quantities.

We introduce the average entropy $S_o$ per oscillator and the average energy $U_o$ per oscillator, and we have

$$S = NS_o$$

$$P = \frac{U}{\varepsilon} = \frac{NU_o}{\varepsilon} \tag{1-72}$$

Hence

$$\begin{aligned} S_o &= k\left[\left(1+\frac{P}{N}\right)\log\left(1+\frac{P}{N}\right) - \frac{P}{N}\log\frac{P}{N}\right] \\ &= k\left[\left(1+\frac{U_o}{\varepsilon}\right)\log\left(1+\frac{U_o}{\varepsilon}\right) - \frac{U_o}{\varepsilon}\log\frac{U_o}{\varepsilon}\right] \end{aligned} \tag{1-73}$$

Planck derived finally

$$\frac{1}{T} = \frac{\partial S_o}{\partial U_o} = \frac{k}{\varepsilon}\left[\log\left(1+\frac{U_o}{\varepsilon}\right) - \log\frac{U_o}{\varepsilon}\right] \tag{1-74}$$

and

$$U_o = \frac{\varepsilon}{\exp(\varepsilon/kT)-1} \tag{1-75}$$

By introducing the new constant $h$ Planck obtained finally

$$U_o = \frac{h\nu}{\exp(h\nu/kT)-1} \tag{1-76}$$

which is identical with the result he derived a few months earlier.

Very few people realized that Planck's work of 1900 constituted a major discovery in physics because his efforts attracted very little attention until 1905. During that year Einstein published a new generalized theoretical description of the radiation field. He showed that a radiation field could be considered as an assembly of energy quanta. In a monochromatic radiation field of frequency $\nu$ the energy quanta have the magnitude $h\nu$; in a more general field we have different quanta with various magnitudes $h\nu$. Later on G. N. Lewis introduced the name *photons* for the energy quanta.

An application of Einstein's general theory was the theoretical explanation of the photoelectric effect, and for this part of the paper Einstein was later awarded the Nobel prize. The photoelectric effect is usually measured from a vacuum tube containing a metallic plate and an electrode a small distance away. If the metallic plate is irradiated by a monochromatic beam of light of sufficiently high frequency $\nu$, then the light ejects electrons out of the metal plate. These electrons may give rise to an electric current between the metallic plate and the receiving electrode if a suitable potential difference is maintained between the plate and the electrode. An interesting feature of the photoelectric effect is that no current is observed if the frequency is below a certain frequency $\nu_o$, no matter how intense the light is.

Einstein explained the photoelectric effect by assuming that the ejection of the electrons out of the metal is due to collisions between the photons and the electrons in the metal. It follows that the maximum amount of energy that can be transferred to an electron by a photon is equal to $h\nu$. Furthermore, a certain amount of energy $eW$ is required to move an electron out of the metal. The maximum amount of kinetic energy $T_m$ of an ejected electron is thus given by

$$T_m = h\nu - eW \tag{1-77}$$

Obviously, no photoelectric current is observed unless the frequency of the incident light is larger than a limiting frequency $\nu_o$ which is given by

$$h\nu_o = eW \tag{1-78}$$

For frequencies larger than $\nu_o$ the photocurrent is proportional to the intensity of the radiation. The kinetic energy of the ejected electrons is proportional to the frequency of the radiation. All these theoretical predictions are in perfect agreement with the experimental results.

The next application of the concept of quantization was in the area of specific heats of solids. If we consider a solid containing $N$ atoms, then there are $3N$ possible modes of vibration in the solid. According to the classical equipartition theorem each mode of vibration has an energy $kT$, so the total vibrational energy $E$ of the solid is given by

$$E = 3NkT \tag{1-79}$$

and its specific head $C_v$ is

$$C_v = \frac{\partial E}{\partial T} = 3Nk \tag{1-80}$$

independent of the temperature. This is the theoretical justification of Dulong and Petit's law which states that the molar heat capacity of a solid is equal to $3R$ (=6 cal/mole degree).

Toward the end of the nineteenth century lower temperatures could be attained, and it was found that the molar heat capacity of solids could be considerably smaller than 6 cal/mole degree and that there were strong indications that $C_v$ approaches zero when the temperature $T$ approaches the absolute zero. Einstein showed in 1907 that this behavior could be explained if it is assumed that the energy of a vibrational mode is described by the quantized expression (1-76) rather than by the classical equipartition theorem. Einstein made the simplifying assumption that all vibrational modes have the same effective frequency $\nu_o$. It follows then that

$$E=3NkT\frac{(h\nu_o/kT)}{\exp(h\nu_o/kT)-1} \tag{1-81}$$

and

$$c_v=3Nk\frac{(h\nu_o/kT)^2\exp(h\nu_o/kT)}{[\exp(h\nu_o/kT)-1]^2} \tag{1-82}$$

Debye incorporated the frequency distribution of the vibrations into the theory and he derived an expression for the specific heat of solids which is in excellent agreement with the experiments.

We conclude this historical sketch by mentioning one of the most important scientific meetings of this century, namely the Solvay Congress, held in 1911. Here, all prominent physicists of that time convened in Brussels in order to discuss the new concepts of quantum theory. The concept of quantization was generally accepted, but at the same time it was realized that much work remained to be done in order to find further justification for the quantization rules and to derive a more comprehensive theoretical foundation for these new ideas. In the following section we describe the search for these generalized new quantum mechanical theories. We should realize that the progress of science was seriously hampered by World War I, which ranged from 1914 to 1918.

## 1-5 Angular Momentum and Hydrogen Atom in Classical Mechanics

As we announced in the first section of this chapter, we do not intend to describe the complete history of the discovery of quantum theory. Instead, we present certain specific developments as separate topics. For instance, in the previous section we discussed the quantization of the radiation field and the harmonic oscillator and in the following section we describe Bohr's atomic theory. The latter theory was of course intended as a general description of atomic structure, but its initial applications dealt mostly with the hydrogen atom and with the properties of angular momentum. It seems appropriate therefore to describe angular momentum and the properties of the hydrogen atom according to classical mechanics first.

In classical mechanics we define the angular momentum $\mathbf{M}$ of a one-particle system as

$$\mathbf{M}=\mathbf{r}\times\mathbf{p} \tag{1-83}$$

and the angular momentum $\mathbf{M}$ of a many-particle system as

$$\mathbf{M}=\sum_j(\mathbf{r}_j\times\mathbf{p}_j) \tag{1-84}$$

In order to understand these definitions, we have to discuss a little more vector analysis. In Section 1-2 we defined the sum of two vectors **u** and **v**. Now we wish to define the product of the two vectors, while realizing that there are two kinds of products. The first product, which is called the scalar product or dot product of **u** and **v**, is a scalar quantity and is defined as

$$\mathbf{u}\cdot\mathbf{v}=u_xv_x+u_yv_y+u_zv_z=uv\cos\theta_{uv} \tag{1-85}$$

where $\theta_{uv}$ is the angle between the vectors **u** and **v**. The second product, which is called the vector product or cross product, is a vector **w** and is defined as

$$\mathbf{w}=\mathbf{u}\times\mathbf{v}$$

$$w_x=u_yv_z-u_zv_y \tag{1-86}$$

$$w_y=u_zv_x-u_xv_z$$

$$w_z=u_xv_y-u_yv_x$$

This definition can be visualized by considering the case where both vectors **u** and **v** are placed in the *XY* plane. It follows then from the definition (1-86) that the vector **w** is then directed along the *Z* axis. Its magnitude *w* is given by

$$w_z=uv\sin\theta_{uv} \tag{1-86a}$$

where $\theta_{uv}$ is the angle between **u** and **v**, going anticlockwise from **u** to **v**. It follows that **w** points in the positive *Z* direction if the angle $\theta_{uv}$ is smaller than 180° and **w** points in the negative *Z* direction if the angle $\theta_{uv}$ is between 180° and 360°.

In the general case, the direction of **w** is perpendicular to the plane formed by the two vectors **u** and **v**, and it points in the direction of motion of a corkscrew being turned from **u** and **v** (see Fig. 1-4) through the smaller angle.

Let us now consider the hydrogen atom according to classical mechanics and investigate the role of the angular momentum. We have two particles: a heavy nucleus with mass *M*, coordinate $\mathbf{r}_n$, and momentum $\mathbf{p}_n$, and a much lighter electron with mass *m*, coordinate $\mathbf{r}_e$, and momentum $\mathbf{p}_e$. The potential energy *U* of this system depends only on the distance $|\mathbf{r}_e-\mathbf{r}_n|$, so that we can

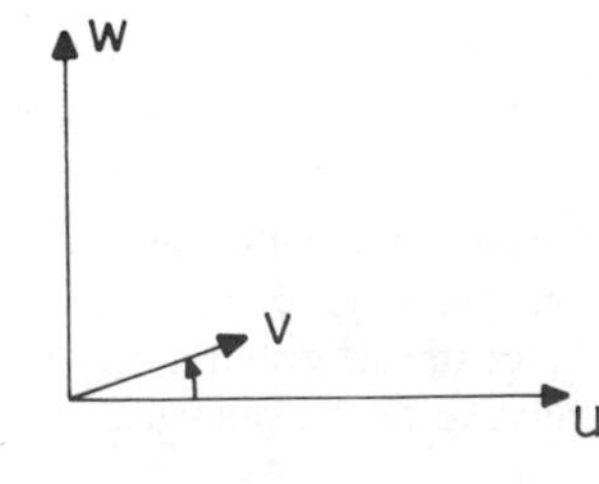


**Fig. 1-4** Representation of the cross product **w** of two vectors **u** and **v** according to Eq. (1-86).

write the Hamiltonian as

$$H=\frac{p_e^2}{2m}+\frac{p_n^2}{2M}+U(|\mathbf{r}_e-\mathbf{r}_n|) \tag{1-87}$$

We begin our discussion by introducing a new set of coordinates (see Fig. 1-5). The position $\mathbf{R}$ of the center of gravity of the two-particle system is defined as

$$(m+M)\mathbf{R}=M\mathbf{r}_n+m\mathbf{r}_e \tag{1-88}$$

and the relative position of the electron with respect to the nucleus is given by

$$\mathbf{r}=\mathbf{r}_e-\mathbf{r}_n \tag{1-89}$$

If we now transform from the coordinates $\mathbf{r}_e$ and $\mathbf{r}_n$ to $\mathbf{r}$ and $\mathbf{R}$, we find that this transformation is given by

$$\mathbf{r}_e=\mathbf{R}+\frac{M}{m+M}\mathbf{r}$$
$$\mathbf{r}_n=\mathbf{R}-\frac{m}{m+M}\mathbf{r} \tag{1-90}$$

It may be recalled that the momenta $\mathbf{p}_e$ and $\mathbf{p}_n$ are defined as

$$\mathbf{p}_e=m\frac{d\mathbf{r}_e}{dt} \qquad \mathbf{p}_n=M\frac{d\mathbf{r}_n}{dt} \tag{1-91}$$

We introduce the new momenta

$$\mathbf{P}=(m+M)\frac{d\mathbf{R}}{dt} \qquad \mathbf{p}=\mu\frac{d\mathbf{r}}{dt} \tag{1-92}$$

conjugate to $\mathbf{R}$ and $\mathbf{r}$, respectively, where

$$\mu=\frac{mM}{m+M} \tag{1-93}$$

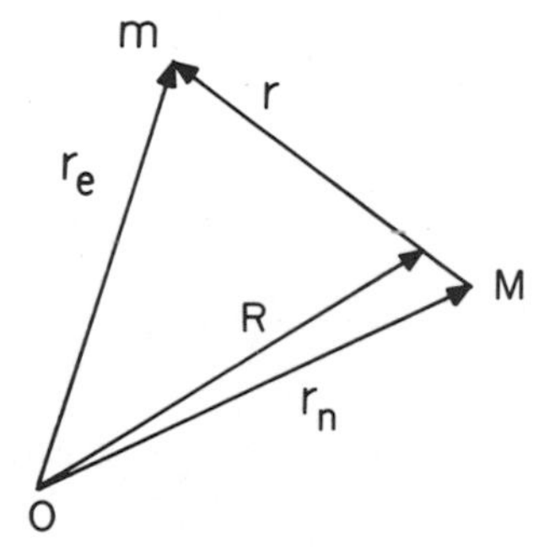


**Fig. 1-5** Coordinate transformation for the description of the hydrogen atom. Here $\mathbf{r}_e$ denotes the electron, $\mathbf{r}_n$ the nucleus, $\mathbf{R}$ the center of gravity, and $\mathbf{r}$ the position of the electron relative to the nucleus.

Substitution of Eqs. (1-89) to (1-92) into Eq. (1-87) for the Hamiltonian gives

$$H=\frac{P^2}{2(M+m)}+\frac{p^2}{2\mu}+U(r) \tag{1-94}$$

Similarly, it is found that the angular momentum

$$\mathbf{M}=(\mathbf{r}_e\times\mathbf{p}_e)+(\mathbf{r}_n\times\mathbf{p}_n) \tag{1-95}$$

is transformed into

$$\mathbf{M}=(\mathbf{R}\times\mathbf{P})+(\mathbf{r}\times\mathbf{p}) \tag{1-96}$$

It is interesting to note that both the old and the new expressions for the angular momentum have similar forms.

The Hamiltonian (1-94) is the sum of a part that depends only on **P** and a part that depends only on **r** and **p**. It can be derived from the considerations in Section 1-2 that the equations of motion can be separated into two independent sets in that case. The first set describes the motion of the center of gravity. Since the center of gravity moves as a free particle (the potential does not depend on **R**), there is not much more to say about it.

The interesting part of the problem is the motion of the electron with respect to the nucleus, which can be considered separately. This motion is described by the Hamiltonian

$$H=\frac{p^2}{2\mu}+U(r) \tag{1-97}$$

and its angular momentum is

$$\mathbf{M}=\mathbf{r}\times\mathbf{p} \tag{1-98}$$

It should be noted that Eq. (1-97) contains the reduced mass $\mu$ instead of the electronic mass $m$. Since $M$ is much larger than $m$, the difference between $m$ and $\mu$ is small, but the difference can be detected in the spectroscopic experiments, which are so extremely accurate.

Let us now show why the angular momentum is such an important quantity in the present problem. One of its components, for example, the $z$ component, is given by

$$M_z=xp_y-yp_x \tag{1-99}$$

whose time derivative is

$$\frac{dM_z}{dt}=p_y\frac{\partial x}{\partial t}-p_x\frac{\partial y}{\partial t}+x\frac{\partial p_y}{\partial t}-y\frac{\partial p_x}{\partial t} \tag{1-100}$$

We use Hamilton's equations (1-21) applied to the Hamiltonian (1-97) to replace the time derivatives:

$$\frac{\partial x}{\partial t}=\frac{\partial H}{\partial p_x}=\frac{p_x}{\mu}$$

$$\frac{\partial y}{\partial t}=\frac{\partial H}{\partial p_y}=\frac{p_y}{\mu} \tag{1-101}$$

$$\frac{\partial p_x}{\partial t}=-\frac{\partial H}{\partial x}=-\frac{x}{r}\frac{\partial U}{\partial r}$$

$$\frac{\partial p_y}{\partial t}=-\frac{\partial H}{\partial y}=-\frac{y}{r}\frac{\partial U}{\partial r}$$

If we substitute these results into Eq. (1-100), we find that

$$\frac{dM_z}{dt}=0 \tag{1-102}$$

In the same way it can be shown that

$$\frac{dM_x}{dt}=\frac{dM_y}{dt}=0 \tag{1-103}$$

Consequently, **M** is a constant of the motion and does not depend on time.

The fact that **M** is time independent makes it possible to solve the equations of motion of the hydrogen atom. We take the $z$ axis along the constant **M**, which now becomes

$$M=\mu\left(x\frac{dy}{dt}-y\frac{dx}{dt}\right) \tag{1-104}$$

It follows from the definition (1-86) that **M** is perpendicular to both **r** and **p**, and since **M** is along the $z$ axis, **r** and **p** have to be in the $xy$ plane.

Let us now introduce polar coordinates in the $xy$ plane,

$$x=r\cos\varphi$$

$$y=r\sin\varphi \tag{1-105}$$

and express $M$ in terms of polar coordinates. We have

$$\frac{dx}{dt}=\frac{dr}{dt}\cos\varphi-r\sin\varphi\frac{d\varphi}{dt}$$

$$\frac{dy}{dt}=\frac{dr}{dt}\sin\varphi+r\cos\varphi\frac{d\varphi}{dt} \tag{1-106}$$

If we define the angular velocity $\omega$ as

$$\omega = \frac{d\varphi}{dt} \tag{1-107}$$

then it follows easily from Eqs. (1-104) and (1-106) that

$$M = \mu r^2 \omega \tag{1-108}$$

Before we proceed to consider the hydrogen atom we derive some general theorems from the time dependence of **M**. We consider a particle in a central force field. We show in Fig. 1-6 that the area $dS$ that is swept out by its position vector **r** during a small time interval $dt$ is given by

$$dS = \tfrac{1}{2} r^2 \sin d\varphi = \tfrac{1}{2} r^2 d\varphi \tag{1-109}$$

It follows from Eq. (1-108) that

$$\frac{dS}{dt} = \frac{1}{2} r^2 \frac{d\varphi}{dt} = \frac{M}{2\mu} \tag{1-110}$$

which means that the derivative of $S$ with respect to the time is a constant.

In the case of a periodic motion with a closed orbit we can easily calculate the time $T$ that is needed to describe one periodic motion. By integrating Eq. (1-110) we find that

$$S = \frac{MT}{2\mu} \tag{1-111}$$

where $S$ is the area contained within the periodic orbital.

Let us now consider the classical theory of the hydrogen atom. We take the $Z$ axis along the direction of the angular momentum **M**. The particle then

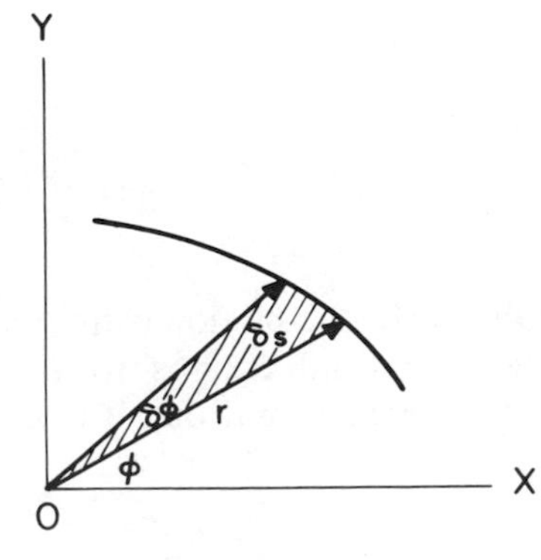


**Fig. 1-6** The area $\delta S$ that is swept out by the position vector **r** of a particle when the polar angle $\varphi$ increases by an amount $\delta\varphi$.

moves in the $x\,y$ plane and its energy is given by

$$E=\frac{\mu}{2}\left[\left(\frac{dx}{dt}\right)^2+\left(\frac{dy}{dt}\right)^2\right]-\frac{e^2}{r} \tag{1-112}$$

according to Eq. (1-97). We use the polar coordinates of Eq. (1-105) and we substitute Eq. (1-106). The result is

$$E=\frac{\mu}{2}\left(\frac{dr}{dt}\right)^2+\frac{\mu r^2}{2}\left(\frac{d\varphi}{dt}\right)^2-\frac{e^2}{r} \tag{1-113}$$

or

$$E=\frac{\mu}{2}\left(\frac{dr}{dt}\right)^2+\frac{M^2}{2\mu r^2}-\frac{e^2}{r} \tag{1-114}$$

Equation (1-114) can be solved exactly, but we prefer to discuss first some general properties of the solution. It is helpful to introduce the effective potential function

$$U_{\text{eff}}(r)=-\frac{e^2}{r}+\frac{M^2}{2\mu r^2} \tag{1-115}$$

because this makes it easier to visualize the behavior of the solution of the equation

$$E=\frac{\mu}{2}\left(\frac{dr}{dt}\right)^2+U_{\text{eff}}(r) \tag{1-116}$$

We have sketched the general form of $U_{\text{eff}}$ in Fig. (1-7). It can be seen that $U$ tends to infinity when $r$ tends to zero; it has a minimum

$$U_{\text{min}}=-\frac{\mu e^4}{2M^2} \tag{1-117}$$

at

$$r_o=\frac{M^2}{\mu e^2} \tag{1-118}$$

and it approaches zero when $r$ tends to infinity. Obviously, the motion is finite and the particle is bound for $E<0$, and the motion is infinite for positive $E$. We will show that the orbit is an ellipse for negative energies and that it is a hyperbola for positive energies.

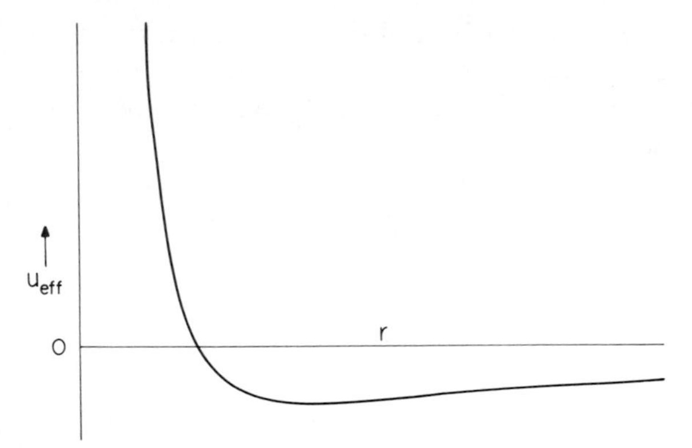


**Fig. 1-7** Sketch of the potential function $U_{eff}(r)$ of Eq. (1-115) for the radial motion of the hydrogen atom.

We can rewrite Eq. (1-113) as

$$\frac{dr}{dt}=\left[\frac{2E}{\mu}+\frac{2e^2}{\mu r}-\frac{M^2}{\mu^2 r^2}\right]^{1/2} \tag{1-119}$$

and we can rewrite Eq. (1-108) as

$$\frac{d\varphi}{dt}=\frac{M}{\mu r^2} \tag{1-120}$$

By combining the two equations we obtain

$$\frac{d\varphi}{dr}=\frac{M}{\mu r^2}\left[\frac{2E}{\mu}+\frac{2e^2}{\mu r}-\frac{M^2}{\mu^2 r^2}\right]^{-1/2} \tag{1-121}$$

The differential equation can be solved by means of an elementary integration. We do not present the details of this integration, we just mention that the solution can be written in the form

$$r=\frac{M^2/\mu e^2}{1+\varepsilon\cos\varphi}$$

$$\varepsilon=\left[1+\left(\frac{2EM^2}{\mu e^4}\right)\right]^{1/2} \tag{1-122}$$

This expression represents an elliptical orbit with the parameter $\varepsilon$ defined as the eccentricity of the ellipse.

The dimensions of the ellipse can be derived from the original Eq. (1-119) in a fairly simple manner. We derive the values of $r$ for which its time derivative is zero; these values describe the turning points of the orbit. The corresponding equation is

$$\frac{2E}{\mu}+\frac{2e^2}{\mu r}-\frac{M^2}{\mu^2 r^2}=0 \tag{1-123}$$

or

$$r^2+\frac{e^2}{E}r-\frac{M^2}{2\mu E}=0 \tag{1-124}$$

The two roots of this equation, $r_1$ and $r_2$, are given by

$$r_{1,2}=-\frac{e^2}{2E}\pm\left(\frac{e^4}{4E^2}+\frac{M^2}{2\mu E}\right)^{1/2} \tag{1-125}$$

We denote the major and the minor semi-axes of the ellipse by $a$ and $b$ and the distance between the two focal points by $2c$ (see Fig. 1-8). It is easily seen that

$$2a=r_1+r_2=-\frac{e^2}{E}$$

$$2c=r_1-r_2=2\left(\frac{e^4}{4E^2}+\frac{M^2}{2\mu E}\right)^{1/2}=\left(\frac{e^4}{E^2}+\frac{2M^2}{\mu E}\right)^{1/2}$$

$$b=(a^2-c^2)^{1/2}=\left(-\frac{M^2}{2\mu E}\right)^{1/2} \tag{1-126}$$

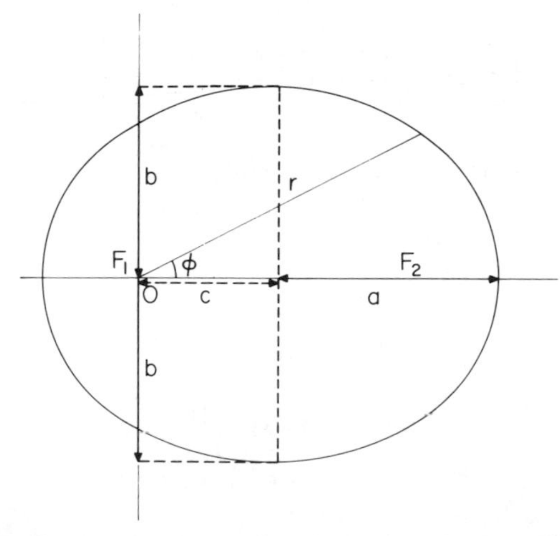


**Fig. 1-8** The classical motion of an electron in a Coulombic force field. When the electron is bound, the orbit is an ellipse with one of the focal points as the center of attraction.

The surface area $S$ within the ellipse is given by

$$S=\pi ab=\frac{\pi\mu e^2 M}{k^3} \qquad k^2=-2\mu E \tag{1-127}$$

Finally, it follows from Eq. (1-111) that the time of revolution $T$ is given by

$$T=\frac{2\mu S}{M}=\frac{2\pi\mu^2 e^2}{k^3} \tag{1-128}$$

This result, where $T$ is proportional to the $(-3/2)$ power of the negative energy $E$, is fairly well known. It was used by Bohr in his early work on the hydrogen atom, as we shall see in the next section.

## 1-6 The Old Quantum Theory

As we mentioned before, we describe the discovery of quantum mechanics by discussing various lines of development separately. The present section deals mainly with Bohr's theory of the structure of the hydrogen atom. Bohr's theory utilized Planck's quantization concepts for a theoretical description of Rutherford's atomic model.

According to Rutherford's model an atom consists of a very small, relatively heavy and positively charged nucleus surrounded by a certain number of negatively charged electrons that orbit around it like the planets around the sun. The diameter of an atom is of the order of $10^{-8}$ cm, and therefore the Ångstrom ($1A=10^{-8}$ cm) is often used as the unit of length in atomic problems. The nuclear diameter is of the order of $10^{-4}$ $A$; its weight is practically equal to the weight of the atom and its positive atomic charge is $Ze$. Here $Z$ is the atomic number, which is about half the atomic weight, and the charge of each electron is $-e$. The simplest atom is hydrogen, where there is only one electron orbiting around a nucleus with charge $e$, which is about two thousand times heavier than the electron. The hydrogen nucleus has a name of its own; it is known as proton.

This model of the atom was first proposed by Ernest Rutherford in 1911. During the nineteenth century most of the experimental information on atomic structure was derived from the atomic spectra. In 1897 J. J. Thomson discovered the existence of the electron and in the following decade various scattering experiments gave additional information. In 1912 Niels Bohr joined Rutherford in Manchester where he set out to derive a theoretical justification for the stability of the atomic model and also to offer a quantitative explanation for the experimental results on atomic spectra. We will first give a brief description of the experimental results on atomic spectra and then we will discuss Bohr's work.

The absorption spectrum of a given sample is measured by placing it in a beam of white light, where the intensity distribution as a function of the wavelength is constant, and by measuring the intensity distribution of the light that has passed through the sample. It is also possible to measure the emission spectrum of the sample; for example, when it is a dilute gas, it can be subjected to an electric discharge, and when the discharge causes light emission, then the intensity distribution of the emitted light is the emission spectrum of the sample. Ordinarily the absorption and emission spectra of the sample are complimentary; this means, as we illustrate in Fig. 1-9, that the valleys of the absorption spectrum (at the top) correspond to the peaks of the emission spectrum (at the bottom). The remarkable feature of an atomic emission spectrum is that it consists of a series of very narrow peaks at certain wavelengths, the spectral lines. Each spectral line is determined by its wavelength and its intensity. An atomic spectrum is therefore described by a set of numbers that describe the wavelengths of the spectral lines, and by a set of corresponding intensity classifications. The wavelengths can be determined with great accuracy (1 part in $10^7$ to $10^8$). The intensities, on the other hand, could not be measured very accurately, but this did not matter much since the intensities never played a very important role in the development of quantum mechanics.

The four hydrogen atom lines in the visible part of the spectrum were measured quite accurately by A. Ångstrom in 1868. In 1885 Johann Jakob Balmer, a numerologist, discovered that the wavelengths $\lambda$ of these lines could all be written as

$$\lambda = \frac{Am^2}{m^2 - 2^2} \qquad m = 3,4,5,6 \tag{1-129}$$

A more general expression for the hydrogen atom spectral lines was proposed a few years later by Johannes Robert Rydberg; in this equation the Rydberg constant $R_H$ was first introduced.

A most important discovery was made by Walter Ritz in 1908. Ritz noted first that we should consider the frequencies $\nu$ of the spectral lines. It is then possible to construct a set of terms $T_n$ so that each spectral wave number $\sigma_{n,m}$ is the difference of two terms:

$$\frac{1}{\lambda_{n,m}} = \sigma_{n,m} = |T_n - T_m| \tag{1-130}$$

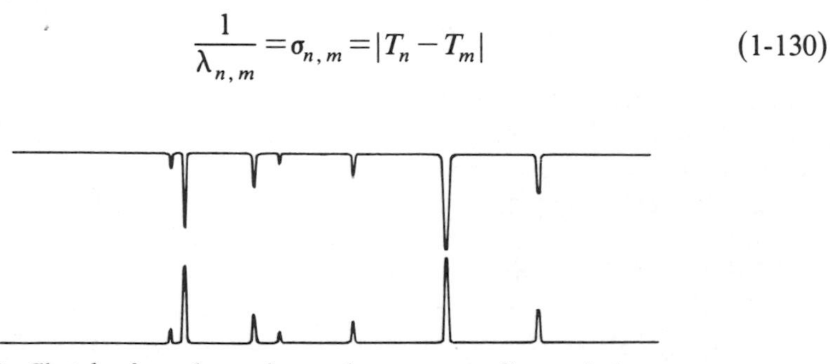

**Fig. 1-9** Sketch of an absorption and a corresponding emission spectrum.

This result is not as trivial as it seems; it means that we need only 10 terms to describe 45 spectral lines, 100 terms for 4950 lines, and so on.

It was found eventually that in the case of the hydrogen atom spectrum the terms are given by

$$T_n = \frac{R_H}{n^2} = n = 1,2,3,\ldots \tag{1-131}$$

and the wave numbers of the hydrogen atom spectrum can all be written as

$$\sigma_{n,m} = R_H\left(\frac{1}{n^2} - \frac{1}{m^2}\right) \qquad n<m \tag{1-132}$$

where $n$ and $m$ are both positive integers. Obviously, Balmer's equation (1-129) corresponds to the case where $m=2$. We should summarize that the most remarkable features of the hydrogen spectrum are (1) the occurrence of sharp discrete lines and (2) the representation by integer numbers.

Apart from the problem in interpreting the form of the atomic spectra, there was a major problem associated with the Rutherford atomic model. In the previous section we described the motion of a particle in a Coulombic force field, but we should realize that this model is inadequate to represent the motion of the electrons around a nucleus according to the Rutherford atomic model. If the motion of the electrons around a nucleus is described according to classical mechanics and according to Maxwell electrodynamic theory, then the motion of the electrons is accompanied by the emission of radiation. This emission of radiation causes a lowering of the potential energy of the system or a contraction. In other words, the theoretical application of classical mechanics in combination with Maxwell electromagnetic theory leads to the prediction of an unstable system.

A well-known theorem in electromagnetic theory states that an assembly of point charges has no stable configuration according to classical theory. In order to avoid this difficulty Bohr postulated first the existence of stationary states or stationary orbits. By definition, if an electron is in a stationary orbit, it does not radiate energy as long as it stays in the same stationary orbit.

Bohr derived the nature of the stationary orbits by making use of Planck's quantization concept. If a hydrogen atom is considered an "atomic oscillator," then it is possible to draw an analogy with the harmonic oscillator. Bohr assumed that the atomic oscillator is capable only of emitting integer amounts of a basic energy quantum $h\nu$. Bohr assumed further that the frequency $\nu_q$ of the energy quantum is half the frequency of revolution $\nu_r$ of the classical motion. The two assumptions can be combined to the equation

$$-E_n = nh\nu_q = \frac{nh\nu_r}{2} = \frac{nh}{2T} \tag{1-133}$$

This equation can be combined with the classical equation (1-128) which relates the energy $E$ to the time of revolution $T$:

$$(-2\mu E)^{3/2}=2\pi\mu^2 e^2 \nu_r \tag{1-134}$$

The energy of the atomic oscillator must be consistent with the quantum condition (1-133). In order to determine its value we eliminate the revolution frequency $\nu_r$ from the two equations. We obtain

$$\frac{(-2\mu E_n)^{3/2}}{2\pi\mu^2 e^2}=\frac{-2E_n}{nh} \tag{1-135}$$

or

$$E_n=\frac{-2\pi^2\mu e^4}{n^2h^2} \tag{1-136}$$

It is found that only certain discrete values of the energy are consistent with Bohr's quantization condition (1-133) and it follows that only those orbits that correspond to the discrete energy values $E_n$ of Eq. (1-136) are allowed.

A spectroscopic transition occurs when the atomic oscillator passes from one stationary state $E(n_1)$ to a different stationary state $E(n_2)$. The amount of energy that is emitted in such a transition is given by

$$\Delta E=E(n_1)-E(n_2)=\frac{-2\pi^2\mu e^4}{h^2}\left(\frac{1}{n_1^2}-\frac{1}{n_2^2}\right) \tag{1-137}$$

The corresponding frequency of the transition is

$$\nu=\frac{\Delta E}{h}=\frac{2\pi^2\mu e^4}{h^3}\left(\frac{1}{n_2^2}-\frac{1}{n_1^2}\right) \tag{1-138}$$

where $n_1$ and $n_2$ are positive integers.

The essential part of Bohr's theory deals with the postulate on stationary states and with the assumption that any spectroscopic transition involves two stationary states. The most arbitrary part of Bohr's theory is the quantization condition (1-133). However, this quantization rule is the part of the theory that leads to quantitative predictions about the hydrogen spectrum. It is ironic that the general acceptance of Bohr's theory was mainly due to these quantitative predictions, which led to a calculation of the Rydberg constant and to predictions of the spectral lines of the helium positive ion.

It should be noted that Paul Ehrenfest had proposed earlier that in an atomic system the angular momentum $M$ should be quantized. It was noted by Bohr that his quantization condition could be introduced also as a quantization condition for the atomic angular momentum:

$$M=n\left(\frac{h}{2\pi}\right) \tag{1-139}$$

consistent with Ehrenfest's assumption. In the case of atomic hydrogen Eq. (1-139) reduces to

$$\mu vr=n\left(\frac{h}{2\pi}\right) \tag{1-140}$$

which is equivalent with Bohr's quantum rule (1-133). A more general formulation of the quantum rule in terms of the phase integral was subsequently presented by Arnold Sommerfeld.

In the following years Bohr added another postulate to his general theory, namely the correspondence principle. In simple terms this principle requires that for large quantum numbers the results of quantum theory should approach those of classical mechanics.

Bohr returned to Copenhagen in 1916, and in 1920 he became director of a newly founded Institute for Theoretical Physics. This institute became one of the centers for the study of the old quantum theory. Important contributions were made, not just by Bohr himself but also by various collaborators who were associated with the institute. In particular we should mention the Dutch physicist Hendrik A. Kramers, who joined Bohr in 1919 and who became Bohr's first assistant, and the German Werner K. Heisenberg, who joined the institute in 1923 and who worked primarily with Kramers.

Looking back, the use of the old quantum theory was based as much on intuition as on rigid rules. Many problems were first solved classically, then the classical solutions were quantized in some fashion, and finally the results were verified by means of the correspondence principle. In most cases this procedure led to the correct answers, but often this was due more to the physical intuition of the practitioners than to the logical consistency of the method.

It became more and more clear that there was a need for a new logical foundation for quantum mechanics from which the various quantization rules could be derived. The discovery of these new fundamental theories started around the year 1924. We discuss three different developments, first the derivation of matrix mechanics by Heisenberg during 1924–1925 in Chapter 2, then in subsequent chapters the work by de Broglie on the wave-particle dualism, and, finally, the discovery of the Schrödinger equation. We should note that Max Born made important contributions to all of these developments and we discuss his work also.

As we mentioned before, we will discuss various topics in mathematics together with the corresponding areas of quantum mechanics. According to this plan the following chapter contains a discussion of the theory of matrices and determinants together with Heisenberg's matrix mechanics. In the subsequent chapter we discuss Fourier analysis and the theory of wave mechanics together with the work of de Broglie. We feel that this interplay of mathematics and physics makes the subject more interesting to the reader.

## Problems

**1** Prove that $\mathbf{u}\cdot(\mathbf{v}\times\mathbf{w})=\mathbf{v}\cdot(\mathbf{w}\times\mathbf{u})=\mathbf{w}\cdot(\mathbf{u}\times\mathbf{v})$.

**2** Express the product $\mathbf{u}\times(\mathbf{v}\times\mathbf{w})$ in terms of scalar products only.

**3** By using the results of Problems 1 and 2, express $(\mathbf{a}\times\mathbf{b})\cdot(\mathbf{c}\times\mathbf{d})$ in terms of scalar products only.

**4** Calculate the products $\mathbf{a}\times(\mathbf{b}\times\mathbf{c})$ and $\mathbf{a}\cdot(\mathbf{b}\times\mathbf{c})$ for $\mathbf{a}=(1,3,2)$, $\mathbf{b}=(2,-1,1)$, and $\mathbf{c}=(-1,2,1)$.

**5** In an astronomy observatory it is found that the position of a star $X$ moves through an arc of 52′20″ (52 minutes, 20 seconds) during a time interval of 1 hour. It is known that at that time the star is at a distance of $1.08\times10^9$ km away from earth. Six months later the same star is observed to move through an arc of 2′37″ (2 minutes, 37 seconds) during a time interval of 48 minutes. What is the distance between that star and earth at the time of the second observation?

**6** Consider a function $F(q,p)$ of the generalized coordinates $q$ and momenta $p$ belonging to a system with the Hamiltonian $H(q,p)$. Prove that

$$\frac{dF}{dt}=\sum_i\left(\frac{\partial F}{\partial q_i}\frac{\partial H}{\partial p_i}-\frac{\partial F}{\partial p_i}\frac{\partial H}{\partial q_i}\right)$$

**7** Determine for the motion of an electron in the hydrogen atom with Hamiltonian

$$H=\left(\frac{p^2}{2m}\right)-\left(\frac{e^2}{r}\right)$$

the time derivative $(dF/dt)$ of the quantity

$$F=\left(p^2+r^2\right)^{1/2}=\left(p_x^2+p_y^2+p_z^2+x^2+y^2+z^2\right)^{1/2}$$

**8** In the classical description of the hydrogen atom motion the orbit becomes a parabola when the energy $E$ is equal to zero. Derive the time dependence $r(t)$ and $\varphi(t)$ for the case $E=0$ and prove that the orbit is a parabola.

**9** The spectra of H and $He^+$ can both be represented as $\nu_{n,m}=(Z_X^2R_X)(n^{-2}-m^{-2})$, where $Z$ is the nuclear charge (atomic number) and $R_X$ is the Rydberg constant of the corresponding nucleus. If the mass ratio between the H and He nuclei is 1.0080 : 4.0021 and the ratio between the electron and proton masses is 1 : 1836.11, determine the Rydberg constant for $He^+$ from $R_H = 109,677.581 \text{ cm}^{-1}$.

**10** If $R_H = 109,677.581 \text{ cm}^{-1}$, which spectral lines of atomic hydrogen are in the visible part of the spectrum? We take the latter as the region with wavelengths between 4000 and 7000 $A$.

**11** By assuming that a harmonic oscillator of frequency $\nu$ can have stationary states with energies $nh\nu$ $(n=0,1,2,3,4,\ldots)$ we can calculate the average energy $\varepsilon(T)$ of the harmonic oscillator by taking a Boltzmann distribution over the stationary states. Show that this approach leads to Einstein's expressions (1-81) and (1-82) for the specific heat of a solid.

## Recommended Reading

Chapter 1 deals with the early history of quantum mechanics and with some classical mechanics. We list the standard text on the history of quantum mechanics, a few biographies of Niels Bohr, and a few textbooks on classical mechanics.

Max Jammer, *The Conceptual Development of Quantum Mechanics*, McGraw-Hill Book Company, New York (1966).

D. ter Haar, *The Old Quantum Theory*, Pergamon Press, Oxford (1967).

R. Moore, *Niels Bohr: The Man, His Science and the World They Changed*, Alfred A. Knopf, New York (1966).

S. Rozental, *Niels Bohr, His Life and Work as Seen by His Friends and Colleagues*, North Holland, Amsterdam (1967).

H. Goldstein, *Classical Mechanics*, Addison-Wesley Publishing Co., Reading, Mass. (1950).

L. D. Landau and E. M. Lifshitz, *Mechanics*, Pergamon Press, Oxford (1960).

CHAPTER TWO

# Matrix Mechanics, Matrices, and Determinants

## 2-1 Introduction

In the present chapter we continue our history of the discovery of quantum mechanics by discussing the development of matrix mechanics and of the uncertainty relations. According to the program that we announced in the previous chapter, we discuss also the relevant mathematics; in the present chapter this is the theory of matrices and determinants. The subsequent historical developments in quantum mechanics, such as the discovery of wave mechanics, wave-particle dualism, and the Schrödinger equation, will be presented in Chapters 3 and 4.

We should realize that mathematics is such a widespread discipline that it is hardly possible for any single person to have a general understanding of all branches of mathematics. There are many specialized areas in mathematics that are known only to a handful of experts. The theory of matrices and determinants was such an area until it became of vital importance in quantum mechanics around 1925. Presently, the theory of linear equations, matrices and determinants, or linear algebra is a standard part of the physics curriculum. Many of our readers are probably familiar with the subject. Nevertheless, it is our policy to discuss any part of mathematics that we use in our discussion of quantum mechanics, and therefore we present a brief but fairly complete account of linear algebra. At the end of the chapter we discuss the discovery of matrix mechanics, which was primarily the work of the German physicist Werner Karl Heisenberg.

A knowledge of linear algebra is necessary for the understanding of many areas in quantum mechanics. For example, the description of permutations in the following section is essential for the formulation of the exclusion principle, the properties of matrices and determinants constitute a basis for an understanding of perturbation theory and variational methods, the properties of Hermitian matrices relate to the fundamentals of quantum mechanics, and so on. We feel that a discussion of linear algebra is an essential part of quantum mechanics. The connection between Heisenberg's work and the theory of matrices offers some justification for presenting linear algebra in the present chapter, but we need the results of linear algebra in many other chapters in this book as well.

We discuss first the description of permutations, next the definitions and properties of matrices, and then the various properties of determinants. Finally, we present the properties of linear equations and the diagonalization of matrices. At the end of the chapter we describe the work by Heisenberg and others on the development of matrix mechanics and on the uncertainty principle.

## 2-2 Permutations

Let us consider the set of numbers $(1,2,3,4,\ldots,N)$ arranged in a monotonic increasing sequence. Any other sequence, for example $(n_1,n_2,n_3,\ldots,n_N)$, of these numbers is called a permutation of the original sequence. The operator that leads to the change from the original sequence $(1,2,3,\ldots,N)$ to the permuted sequence $(n_1,n_2,\ldots,n_N)$ is called a permutation operator $P$, and we write

$$(n_1,n_2,n_3,\ldots,n_N)=P(1,2,3,4,\ldots,N) \tag{2-1}$$

The specific nature of $P$ is determined by the two sequences that occur on the left and right. In the case that these two sequences are identical, $P$ is called the identity permutation $P_o$.

A permutation operator can be described by means of its cycles. Let us illustrate this by writing a permuted sequence of nine numbers under its original sequence:

$$\begin{matrix} 1 & 2 & 3 & 4 & 5 & 6 & 7 & 8 & 9 \\ 8 & 3 & 1 & 7 & 2 & 6 & 9 & 5 & 4 \end{matrix} \tag{2-2}$$

In order to describe this permutation, we start with the number 1 of the upper sequence, and we note that in the lower sequence its place has been taken by the number 8. We now look at this number 8 in the upper sequence, and we find number 5 below it. Under number 5 we find number 2; under number 2 we find number 3; under number 3, we find our starting point, number 1. We write this succession of replacements as the cycle $(1,8,5,2,3)$. The first number that does not occur in this cycle is 4, and we take this as the starting point of a second cycle, which is $(4,7,9)$. Finally, we have number 6, which has remained in the same position. We denote this cycle by $(6)$. The permutation (2-2) can thus be represented as $(1,8,5,2,3)$ $(4,7,9)$ $(6)$, that is, as three cycles of five, three, and one elements, respectively.

Each permutation can also be obtained by a succession of pairwise interchanges. As an example we show in (2-3) how the lower sequence of (2-2) is obtained from the upper sequence in this way. On each line we have underlined the two elements that we have interchanged to obtain the subsequent line:

$$\begin{array}{ccccccccc}
\underline{1} & 2 & 3 & 4 & 5 & 6 & 7 & \underline{8} & 9 \\
8 & 2 & 3 & 4 & \underline{5} & 6 & 7 & \underline{1} & 9 \\
8 & \underline{2} & 3 & 4 & \underline{1} & 6 & 7 & 5 & 9 \\
8 & \underline{1} & \underline{3} & 4 & 2 & 6 & 7 & 5 & 9 \\
8 & 3 & 1 & \underline{4} & 2 & 6 & \underline{7} & 5 & 9 \\
8 & 3 & 1 & 7 & 2 & 6 & \underline{4} & 5 & \underline{9} \\
8 & 3 & 1 & 7 & 2 & 6 & 9 & 5 & 4
\end{array} \tag{2-3}$$

Obviously we could also have obtained this result by means of a different succession of pairwise interchanges. In each case the total number of interchanges can vary but its parity cannot; for example, the permutation (2-2) can be obtained only by an even number of interchanges. It is therefore possible to use this number of interchanges as a criterion for the division of permutations into even and odd permutations. If a permutation can be achieved by an even number of pairwise interchanges, it is called even, and if it is achieved by an odd number of interchanges, it is called odd. In this connection we introduce the symbol $\delta_P$, which is equal to plus unity when $P$ is even and minus unity when $P$ is odd. It is easily shown that for a permutation $P$ of $N$ elements that can be represented by $C$ cycles, $\delta_P$ is given by

$$\delta_P = (-1)^{N+C} \tag{2-4}$$

In this discussion we have taken a sequence of numbers as the basis for our argument, but we can consider permutations of any other set of quantities. In general, we speak of the elements of a permutation. Often we consider permutations of a set of variables $q_1, q_2, \ldots, q_N$ of a function $\Psi$. As an example we take a function $\Psi$ of four variables $q_1$, $q_2$, $q_3$, and $q_4$. A permutation of the variables, in general, leads to a different function. For instance, if $P$ is described by

$$P = (13)(24) \tag{2-5}$$

then

$$P\Psi(q_1, q_2, q_3, q_4) = \Psi(q_3, q_4, q_1, q_2) \tag{2-6}$$

Any function $\Psi(q_1, q_2, \ldots, q_N)$ that satisfies the condition

$$\Psi(q_1, q_2, \ldots, q_N) = P\Psi(q_1, q_2, \ldots, q_N) \tag{2-7}$$

is called symmetric with respect to permutations. If the function satisfies the condition

$$\Psi(q_1, q_2, \ldots, q_N) = P\delta_P\Psi(q_1, q_2, \ldots, q_N) \tag{2-8}$$

then it is called antisymmetric with respect to permutations. Equation (2-8) means that $\Psi$ remains unchanged if its variables are subjected to an even permutation and that $\Psi$ changes sign if it is subjected to an odd permutation.

An arbitrary function $\Psi(q_1, q_2, \ldots, q_N)$ is, in general, neither symmetric nor antisymmetric with respect to permutations. However, we can construct a symmetric function $\Phi$ from it by taking

$$\Phi = \sum_P P\Psi(q_1, q_2, \ldots, q_N) \tag{2-9}$$

and an antisymmetric function $\Psi$ by

$$\Psi = \sum_P P\delta_P\Psi(q_1, q_2, \ldots, q_N) \tag{2-10}$$

The summations in Eqs. (2-9) and (2-10) are to be taken over all possible permutations, including the identity permutation.

Finally, we derive the total number $S(N)$ of possible permutations for $N$ elements. If there is only one element, then the only possible permutation is the identity permutation, so that

$$S(1) = 1 \tag{2-11}$$

For a system of two elements there are two permutations, namely the identity permutation and the exchange of the two elements, and therefore

$$S(2) = 2 \tag{2-12}$$

The general expression for $S(N)$ can be obtained by deriving a relation between $S(N)$ and $S(N+1)$. Let us consider a particular permutation of $N$ numbers that is described by a particular sequence

$$n_1, n_2, n_3, n_4, \ldots, n_N \tag{2-13}$$

of the numbers. If we now try to fit an additional number $(N+1)$ into this sequence, then we see that there are $(N+1)$ different ways of doing this, and each of these arrangements leads to a different permutation for the $(N+1)$ numbers. We find that an arbitrary permutation of $N$ elements gives $(N+1)$ different permutations for $(N+1)$ elements, and therefore

$$S(N+1) = (N+1)S(N) \tag{2-14}$$

From Eqs. (2-11), (2-12), and (2-14) it is easily found that

$$S(N) = 1 \cdot 2 \cdot 3 \cdot 4 \cdots N = N! \tag{2-15}$$

## 2-3 Matrices

A matrix is defined as a two-dimensional, rectangular array of numbers or functions. It is customarily represented as

$$\begin{bmatrix} a_{1,1} & a_{1,2} & a_{1,3} & \cdots & \cdots & \cdots & a_{1,N} \\ a_{2,1} & a_{2,2} & a_{2,3} & \cdots & \cdots & \cdots & a_{2,N} \\ a_{3,1} & a_{3,2} & a_{3,3} & \cdots & \cdots & \cdots & a_{3,N} \\ \cdots & \cdots & \cdots & \cdots & \cdots & \cdots & \cdots \\ \cdots & \cdots & \cdots & \cdots & \cdots & \cdots & \cdots \\ a_{M,1} & a_{M,2} & a_{M,3} & \cdots & \cdots & \cdots & a_{M,N} \end{bmatrix} \tag{2-16}$$

or briefly as

$$[a_{i,j}] \qquad \begin{matrix} i=1,2,3,\ldots, M \\ j=1,2,3,\ldots, N \end{matrix} \tag{2-17}$$

All the elements that are on the same horizontal line are said to form a row, and all elements that are on the same vertical line are said to form a column. The $m$th row of the matrix (2-16) is composed of the elements $(a_{m,1}, a_{m,2}, a_{m,3},\ldots, a_{m,N})$ and the $n$th column is $(a_{1,n}, a_{2,n}, a_{3,n},\ldots, a_{M,n})$. The sum of two matrices $[a_{i,j}]$ and $[b_{i,j}]$ is defined as

$$[c_{i,j}]=[a_{i,j}]+[b_{i,j}] \tag{2-18}$$

where each element $c_{i,j}$ is the sum of the corresponding two elements $a_{i,j}$ and $b_{i,j}$:

$$c_{i,j}=a_{i,j}+b_{i,j} \tag{2-19}$$

This definition is logical only if the dimensions of the matrices $[a_{i,j}]$ and $[b_{i,j}]$ are the same or, as the mathematicians say, if they are conformable for addition or subtraction. We conclude, therefore, that two matrices can be added or subtracted only if they are conformable for addition or subtraction. It is easily verified from the definition that

$$[a_{i,j}]+[b_{i,j}]=[b_{i,j}]+[a_{i,j}] \tag{2-20}$$

The product matrix $[u_{i,j}]$ of two matrices $[a_{i,j}]$ and $[b_{i,j}]$,

$$[u_{i,j}]=[a_{i,j}]\times[b_{i,j}] \tag{2-21}$$

is defined in such a way that the elements $u_{i,j}$ are given by

$$u_{i,j}=\sum_k a_{i,k}b_{k,j} \tag{2-22}$$

It follows from this definition that $u_{i,j}$ is obtained by taking the $i$th row of the first matrix and the $j$th column of the second matrix, by multiplication of the corresponding elements, and finally by summation. This procedure can be followed only if the rows of the first matrix and the columns of the second matrix are of equal length, that is, if the horizontal dimension of the first matrix and the vertical dimension of the second matrix are equal or if they are conformable for multiplication. An example of the multiplication of two matrices is

$$\begin{bmatrix} 1 & 0 & 3 & 4 \\ 2 & 1 & 0 & 1 \end{bmatrix} \times \begin{bmatrix} 3 & 0 \\ 1 & 2 \\ 0 & 5 \\ 1 & 3 \end{bmatrix} = \begin{bmatrix} 7 & 27 \\ 8 & 5 \end{bmatrix} \tag{2-23}$$

It follows from the definitions (2-21) and (2-22) that, in general,

$$[a_{i,j}] \times [b_{i,j}] \neq [b_{i,j}] \times [a_{i,j}] \tag{2-24}$$

For example, if we change the order of multiplication of the two matrices of Eq. (2-23), we obtain

$$\begin{bmatrix} 3 & 0 \\ 1 & 2 \\ 0 & 5 \\ 1 & 3 \end{bmatrix} \times \begin{bmatrix} 1 & 0 & 3 & 4 \\ 2 & 1 & 0 & 1 \end{bmatrix} = \begin{bmatrix} 3 & 0 & 9 & 12 \\ 5 & 2 & 3 & 6 \\ 10 & 5 & 0 & 5 \\ 7 & 3 & 3 & 7 \end{bmatrix} \tag{2-25}$$

which is quite different from Eq. (2-23).

A vector $\mathbf{v} = (v_x, v_y, v_z)$ can also be written in the form of a matrix. We can do this in two different ways, namely as

$$\mathbf{v} = [v_x \, v_y \, v_z] \tag{2-26}$$

or as

$$\mathbf{v} = \begin{bmatrix} v_x \\ v_y \\ v_z \end{bmatrix} \tag{2-27}$$

The quantities in Eqs. (2-26) and (2-27) are called a row vector and a column vector, respectively. In these definitions we do not have to restrict ourselves to three-dimensional vectors. In general, we have row vectors and column vectors of order $N$. The word "order" is used to describe the dimensions of a matrix, and generally we say that a matrix of $m$ rows and $n$ columns is of order "$m$ by $n$." If the matrix is square and has $n$ rows and $n$ columns, we call it a square matrix of order $n$. The matrix (2-26) is of order "1 by 3," and the matrix (2-27) is of order "3 by 1"; but in the case of row or column vectors this is usually abbreviated as above.

The inner product of two vectors **u** and **v**, each of order $N$, can be written in matrix form as

$$\mathbf{u}\cdot\mathbf{v}=[u_1 u_2 \cdots u_N]\times\begin{bmatrix} v_1 \\ v_2 \\ \cdots \\ \cdots \\ v_N \end{bmatrix} \tag{2-28}$$

Let us now consider some properties and definitions of square matrices. If we multiply a column vector **u** of order $N$ by a square matrix $[a_{i,j}]$ of the same order, we obtain another column vector **v**:

$$\begin{bmatrix} a_{1,1} & a_{1,2} & \cdots & \cdots & a_{1,N} \\ a_{2,1} & a_{2,2} & \cdots & \cdots & a_{2,N} \\ \cdots & \cdots & \cdots & \cdots & \cdots \\ \cdots & \cdots & \cdots & \cdots & \cdots \\ a_{N,1} & a_{N,2} & \cdots & \cdots & a_{N,N} \end{bmatrix}\times\begin{bmatrix} u_1 \\ u_2 \\ \cdots \\ \cdots \\ u_N \end{bmatrix}=\begin{bmatrix} v_1 \\ v_2 \\ \cdots \\ \cdots \\ v_N \end{bmatrix} \tag{2-29}$$

The elements of **v** are given by

$$v_k=\sum_i a_{k,i}u_i \tag{2-30}$$

and we see that Eq. (2-29) represents a linear transformation between two vectors of order $N$.

There are various types of square matrices whose elements have certain properties, and we will define some of these types. The identity matrix is defined as

$$\begin{bmatrix} 1 & 0 & 0 & \cdots & \cdots & 0 \\ 0 & 1 & 0 & \cdots & \cdots & 0 \\ 0 & 0 & 1 & \cdots & \cdots & 0 \\ \cdots & \cdots & \cdots & \cdots & \cdots & \cdots \\ \cdots & \cdots & \cdots & \cdots & \cdots & \cdots \\ 0 & 0 & 0 & \cdots & \cdots & 1 \end{bmatrix} \tag{2-31}$$

and its elements are given by

$$a_{i,j}=\delta_{i,j} \tag{2-32}$$

Here we have introduced the Kronecker symbol $\delta_{i,j}$, which is defined as equal to one if $i=j$ and as equal to zero if $i\neq j$. The elements $a_{i,i}$ are defined as the diagonal of the matrix, and in the identity matrix all diagonal elements are equal to unity and all off-diagonal elements are equal to zero. A diagonal matrix is defined by

$$a_{i,j}=a_i\delta_{i,j} \tag{2-33}$$

Here all off-diagonal elements are again zero, but the diagonal elements can now be different from one another.

A symmetric matrix has elements that satisfy the condition

$$a_{i,j}=a_{j,i} \tag{2-34}$$

If

$$a_{i,j}=-a_{j,i} \tag{2-35}$$

we speak of a skew-symmetric matrix. In this case we have

$$a_{i,i}=-a_{i,i} \tag{2-36}$$

and it follows that the diagonal elements of a skew-symmetric matrix are all zero. The elements of a matrix can, in general, be complex quantities, but if they are all real or imaginary, we have a real or an imaginary matrix, respectively. A very important type of matrix in quantum mechanics is the Hermitian matrix, which is defined by the property

$$a_{i,j}=a^*_{j,i} \tag{2-37}$$

Finally, we have the unitary matrix, which is defined by

$$\sum_k a_{i,k}a^*_{j,k}=\delta_{i,j}$$

$$\sum_k a_{k,i}a^*_{k,j}=\delta_{i,j} \tag{2-38}$$

Here the rows from a mutually orthogonal set of unit vectors, and the same is true of the columns. A coordinate transformation that results from a rotation of the coordinate axes is described by a unitary transformation.

Let us now consider two square matrices $[a_{i,j}]$ and $[b_{i,j}]$ of the same order $N$ and the possible relations between them. If

$$b_{i,j}=a_{j,i} \tag{2-39}$$

we call $[b_{i,j}]$ the transpose of $[a_{i,j}]$. If the matrix $[b_{i,j}]$ is denoted by the symbol **B** and the matrix $[a_{i,j}]$ by the symbol **A**, as is often done in the mathematical literature, we can use the notation

$$\mathbf{B}=\tilde{\mathbf{A}} \tag{2-40}$$

in order to describe **B** as the transpose of **A**. If the elements of **B** are the complex conjugates of the elements of **A**,

$$b_{i,j}=a^*_{i,j} \tag{2-41}$$

we say that the matrix **B** is the complex conjugate of **A**. The notation for this is

$$\mathbf{B}=\mathbf{A}^* \tag{2-42}$$

We call **B** the associate matrix of **A**, or

$$\mathbf{B}=\mathbf{A}^\dagger \tag{2-43}$$

if their elements are related by

$$b_{i,j}=a_{j,i}^* \tag{2-44}$$

It is easily seen that

$$\mathbf{A}^\dagger=(\tilde{\mathbf{A}})^* \tag{2-45}$$

We saw in Eq. (2-24) that, in general, the product of two matrices **A** and **B** depends on the order of multiplication. However, if the matrices satisfy the condition

$$\mathbf{A}\times\mathbf{B}=\mathbf{B}\times\mathbf{A} \tag{2-46}$$

then we say that they commute. Similarly, **A** and **B** anticommute when

$$\mathbf{A}\times\mathbf{B}=-\mathbf{B}\times\mathbf{A} \tag{2-47}$$

If

$$\mathbf{A}\times\mathbf{B}=\mathbf{I} \tag{2-48}$$

where **I** is an identity matrix, then **B** is the inverse of **A**, which is written as

$$\mathbf{B}=(\mathbf{A})^{-1} \tag{2-49}$$

It can be shown that for a unitary matrix **U** we have

$$\mathbf{U}\times\mathbf{U}^\dagger=\mathbf{U}^\dagger\times\mathbf{U}=\mathbf{I} \tag{2-50}$$

If **U** has the elements $u_{i,j}$, then $\mathbf{U}^+$ has the elements $u_{j,i}^*$, and Eq. (2-50) states that

$$\sum_k u_{i,k}u_{j,k}^*=\delta_{i,j} \tag{2-51}$$

$$\sum_k u_{k,i}^*u_{k,j}=\delta_{i,j} \tag{2-52}$$

which is identical to Eq. (2-38).

## 2-4 Determinants

Let us consider an arbitrary square matrix of order $N$:

$$\mathsf{A}=\begin{bmatrix} a_{1,1} & a_{1,2} & a_{1,3} & \cdots & \cdots & a_{1,N} \\ a_{2,1} & a_{2,2} & a_{2,3} & \cdots & \cdots & a_{2,N} \\ a_{3,1} & a_{3,2} & a_{3,3} & \cdots & \cdots & a_{3,N} \\ \cdots & \cdots & \cdots & \cdots & \cdots & \cdots \\ \cdots & \cdots & \cdots & \cdots & \cdots & \cdots \\ a_{N,1} & a_{N,2} & a_{N,3} & \cdots & \cdots & a_{N,N} \end{bmatrix} \tag{2-53}$$

By following a rather complex procedure, which will shortly be outlined, we can derive a number $A$ from this matrix, and this number is known as the determinant corresponding to the matrix. We write this as

$$A=\begin{vmatrix} a_{1,1} & a_{1,2} & \cdots & \cdots & a_{1,N} \\ a_{2,1} & a_{2,2} & \cdots & \cdots & a_{2,N} \\ \cdots & \cdots & \cdots & \cdots & \cdots \\ \cdots & \cdots & \cdots & \cdots & \cdots \\ a_{N,1} & a_{N,2} & \cdots & \cdots & a_{N,N} \end{vmatrix} \tag{2-54}$$

Often we do not take the trouble to evaluate the number $A$; instead we discuss its features from the array of numbers (2-54) by making use of the properties of determinants. We should remember that a determinant is always a single number even when it is represented as an array of numbers.

In order to evaluate $A$, we first select one element of each row of the matrix in such a way that each column is represented by only one element. A group of such elements is

$$(a_{1,n_1}; a_{2,n_2}; a_{3,n_3}; \ldots; a_{N,n_N}) \tag{2-55}$$

The second indices of the elements $(n_1, n_2, n_3, \ldots, n_N)$ are one of the permutations of the numbers $(1,2,3,\ldots, N)$. All possible groups of the type of Eq. (2-55) are obtained by taking all possible permutations of the $N$ numbers; their number is $N!$ according to Eq. (2-15). The determinant $A$ is now defined as

$$A=\sum_P \delta_P P(n_1, n_2, \ldots, n_N) a_{1,n_1} a_{2,n_2} a_{3,n_3} \cdots a_{N,n_N} \tag{2-56}$$

where the permutations have to be taken with respect to the natural order $(1,2,3,\ldots, N)$. This means that we have to take all possible groups of the type of Eq. (2-55), and in each group we take the product of the elements and the corresponding $\delta_P$ (plus unity when $P$ is even and minus unity when $P$ is odd) and, finally, add all $N!$ terms.

According to this definition the determinants of order two and three are obtained as

$$\begin{vmatrix} a_{1,1} & a_{1,2} \\ a_{2,1} & a_{2,2} \end{vmatrix} = a_{1,1}a_{2,2} - a_{1,2}a_{2,1} \tag{2-57}$$

and

$$\begin{vmatrix} a_{1,1} & a_{1,2} & a_{1,3} \\ a_{2,1} & a_{2,2} & a_{2,3} \\ a_{3,1} & a_{3,2} & a_{3,3} \end{vmatrix} = \begin{aligned} & a_{1,1}a_{2,2}a_{3,3} + a_{1,2}a_{2,3}a_{3,1} \\ & + a_{1,3}a_{2,1}a_{3,2} - a_{1,3}a_{2,2}a_{3,1} \\ & - a_{1,1}a_{2,3}a_{3,2} - a_{1,2}a_{2,1}a_{3,3} \end{aligned} \tag{2-58}$$

A determinant of order four has 24 terms, and it becomes impractical to evaluate it directly. We will see that there are less laborious methods for evaluating determinants of higher order.

It can be derived from the definition (2-56) that the exchange of two rows in Eq. (2-54) leads to a determinant value of $-A$. The exchange leads to the same group of products as in Eq. (2-56), but each $\delta_P$ changes sign as a result of the exchange, hence the minus sign. Since the definition of the determinant is symmetric in the columns and rows, we also obtain $-A$ if we exchange two columns. It follows, therefore, that a determinant is zero when two of its rows or two of its columns are identical. In addition, we have seen that a determinant remains unchanged if we transpose the matrix from which it is derived.

Let us now derive some properties of determinants that are helpful in their evaluation. We start by considering the determiniant

$$A = \begin{vmatrix} a_{1,1} & a_{1,2} & \cdots & a_{1,k} & \cdots & a_{1,N} \\ a_{2,1} & a_{2,2} & \cdots & a_{2,k} & \cdots & a_{2,N} \\ a_{3,1} & a_{3,2} & \cdots & a_{3,k} & \cdots & a_{3,N} \\ \cdots & \cdots & \cdots & \cdots & \cdots & \cdots \\ \cdots & \cdots & \cdots & \cdots & \cdots & \cdots \\ a_{N,1} & a_{N,2} & \cdots & a_{N,k} & \cdots & a_{N,N} \end{vmatrix} \tag{2-59}$$

where $k$ is an arbitrary column. It is easily seen that

$$\lambda A = \begin{vmatrix} a_{1,1} & a_{1,2} & \cdots & \lambda a_{1,k} & \cdots & a_{1,N} \\ a_{2,1} & a_{2,2} & \cdots & \lambda a_{2,k} & \cdots & a_{2,N} \\ a_{3,1} & a_{3,2} & \cdots & \lambda a_{3,k} & \cdots & a_{3,N} \\ \cdots & \cdots & \cdots & \cdots & \cdots & \cdots \\ \cdots & \cdots & \cdots & \cdots & \cdots & \cdots \\ a_{N,1} & a_{N,2} & \cdots & \lambda a_{N,k} & \cdots & a_{N,N} \end{vmatrix} \tag{2-60}$$

since each product in Eq. (2-56) acquires a factor $\lambda$. It follows, therefore, that a determinant is not only zero when two of its columns or rows are identical but also when their elements are proportional to each other.

Let us now consider a second determinant $A'$:

$$A' = \begin{vmatrix} a_{1,1} & a_{1,2} & \cdots & a'_{1,k} & \cdots & a_{1,N} \\ a_{2,1} & a_{2,2} & \cdots & a'_{2,k} & \cdots & a_{2,N} \\ a_{3,1} & a_{3,2} & \cdots & a'_{3,k} & \cdots & a_{3,N} \\ \cdots & \cdots & \cdots & \cdots & \cdots & \cdots \\ \cdots & \cdots & \cdots & \cdots & \cdots & \cdots \\ a_{N,1} & a_{N,2} & \cdots & a'_{N,k} & \cdots & a_{N,N} \end{vmatrix} \tag{2-61}$$

which differs from $A$ only in its $k$th column. It again follows from the definition (2-56) that

$$A+\lambda A' = \begin{vmatrix} a_{1,1} & a_{1,2} & \cdots & (a_{1,k}+\lambda a'_{1,k}) & \cdots & a_{1,N} \\ a_{2,1} & a_{2,2} & \cdots & (a_{2,k}+\lambda a'_{2,k}) & \cdots & a_{2,N} \\ a_{3,1} & a_{3,2} & \cdots & (a_{3,k}+\lambda a'_{3,k}) & \cdots & a_{3,N} \\ \cdots & \cdots & \cdots & \cdots & \cdots & \cdots \\ \cdots & \cdots & \cdots & \cdots & \cdots & \cdots \\ a_{N,1} & a_{N,2} & \cdots & (a_{N,k}+\lambda a'_{N,k}) & \cdots & a_{N,N} \end{vmatrix} \tag{2-62}$$

We now take the column $a'_{i,k}$ equal to another column $a_{i,l}$. Then $A'=0$, since two of its columns are identical. Consequently,

$$A = \begin{vmatrix} a_{1,1} & a_{1,2} & \cdots & a_{1,k}+\lambda a_{1,l} & \cdots & a_{1,N} \\ a_{2,1} & a_{2,2} & \cdots & a_{2,k}+\lambda a_{2,l} & \cdots & a_{2,N} \\ a_{3,1} & a_{3,2} & \cdots & a_{3,k}+\lambda a_{3,l} & \cdots & a_{3,N} \\ \cdots & \cdots & \cdots & \cdots & \cdots & \cdots \\ \cdots & \cdots & \cdots & \cdots & \cdots & \cdots \\ a_{N,1} & a_{N,2} & \cdots & a_{N,k}+\lambda a_{N,l} & \cdots & a_{N,N} \end{vmatrix} \tag{2-63}$$

We see that a determinant remains unchanged if we add one of its columns, multiplied by an arbitrary parameter $\lambda$, to another column. Since this procedure can be repeated a number of times, we conclude that a determinant remains unchanged if we add an arbitrary linear combination of a number of columns or rows to another column or row, respectively.

Let us next consider the expansion of a determinant along one of its rows and columns. For this purpose we first define the minors and cofactors of the elements of the determinant (2-53). The minor of the element $a_{i,j}$ of $A$ is defined as the determinant that is obtained by deleting the $i$th row and $j$th column of $A$. The cofactor of $a_{i,j}$, which we denote by $A_{i,j}$, is the minor of $a_{i,j}$ multiplied by $(-1)^{i+j}$. It can now be shown that

$$A = \sum_j a_{i,j}A_{i,j} = \sum_i a_{i,j}A_{i,j} \tag{2-64}$$

This is known as the expansion of the determinant along its $i$th row or along its $j$th column, respectively.

In order to prove Eq. (2-64), we start from the definition (2-56), and we first consider the expansion of $A$ along the first row. We select all terms that contain the element $a_{1,1}$, and we note that in each of these terms we have chosen an element from the first column and the first row so that none of the other elements can belong to the first row and the first column. According to the definition (2-56) the sum of the terms containing $a_{1,1}$ can be represented as

$$a_{1,1}\sum_{P}\delta_P P(n_2, n_3, \ldots, n_N) a_{2,n_2} a_{3,n_3} \cdots a_{N,n_N} = a_{1,1}A_{1,1} \qquad (2\text{-}65)$$

Next we consider all terms that contain the element $a_{1,j}$. We can evaluate their sum by moving the $j$th column of the determinant to the left side of the determinant, since we can then use Eq. (2-65). This move is achieved by first exchanging column $j$ with column $(j-1)$, then exchanging column $(j-1)$ with column $(j-2)$, and so on, until the original column $j$ has become the first column. As a result of these interchanges the determinant should be multiplied by a factor $(-1)^{j-1}$. If we now apply Eq. (2-65) to this new determinant, we find that for the original determinant the sum of the terms that contains $a_{1,j}$ is given by

$$a_{1,j}(-)^{j-1}\sum_{P}\delta_P P(n_1, \ldots, n_{j-1}, n_{j+1}, \ldots, n_N) a_{n_1 1} \cdots$$

$$a_{n_{j-1}, j-1} a_{n_{j+1}, j+1} \cdots a_{n_N, N} = a_{1,j}A_{1,j} \qquad (2\text{-}66)$$

The value of $A$ is now obtained by summing Eq. (2-66) over all values of $j$:

$$A = \sum_{j=1}^{N} a_{1,j}A_{1,j} \qquad (2\text{-}67)$$

The expansion along the $i$th row can be described if we move the $i$th row to the top of the determinant by first exchanging rows $i$ and $i-1$, then exchanging rows $i-1$ and $i-2$, and so on. We then find from Eq. (2-66) that

$$(-1)^{i-1}A = \sum_{j=1}^{N} a_{i,j}(-1)^{i-1}A_{i,j} \qquad (2\text{-}68)$$

or

$$A = \sum_{j} a_{i,j}A_{i,j} \qquad (2\text{-}69)$$

The second Eq. (2-64) is easily derived from Eq. (2-69) if we take the transpose of the matrix **A**, since this leaves the determinant unchanged.

We now consider the determinant

$$\begin{vmatrix} a_{1,1} & a_{1,2} & \cdots & \cdots & a_{1,j-1} & a_{1,k} & a_{1,j+1} & \cdots & a_{1,N} \\ a_{2,1} & a_{2,2} & \cdots & \cdots & a_{2,j-1} & a_{2,k} & a_{2,j+1} & \cdots & a_{2,N} \\ a_{3,1} & a_{3,2} & \cdots & \cdots & a_{3,j-1} & a_{3,k} & a_{3,j+1} & \cdots & a_{3,N} \\ \cdots & \cdots & \cdots & \cdots & \cdots & \cdots & \cdots & \cdots & \cdots \\ \cdots & \cdots & \cdots & \cdots & \cdots & \cdots & \cdots & \cdots & \cdots \\ a_{N,1} & a_{N,2} & \cdots & \cdots & a_{N,j-1} & a_{N,k} & a_{N,j+1} & \cdots & a_{N,N} \end{vmatrix} = 0 \tag{2-70}$$

which is obtained from Eq. (2-53) by replacing column $j$ by one of the other columns $k$. This determinant (2-70) is equal to zero, since two of its columns are identical. If we expand along the $j$th column, we get

$$\sum_i a_{i,k} A_{i,j} = 0 \qquad j \neq k \tag{2-71}$$

where the cofactors $A_{i,j}$ are identical to the cofactors of the determinant (2-53). Combination of Eqs. (2-64) and (2-71) yields, finally,

$$\sum_i a_{i,j} A_{i,k} = A\delta_{j,k}$$

$$\sum_i a_{j,i} A_{k,i} = A\delta_{j,k} \tag{2-72}$$

We can use Eq. (2-72) to construct the inverse of a square matrix $[a_{i,j}]$. We define the matrix $[b_{i,j}]$ by means of

$$b_{i,j} = \frac{A_{j,i}}{A} \tag{2-73}$$

Since the elements of the product matrix

$$[c_{i,j}] = [a_{i,j}] \times [b_{i,j}] \tag{2-74}$$

are

$$c_{i,j} = \sum_k a_{i,k} b_{k,j} = \frac{1}{A} \sum_k a_{i,k} A_{j,k} = \delta_{i,j} \tag{2-75}$$

it follows that **B** is the inverse of **A**.

It may be useful to illustrate the evaluation of some simple determinants by making use of the properties above. The determinant

$$D_5 = \begin{vmatrix} x & 1 & 0 & 0 & 0 \\ 1 & x & 1 & 0 & 0 \\ 0 & 1 & x & 1 & 0 \\ 0 & 0 & 1 & x & 1 \\ 0 & 0 & 0 & 1 & x \end{vmatrix} \tag{2-76}$$

can be expanded along its first row:

$$D_5 = x\begin{vmatrix} x & 1 & 0 & 0 \\ 1 & x & 1 & 0 \\ 0 & 1 & x & 1 \\ 0 & 0 & 1 & x \end{vmatrix} - \begin{vmatrix} 1 & 1 & 0 & 0 \\ 0 & x & 1 & 0 \\ 0 & 1 & x & 1 \\ 0 & 0 & 1 & x \end{vmatrix} \tag{2-77}$$

Expansion of the two determinants of Eq. (2-77) along their first column gives

$$D_5 = x^2\begin{vmatrix} x & 1 & 0 \\ 1 & x & 1 \\ 0 & 1 & x \end{vmatrix} - x\begin{vmatrix} 1 & 0 & 0 \\ 1 & x & 1 \\ 0 & 1 & x \end{vmatrix} - \begin{vmatrix} x & 1 & 0 \\ 1 & x & 1 \\ 0 & 1 & x \end{vmatrix} \tag{2-78}$$

From Eq. (2-58) we find

$$\begin{vmatrix} x & 1 & 0 \\ 1 & x & 1 \\ 0 & 1 & x \end{vmatrix} = x^3 - 2x \qquad \begin{vmatrix} 1 & 0 & 0 \\ 1 & x & 1 \\ 0 & 1 & x \end{vmatrix} = x^2 - 1 \tag{2-79}$$

Hence we have

$$D_5 = (x^2-1)(x^3-2x) - x(x^2-1) = x(x^2-1)(x^2-3) \tag{2-80}$$

The following is an example of the evaluation of a determinant by adding rows and columns and subsequent expansion:

$$\begin{vmatrix} 1 & 2 & 3 & 4 & 0 \\ 0 & 1 & 3 & -1 & 2 \\ -1 & 4 & 3 & 2 & 1 \\ 0 & 0 & 2 & 1 & 1 \\ 1 & 2 & 3 & 4 & 1 \end{vmatrix} = \begin{vmatrix} 1 & 2 & 3 & 4 & 0 \\ 0 & 1 & 3 & -1 & 2 \\ 0 & 6 & 6 & 6 & 1 \\ 0 & 0 & 2 & 1 & 1 \\ 0 & 0 & 0 & 0 & 1 \end{vmatrix}$$

$$= \begin{vmatrix} 1 & 3 & -1 & 2 \\ 6 & 6 & 6 & 1 \\ 0 & 2 & 1 & 1 \\ 0 & 0 & 0 & 1 \end{vmatrix} = \begin{vmatrix} 1 & 3 & -1 \\ 6 & 6 & 6 \\ 0 & 2 & 1 \end{vmatrix} = \begin{vmatrix} 1 & 2 & -2 \\ 6 & 0 & 0 \\ 0 & 2 & 1 \end{vmatrix}$$

$$= -6\begin{vmatrix} 2 & -2 \\ 2 & 1 \end{vmatrix} = -36 \tag{2-81}$$

Finally, we wish to show that the determinant of a product matrix is equal to the product of the determinants of the original matrices. We consider two matrices $[a_{i,j}]$ and $[b_{i,j}]$, both of order $N$, and their product matrix $[c_{i,j}]$, which is defined by

$$c_{i,j} = \sum_k a_{i,k} b_{k,j} \tag{2-82}$$

The corresponding determinants are $A$, $B$, and $C$, respectively. From the definition of a determinant it follows that

$$\begin{vmatrix} a_{1,1} & a_{1,2} & \cdots & \cdots & a_{1,N} & 0 & 0 & \cdots & \cdots & 0 \\ a_{2,1} & a_{2,2} & \cdots & \cdots & a_{2,N} & 0 & 0 & \cdots & \cdots & 0 \\ \cdots & \cdots & \cdots & \cdots & \cdots & \cdots & \cdots & \cdots & \cdots & \cdots \\ \cdots & \cdots & \cdots & \cdots & \cdots & \cdots & \cdots & \cdots & \cdots & \cdots \\ a_{N,1} & a_{N,2} & \cdots & \cdots & a_{N,N} & 0 & 0 & \cdots & \cdots & 0 \\ -1 & 0 & \cdots & \cdots & 0 & b_{1,1} & b_{1,2} & \cdots & \cdots & b_{1,N} \\ 0 & -1 & \cdots & \cdots & 0 & b_{2,1} & b_{2,2} & \cdots & \cdots & b_{2,N} \\ \cdots & \cdots & \cdots & \cdots & \cdots & \cdots & \cdots & \cdots & \cdots & \cdots \\ \cdots & \cdots & \cdots & \cdots & \cdots & \cdots & \cdots & \cdots & \cdots & \cdots \\ 0 & 0 & \cdots & \cdots & -1 & b_{N,1} & b_{N,2} & \cdots & \cdots & b_{N,N} \end{vmatrix} = AB \tag{2-83}$$

We add the first column, multiplied by a factor $b_{1,1}$, to the $(N+1)$st column; then we add the second column, multiplied by a factor $b_{2,1}$, to the $(N+1)$st column, and so on. In general, we add the $j$th column $(1 \leqslant j \leqslant N)$, multiplied by a factor $b_{j,1}$, to the $(N+1)$st column. We obtain

$$\begin{vmatrix} a_{1,1} & a_{1,2} & \cdots & \cdots & a_{1,N} & c_{1,1} & 0 & \cdots & \cdots & 0 \\ a_{2,1} & a_{2,2} & \cdots & \cdots & a_{2,N} & c_{2,1} & 0 & \cdots & \cdots & 0 \\ \cdots & \cdots & \cdots & \cdots & \cdots & \cdots & \cdots & \cdots & \cdots & \cdots \\ \cdots & \cdots & \cdots & \cdots & \cdots & \cdots & \cdots & \cdots & \cdots & \cdots \\ a_{N,1} & a_{N,2} & \cdots & \cdots & a_{N,N} & c_{N,1} & 0 & \cdots & \cdots & 0 \\ -1 & 0 & \cdots & \cdots & 0 & 0 & b_{1,2} & \cdots & \cdots & b_{1,N} \\ 0 & -1 & \cdots & \cdots & 0 & 0 & b_{2,2} & \cdots & \cdots & b_{2,N} \\ \cdots & \cdots & \cdots & \cdots & \cdots & \cdots & \cdots & \cdots & \cdots & \cdots \\ \cdots & \cdots & \cdots & \cdots & \cdots & \cdots & \cdots & \cdots & \cdots & \cdots \\ 0 & 0 & \cdots & \cdots & -1 & 0 & b_{N,2} & \cdots & \cdots & b_{N,N} \end{vmatrix} = AB \tag{2-84}$$

We treat all columns $N+1, N+2, N+3, \ldots, 2N$ in this way. For example, we add the first column, multiplied by a factor $b_{1,k}$, to column $N+k$, the second column, multiplied by a factor $b_{2,k}$, to column $N+k$, and so on. This procedure gives

$$\begin{vmatrix} a_{1,1} & a_{1,2} & \cdots & \cdots & a_{1,N} & c_{1,1} & c_{1,2} & \cdots & \cdots & c_{1,N} \\ a_{2,1} & a_{2,2} & \cdots & \cdots & a_{2,N} & c_{2,1} & c_{2,2} & \cdots & \cdots & c_{2,N} \\ \cdots & \cdots & \cdots & \cdots & \cdots & \cdots & \cdots & \cdots & \cdots & \cdots \\ \cdots & \cdots & \cdots & \cdots & \cdots & \cdots & \cdots & \cdots & \cdots & \cdots \\ a_{N,1} & a_{N,2} & \cdots & \cdots & a_{N,N} & c_{N,1} & c_{N,2} & \cdots & \cdots & c_{N,N} \\ -1 & 0 & \cdots & \cdots & 0 & 0 & 0 & \cdots & \cdots & 0 \\ 0 & -1 & \cdots & \cdots & 0 & 0 & 0 & \cdots & \cdots & 0 \\ \cdots & \cdots & \cdots & \cdots & \cdots & \cdots & \cdots & \cdots & \cdots & \cdots \\ \cdots & \cdots & \cdots & \cdots & \cdots & \cdots & \cdots & \cdots & \cdots & \cdots \\ 0 & 0 & \cdots & \cdots & -1 & 0 & 0 & \cdots & \cdots & 0 \end{vmatrix} = AB \tag{2-85}$$

It is easily seen now that the determinant on the left side of Eq. (2-85) is equal to $C$, so that

$$AB=C \tag{2-86}$$

which is what we wished to prove.

## 2-5 Homogeneous Linear Equations

The equations

$$\begin{aligned}
a_{1,1}x_1 + a_{1,2}x_2 + a_{1,3}x_3 + \cdots + a_{1,N}x_N &= \lambda_1 \\
a_{2,1}x_1 + a_{2,2}x_2 + a_{2,3}x_3 + \cdots + a_{2,N}x_N &= \lambda_2 \\
\cdots\cdots\cdots\cdots\cdots\cdots\cdots\cdots\cdots\cdots& \\
a_{M,1}x_1 + a_{M,2}x_2 + a_{M,3}x_3 + \cdots + a_{M,N}x_N &= \lambda_M
\end{aligned} \tag{2-87}$$

form a set of $M$ linear equations in the $N$ unknowns $x_i$. We speak of homogeneous equations if all the $\lambda_i$ are equal to zero, otherwise we have a set of inhomogeneous equations. In dealing with linear equations, there are a number of different cases to be considered, depending on the relative magnitudes of $N$ and $M$ and whether the equations are homogeneous or inhomogeneous. We start with the homogeneous equations, where the number of equations is equal to the number of variables:

$$\sum_{j=1}^{N} a_{i,j}x_j = 0 \qquad i=1,2,3,\ldots, N \tag{2-88}$$

Before attempting to solve these equations, let us first discuss what we consider by its solution. First we note that, if we take every variable $x_j$ equal to zero, we obtain a solution for every type of homogeneous equation. This is known as the zero solution, but we discount it since it is trivial and generally of no use to us. Let us now imagine that we have obtained a specific solution $(x'_1, x'_2, x'_3, \ldots, x'_N)$ of a homogeneous equation. It is easily verified that $(\rho x'_1, \rho x'_2, \ldots, \rho x'_N)$ is also a solution if $\rho$ is an arbitrary nonzero parameter. What counts, however, is the ratio between the variables $x_j$, so that we should look upon $(x'_j)$ and $(\rho x'_j)$ as a single solution of the equations. In general, we count only solutions that are linearly independent; that is, we consider that we have $s$ independent solutions $(x_j^1), \ldots, (x_j^2), \ldots, (x_j^s)$ if there exists no relationship

$$\sum_{\alpha} \rho_\alpha x_j^\alpha = 0 \qquad j=1,2,\ldots, N \tag{2-89}$$

between the solutions. Naturally, if we have $s$ linearly independent solutions, then each linear combination of them is also a solution of the equations. We have then actually $(\infty)^s$ solutions, but we count them only as $s$ solutions.

We return now to Eq. (2-88). Apparently these equations are completely determined by the matrix $[a_{i,j}]$ of the coefficients, and we can therefore derive the solutions from this matrix only. We first show that the equations have no solution other than the zero solution if the determinant $A$, belonging to the matrix $[a_{i,j}]$, is different from zero. If we multiply the first Eq. (2-88) by $A_{1,1}$, the second by $A_{2,1}$, and so on, and then add them, we obtain

$$\sum_j a_{j,1}A_{j,1}x_1 + \sum_j a_{j,2}A_{j,1}x_2 + \cdots + \sum_j a_{j,N}A_{j,1}x_N = 0 \qquad (2\text{-}90)$$

or, according to Eq. (2-72),

$$Ax_1 = 0 \qquad (2\text{-}91)$$

Since $A$ is different from zero, $x_1 = 0$. In the same way we prove that every $x_j = 0$, and since we discount the zero solution, we conclude that the equations have no solution if $A$ is different from zero.

If $A = 0$ we have, according to Eq. (2-72),

$$\sum_j a_{i,j}A_{k,j} = 0 \qquad i = 1, 2, 3, \ldots, N \qquad (2\text{-}92)$$

Hence

$$x_j = A_{k,j} \qquad (2\text{-}93)$$

is a solution of the equations. We see that we can obtain a valid solution of the equations if at least one of the cofactors $A_{i,j}$ is different from zero.

At first sight it seems as if we have obtained $N$ different solutions, but this is not true. We will prove that each solution is either zero or proportional to the others if at least one of the $A_{k,j}$ is different from zero. Let us assume that $A_{1,1}$ is different from zero. Then we omit the first Eq. (2-88), and we write the others as

$$\begin{aligned} a_{2,2}x_2 + a_{2,3}x_3 + \cdots + a_{2,N}x_N &= -a_{2,1}x_1 \\ a_{3,2}x_2 + a_{3,3}x_3 + \cdots + a_{3,N}x_N &= -a_{3,1}x_1 \\ \cdots\cdots\cdots\cdots\cdots\cdots\cdots\cdots \\ a_{N,2}x_2 + a_{N,3}x_3 + \cdots + a_{N,N}x_N &= -a_{N,1}x \end{aligned} \qquad (2\text{-}94)$$

The coefficients on the left side of Eq. (2-94) form a matrix $[b_{i,j}]$, with $b_{i,j} = a_{i-1,j-1}$. We denote the cofactors of $b_{i,j}$ by $B_{i,j}$. Let us now multiply the first Eq. (2-94) by $B_{1,1}$, the second Eq. (2-94) by $B_{2,1}$, and so on, and add all the equations. The result is

$$-Bx_2 = x_1(a_{2,1}B_{1,1} + a_{3,1}B_{2,1} + a_{4,1}B_{3,1} + \cdots + a_{N,1}B_{N-1,1}) \qquad (2\text{-}95)$$

We observe that the expression between parentheses on the right side of Eq.

(2-95) can be written as

$$\begin{vmatrix} a_{2,1} & a_{2,3} & a_{2,4} & \cdots & a_{2,N} \\ a_{3,1} & a_{3,3} & a_{3,4} & \cdots & a_{3,4} \\ \cdots & \cdots & \cdots & \cdots & \cdots \\ a_{N,1} & a_{N,3} & a_{N,4} & \cdots & a_{N,N} \end{vmatrix} = -A_{1,2} \tag{2-96}$$

and that $B = A_{1,1}$, where $A_{i,j}$ are cofactors of the determinant $|a_{i,j}|$. We thus have

$$A_{1,1}x_2 = A_{1,2}x_1 \tag{2-97}$$

for all values of $x_2$ and $x_1$ that satisfy Eqs. (2-94). However, all solutions of Eqs. (2-88) satisfy Eqs. (2-94) also, and according to Eq. (2-93) we have

$$A_{1,1}A_{k,2} = A_{1,2}A_{k,1} \tag{2-98}$$

or if $A_{1,2}$ and $A_{1,1}$ are both different from zero,

$$\frac{A_{k,1}}{A_{1,1}} = \frac{A_{k,2}}{A_{1,2}} \tag{2-99}$$

Along the same lines we can prove that

$$\frac{A_{k,1}}{A_{1,1}} = \frac{A_{k,2}}{A_{1,2}} = \frac{A_{k,3}}{A_{1,3}} \cdots = \frac{A_{k,N}}{A_{1,N}} \tag{2-100}$$

If any of the $A_{1,i}$ is zero, it follows from Eq. (2-98) or its analogue that the corresponding $A_{k,i}$ is also zero. It follows, therefore, that Eq. (2-93) represents only one solution for the $N$ homogeneous equations with $N$ unknowns.

The reader may wonder what happens when all cofactors $A_{i,j}$ are zero, but we bypass this question for the time being and instead consider first the case of $N$ homogeneous equations with $(N+1)$ variables:

$$\begin{aligned} a_{1,1}x_1 + a_{1,2}x_2 + a_{1,3}x_3 + \cdots + a_{1,N}x_N + a_{1,N+1}x_{N+1} &= 0 \\ a_{2,2}x_1 + a_{2,2}x_2 + a_{2,3}x_3 + \cdots + a_{2,N}x_N + a_{2,N+1}x_{N+1} &= 0 \\ \cdots\cdots\cdots\cdots\cdots\cdots\cdots\cdots\cdots\cdots\cdots\cdots \\ a_{N,1}x_1 + a_{N,2}x_2 + a_{N,3}x_3 + \cdots + a_{N,N}x_N + a_{N,N+1}x_{N+1} &= 0 \end{aligned} \tag{2-101}$$

This case is easily transformed to the previous situation if we add an additional equation, for example, if we write the first equation twice. We then obtain $(N+1)$ equations with $(N+1)$ unknowns, and the solution can be written as the cofactor of the first row, since the determinant of the coefficients is zero. The solution of Eq. (2-101) is therefore obtained as follows. First we write the matrix of the coefficients:

$$\begin{bmatrix} a_{1,1} & a_{1,2} & \cdots & \cdots & a_{1,N+1} \\ a_{2,1} & a_{2,2} & \cdots & \cdots & a_{2,N+1} \\ \cdots & \cdots & \cdots & \cdots & \cdots \\ a_{N,1} & a_{N,2} & \cdots & \cdots & a_{N,N+1} \end{bmatrix} \tag{2-102}$$

We then define $\Delta_n$ as the determinant corresponding to the square matrix of order $N$, which we obtain by deleting the $n$th column from the matrix (2-102). The solution of Eq. (2-101) is now

$$x_n = (-1)^{n-1}\Delta_n \tag{2-103}$$

This method is known as Cramer's rule, and it assumes that at least one of the determinants $\Delta_n$ is different from zero.

We will not derive the solution of every type of homogeneous equation in detail, but we will state some general rules. For this purpose it is useful to introduce the concept of the rank of a matrix. If we consider an arbitrary rectangular matrix of $N$ columns and $M$ rows, then we can construct square matrices by deleting a certain number of columns and rows from the original matrix. If the largest such matrix that has a nonzero determinant is of the $R$th order, we say that the original $N \times M$ matrix has the rank $R$. It is easily seen that the largest possible value of $R$ is equal to $N$ if $N \leq M$ and equal to $M$ when $M \leq N$. The general rule is that $M$ linear homogeneous equations in $N$ unknowns have $(N-R)$ linearly independent solutions if $R$ is the rank of the matrix of the coefficients. As we have mentioned, it is not our intention to prove this general rule, but it is easily verified that it applies to the cases discussed above.

## 2-6 Inhomogeneous Linear Equations

The solution of a set of inhomogeneous equations can, in general, be derived by making use of the theory of homogeneous equations. As an example we take the case of $N$ equations with $N$ unknowns:

$$\begin{aligned}
a_{1,1}x_1 + a_{1,2}x_2 + \cdots + a_{1,N}x_N + \lambda_1 &= 0 \\
a_{2,1}x_1 + a_{2,2}x_2 + \cdots + a_{2,N}x_N + \lambda_2 &= 0 \\
\cdots\cdots\cdots\cdots\cdots\cdots\cdots\cdots\cdots \\
a_{N,1}x_1 + a_{N,2}x_2 + \cdots + a_{N,N}x_N + \lambda_N &= 0
\end{aligned} \tag{2-104}$$

The solution is obtained by considering the set of homogeneous equations

$$\begin{aligned}
a_{1,1}x_1 + a_{1,2}x_2 + \cdots + a_{1,N}x_N + \lambda_1 y &= 0 \\
a_{2,1}x_2 + a_{2,2}x_2 + \cdots + a_{2,N}x_N + \lambda_2 y &= 0 \\
\cdots\cdots\cdots\cdots\cdots\cdots\cdots\cdots\cdots \\
a_{N,1}x_1 + a_{N,2}x_2 + \cdots + a_{N,N}x_N + \lambda_N y &= 0
\end{aligned} \tag{2-105}$$

The rank of the matrix $\mathbf{M}$,

$$\mathbf{M} = \begin{bmatrix} \lambda_1 & a_{1,1} & a_{1,2} & \cdots & a_{1,N} \\ \lambda_2 & a_{2,1} & a_{2,2} & \cdots & a_{2,N} \\ \cdots & \cdots & \cdots & \cdots & \cdots \\ \lambda_N & a_{N,1} & a_{N,2} & \cdots & a_{N,N} \end{bmatrix} \tag{2-106}$$

is always smaller than $N+1$, so that Eqs. (2-105) always have at least one solution. If the rank $R_M$ of $\mathbf{M}$ is $N$, then the solution is, according to Eq. (2-103),

$$y=\begin{vmatrix} a_{1,1} & a_{1,2} & \cdots & a_{1,N} \\ a_{2,1} & a_{2,2} & \cdots & a_{2,N} \\ \cdots & \cdots & \cdots & \cdots \\ a_{N,1} & a_{N,2} & \cdots & a_{N,N} \end{vmatrix}=\Delta \tag{2-107}$$

and

$$x_n=-\Delta_n \tag{2-108}$$

where $\Delta_n$ is obtained from $\Delta$ by replacing its $n$th column by the column $\lambda_i$. Since only the ratio of the unknowns is determined by Eqs. (2-105), we now find the solution of the inhomogeneous Eqs. (2-104) by setting $y$ equal to unity. Hence the solution of Eq. (2-104) is

$$x_n=-\frac{\Delta_n}{\Delta} \tag{2-109}$$

This procedure leads to the solution of the set of inhomogeneous equations as long as $\Delta$ is different from zero. If $\Delta=0$, the inhomogeneous equations have no solution, and we call them inconsistent. An example of such a case is

$$\begin{aligned} x+y&=1 \\ 2x+2y&=3 \end{aligned} \tag{2-110}$$

It is easily seen that these two equations are not consistent with one another, and it is therefore impossible to find a solution for them.

We finally state the general rule for establishing the number of possible solutions for a set of $k$ linear, inhomogeneous equations in $N$ unknowns:

$$\begin{aligned} a_{1,1}x_1+a_{1,2}x_2+\cdots+a_{1,N}x_N+\lambda_1&=0 \\ a_{2,1}x_1+a_{2,2}x_2+\cdots+a_{2,N}x_N+\lambda_2&=0 \\ \cdots\cdots\cdots\cdots\cdots\cdots\cdots\cdots& \\ a_{k,1}x_1+a_{k,2}x_2+\cdots+a_{k,N}x_N+\lambda_k&=0 \end{aligned} \tag{2-111}$$

We write the two matrices:

$$\mathbf{M}=\begin{bmatrix} a_{1,1} & a_{1,2} & \cdots & \cdots & a_{1,N} & \lambda_1 \\ a_{2,1} & a_{2,2} & \cdots & \cdots & a_{2,N} & \lambda_2 \\ \cdots & \cdots & \cdots & \cdots & \cdots & \cdots \\ a_{k,1} & a_{k,2} & \cdots & \cdots & a_{k,N} & \lambda_k \end{bmatrix} \tag{2-112}$$

and

$$\mathbf{A}=\begin{bmatrix} a_{1,1} & a_{1,2} & \cdots & \cdots & a_{1,N} \\ a_{2,1} & a_{2,2} & \cdots & \cdots & a_{2,N} \\ \cdots & \cdots & \cdots & \cdots & \cdots \\ a_{k,1} & a_{k,2} & \cdots & \cdots & a_{k,N} \end{bmatrix} \tag{2-113}$$

and we denote their ranks by $R_M$ and $R_A$, respectively. Equations (2-111) are inconsistent and have, therefore, no solutions if $R_M > R_A$. If $R_M = R_A$, the number of solutions is given by $N - R_A + 1$.

## 2-7 Eigenvalue Problems

We consider the set of $N$ homogeneous linear equations in $N$ unknowns:

$$\begin{aligned} (a_{1,1}-\lambda)x_1 + a_{1,2}x_2 + a_{1,3}x_3 + \cdots + a_{1,N}x_N &= 0 \\ a_{2,1}x_1 + (a_{2,2}-\lambda)x_2 + a_{2,3}x_3 + \cdots + a_{2,N}x_N &= 0 \\ a_{3,1}x_1 + a_{3,2}x_2 + (a_{3,3}-\lambda)x_3 + \cdots + a_{3,N}x_N &= 0 \\ \cdots\cdots\cdots\cdots\cdots\cdots\cdots\cdots\cdots\cdots \\ a_{N,1}x_1 + a_{N,2}x_2 + a_{N,3}x_3 + \cdots + (a_{N,N}-\lambda)x_N &= 0 \end{aligned} \tag{2-114}$$

where $\lambda$ is an undetermined parameter. We ask for which values of $\lambda$ this system has nonzero solutions and what these solutions are. Any value $\lambda_n$ for which Eq. (2-114) has a solution is called an eigenvalue, and the corresponding solution $(x_1^n, x_2^n, \ldots, x_N^n)$ is called an eigenvector. The entire problem is known as the eigenvalue problem of the matrix $[a_{i,j}]$.

From the considerations of Section 2-5 it is easily seen how we can solve an eigenvalue problem. A set of $N$ homogeneous equations has solutions only when the determinant of the coefficients is zero, and the eigenvalues of Eq. (2-114) are therefore derived from the equation

$$|a_{i,j} - \lambda\delta_{i,j}| = 0 \tag{2-115}$$

This is a polynomial of the $N$th degree in $\lambda$, and we can write it as

$$(\lambda-\lambda_1)(\lambda-\lambda_2)\cdots(\lambda-\lambda_n)\cdots(\lambda-\lambda_N) = 0 \tag{2-116}$$

where the $\lambda_n$ are the roots. If all $\lambda_n$ are different, we have $N$ single roots. If one or more of the $\lambda_n$ are equal, for example, if the root $\lambda_k$ occurs as a factor $(\lambda-\lambda_k)^r$, we call $\lambda_k$ a multiple root and its multiplicity is $r$.

It can be shown that if $\lambda_n$ is a single root of Eq. (2-115), the matrix

$$[a_{i,j} - \lambda_n\delta_{i,j}] \tag{2-117}$$

has the rank $(N-1)$. Equations (2-114) have, then, one solution $(x_i^n)$ corresponding to $\lambda_n$. By making use of the rules of matrix multiplication, we can write this situation as

$$\begin{bmatrix} a_{1,1}-\lambda_n & a_{1,2} & \cdots & a_{1,N} \\ a_{2,1} & a_{2,2}-\lambda_n & \cdots & a_{2,N} \\ \cdots & \cdots & \cdots & \cdots \\ a_{N,1} & a_{N,2} & \cdots & a_{N,N}-\lambda_n \end{bmatrix} \begin{bmatrix} x_1^n \\ x_2^n \\ \cdots \\ x_N^n \end{bmatrix} = \begin{bmatrix} 0 \\ 0 \\ \cdots \\ 0 \end{bmatrix} \tag{2-118}$$

or as

$$[a_{i,j}]\mathbf{x}^n = \lambda_n \mathbf{x}^n \tag{2-119}$$

We call $\lambda_n$ a nondegenerate eigenvalue of the matrix **A**, and $\mathbf{x}^n$ is its corresponding eigenvector.

If $\lambda_m$ is a multiple root of Eq. (2-115) and its multiplicity is $r$, it can be shown that the matrix

$$\left[a_{i,j} - \lambda_m \delta_{i,j}\right] \tag{2-120}$$

has the rank $N-r$. Hence the substitution of $\lambda_m$ into Eqs. (2-114) leads to $r$ different, linearly independent solutions $\mathbf{x}^{m,\delta}$, each of which satisfies the equation

$$[a_{i,j}]\mathbf{x}^{m,\delta} = \lambda_m \mathbf{x}^{m,\delta} \qquad \delta = 1,2,\ldots,r \tag{2-121}$$

We now say that $\lambda_m$ is an $r$-fold degenerate eigenvalue of the matrix **A**, since it has $r$ different eigenvectors. Each linear combination of $\mathbf{x}^{m,\delta}$ is also an eigenvector of $\lambda_m$.

The evaluation of the eigenvalues and eigenvectors of a large matrix by means of the direct approach that we outlined above can be very laborious. It is therefore not surprising that various alternate methods for solving eigenvalue problems have been suggested. Today these methods usually involve the use of an electronic computer, and very few scientists endeavor the solution of an eigenvalue problem of any matrix that is larger than $3\times 3$ without using the computer. We first discuss some general properties of eigenvalues and eigenfunctions, and then, in Section 2-8, we outline some of the principles that are the basis of the computer programs. We limit ourselves only to eigenvalue problems of Hermitian matrices, since practically all matrices that we encounter in quantum mechanical eigenvalue problems are Hermitian. We may recall from Eq. (2-37) that such a Hermitian matrix is defined by

$$a_{i,j} = a_{j,i}^* \qquad \mathbf{A} = \mathbf{A}^\dagger \tag{2-122}$$

We will first show that the eigenvalues of a Hermitian matrix are all real. If **A** is a Hermitian matrix, $\lambda_n$ is one of its eigenvalues, and $\mathbf{x}^n$ is the corresponding eigenvector, written as a column vector, then we have the equation

$$\mathbf{A}\mathbf{x}^n = \lambda_n \mathbf{x}^n \tag{2-123}$$

Since $\mathbf{x}^n$ is defined as a column vector, the transpose of $\mathbf{x}^n, \tilde{\mathbf{x}}^n$ is the row vector

$$\tilde{\mathbf{x}}^n = [x_1^n x_2^n x_3^n \cdots x_N^n] \tag{2-124}$$

It is easily verified that

$$(\tilde{\mathbf{x}}^n)^* \cdot \mathbf{x}^n = \sum_k (x_k^n)^* \cdot x_k^n = N_n \tag{2-125}$$

where $N_n$ is a positive real number. We have, therefore,

$$(\mathbf{x}^n)^\dagger \mathbf{A}\mathbf{x}^n = \lambda_n N_n \tag{2-126}$$

The complex conjugate of this equation is

$$\tilde{\mathbf{x}}^n \mathbf{A}^* (\mathbf{x}^n)^* = \lambda_n^* N_n \tag{2-127}$$

It follows from the definition of matrix multiplication that

$$\widetilde{(\mathbf{A} \cdot \mathbf{B} \cdot \mathbf{C})} = \tilde{\mathbf{C}} \cdot \tilde{\mathbf{B}} \cdot \tilde{\mathbf{A}} \tag{2-128}$$

Hence we can write Eq. (2-127) as

$$\lambda_n^* N_n = \tilde{\mathbf{x}}^n \mathbf{A}^* (\mathbf{x}^n)^* = (\tilde{\mathbf{x}}^n)^* \tilde{\mathbf{A}}^* \mathbf{x}^n = (\mathbf{x}^n)^\dagger \mathbf{A}^\dagger \mathbf{x}^n \tag{2-129}$$

or, since $\mathbf{A}$ is Hermitian, as

$$(\mathbf{x}^n)^\dagger \mathbf{A}\mathbf{x}^n = \lambda_n^* N_n \tag{2-130}$$

Subtraction of Eq. (2-130) from Eq. (2-126) gives

$$(\lambda_n - \lambda_n^*) N_n = 0 \tag{2-131}$$

and it follows, therefore, that $\lambda_n$ is real.

Two $N$-dimensional vectors $\mathbf{u}$ and $\mathbf{v}$ are defined to be orthogonal if

$$\sum_k u_k^* v_k = 0 \tag{2-132}$$

We can also write this definition in the matrix notation

$$\mathbf{u}^\dagger \cdot \mathbf{v} = 0 \tag{2-133}$$

if we take **u** and **v** as column vectors. We now prove that the two eigenvectors $\mathbf{x}^m$ and $\mathbf{x}^n$ of a Hermitian matrix **A** are orthogonal if they belong to different eigenvalues $\lambda_m$ and $\lambda_n$, respectively.

Since $\mathbf{x}^m$ and $\mathbf{x}^n$ are eigenvectors, they satisfy the equations

$$\mathbf{A}\mathbf{x}^n = \lambda_n \mathbf{x}^n \tag{2-134}$$

and

$$\mathbf{A}^*(\mathbf{x}^m)^* = \lambda_m (\mathbf{x}^m)^* \tag{2-135}$$

We multiply Eq. (2-134) on the left by $(\mathbf{x}^m)^\dagger$ and Eq. (2-135) by $\tilde{\mathbf{x}}^n$. This gives

$$(\mathbf{x}^m)^\dagger \cdot \mathbf{A} \cdot \mathbf{x}^n = \lambda_n (\mathbf{x}^m)^\dagger \cdot (\mathbf{x}^n)$$

$$(\tilde{\mathbf{x}}^n) \cdot \mathbf{A}^* \cdot (\mathbf{x}^m)^* = \lambda_m (\tilde{\mathbf{x}}^n) \cdot (\mathbf{x}^m)^* \tag{2-136}$$

The second equation can also be written as

$$(\tilde{\mathbf{x}}^m)^\dagger \cdot \mathbf{A} \cdot (\mathbf{x}^n) = \lambda_m (\mathbf{x}^m)^\dagger \cdot (\mathbf{x}^n) \tag{2-137}$$

and if we subtract it from the first Eq. (2-136), we obtain

$$(\mathbf{x}^m)^\dagger \cdot (\mathbf{x}^n) = 0 \tag{2-138}$$

since $\lambda_m$ and $\lambda_n$ are different.

Two different eigenvectors that belong to the same eigenvalue do not have to be orthogonal, but by choosing suitable linear combinations of degenerate eigenvectors, we can make them orthogonal. It is always possible, therefore, to obtain the eigenvectors of a Hermitian matrix as an orthonormal set of vectors.

## 2-8 Diagonalization of Hermitian Matrices

Two matrices **B** and **A** are connected by a similarity transformation if **B** can be written as

$$\mathbf{B} = \mathbf{M}^{-1}\mathbf{A}\mathbf{M} \tag{2-139}$$

If **M** is unitary, that is, if

$$\mathbf{B} = \mathbf{U}^{-1}\mathbf{A}\mathbf{U} = \mathbf{U}^\dagger\mathbf{A}\mathbf{U} \tag{2-140}$$

we speak of a unitary transformation. If we start with an arbitrary matrix **A**, the problem of finding a unitary matrix **U** so that the transformed matrix **B** is diagonal is called the diagonalization of the matrix **A**.

If **A** is a Hermitian matrix **H**, it is always possible to diagonalize it. Let $\lambda_n$ be the eigenvalues of **H** and $\mathbf{x}^n$ be its eigenvectors, which have been selected in such a way that they form an orthonormal set. We allow for degeneracies so that several of the eigenvalues $\lambda_n$ can be equal to one another. We now construct a matrix **V** as

$$\mathbf{V} = \begin{bmatrix} x_1^1 & x_1^2 & \cdots & x_1^N \\ x_2^1 & x_2^2 & \cdots & x_2^N \\ \cdots & \cdots & \cdots & \cdots \\ x_N^1 & x_N^2 & \cdots & x_N^N \end{bmatrix} \tag{2-141}$$

It is easily verified that

$$\mathbf{AV} = \begin{bmatrix} \lambda_1 x_1^1 & \lambda_2 x_1^2 & \cdots & \lambda_N x_1^N \\ \lambda_1 x_2^1 & \lambda_2 x_2^2 & \cdots & \lambda_N x_2^N \\ \cdots & \cdots & \cdots & \cdots \\ \lambda_1 x_N^1 & \lambda_2 x_N^2 & \cdots & \lambda_N x_N^N \end{bmatrix} \tag{2-142}$$

and

$$\mathbf{V}^\dagger\mathbf{AV} = \begin{bmatrix} \lambda_1 & 0 & \cdots & \cdots & \cdots & 0 \\ 0 & \lambda_2 & 0 & \cdots & \cdots & 0 \\ 0 & 0 & \lambda_3 & \cdots & \cdots & 0 \\ \cdots & \cdots & \cdots & \cdots & \cdots & \cdots \\ \cdots & \cdots & \cdots & \cdots & \cdots & \cdots \\ 0 & 0 & 0 & \cdots & \cdots & \lambda_N \end{bmatrix} \tag{2-143}$$

since the eigenvectors are orthonormal. We see that we can diagonalize **A** if we know its eigenvalues and eigenvectors.

If, on the other hand, we know the unitary matrix **U** by which **A** is diagonalized, then it is possible to derive the eigenvalues and eigenvectors of **A** from the matrix **U**. First we show that the two matrices **A** and $(\mathbf{U}^\dagger \cdot \mathbf{A} \cdot \mathbf{U})$ have the same eigenvalues. The eigenvalues of **A** are obtained as the roots of the determinant

$$|\mathbf{A} - \lambda \mathbf{I}| \tag{2-144}$$

written in the form of a polynomial. We now have

$$|\mathbf{A} - \lambda\mathbf{I}| = |\mathbf{U}^\dagger|\,|\mathbf{A} - \lambda\mathbf{I}|\,|\mathbf{U}| = |\mathbf{U}^\dagger\mathbf{AU} - \lambda\mathbf{U}^\dagger\mathbf{IU}| = |\mathbf{U}^\dagger\mathbf{AU} - \lambda\mathbf{I}| \tag{2-145}$$

The two polynomials are the same, and they have the same roots; consequently, **A** and $(\mathbf{U}^\dagger \cdot \mathbf{A} \cdot \mathbf{U})$ have the same eigenvalues.

Although the eigenvectors of **A** and $\mathbf{U}^\dagger \cdot \mathbf{A} \cdot \mathbf{U}$ are not the same, they are easily derived from each other. The eigenvalues and eigenvectors of **A** satisfy the equations

$$\mathbf{A}\mathbf{x}^n = \lambda_n \mathbf{x}^n \tag{2-146}$$

We introduce a new set of vectors $\mathbf{y}^n$ that are derived from the $\mathbf{x}^n$ by means of the transformation

$$\mathbf{x}^n = \mathbf{U}\mathbf{y}^n \tag{2-147}$$

where $\mathbf{U}$ is the unitary matrix by which $\mathbf{A}$ is diagonalized. Substitution into Eq. (2-146) gives

$$\mathbf{A}\mathbf{U}\mathbf{y}^n = \lambda_n \mathbf{U}\mathbf{y}^n \tag{2-148}$$

and multiplication on the left by $\mathbf{U}^\dagger$ yields

$$(\mathbf{U}^\dagger\mathbf{A}\mathbf{U})\mathbf{y}^n = \lambda_n \mathbf{y}^n \tag{2-149}$$

We conclude that $\mathbf{A}$ and $(\mathbf{U}^\dagger \cdot \mathbf{A} \cdot \mathbf{U})$ have the same eigenvalues and that their eigenvectors are related to each other through Eq. (2-147).

Let us imagine that we have found a unitary transformation $\mathbf{U}$ that diagonalizes the matrix $\mathbf{A}$:

$$\mathbf{U}^\dagger\mathbf{A}\mathbf{U} = \left[b_i \delta_{i,j}\right] \tag{2-150}$$

The eigenvalues and eigenvectors of $[b_i \delta_{i,j}]$ are easily obtained. It follows immediately from Eq. (2-114) that the eigenvalues are $\lambda_n = b_n$ and that the eigenvectors are

$$y_i^n = \delta_{n,i} \tag{2-151}$$

By making use of Eq. (2-147), we obtain the eigenvectors of $\mathbf{A}$. The eigenvalues of $\mathbf{A}$ are $\lambda_n$, since they are the same as for $\mathbf{B}$. It follows, therefore, that the diagonalization of a matrix and finding its eigenvalues and eigenfunctions are equivalent problems.

It is now possible to show that two matrices $\mathbf{A}$ and $\mathbf{B}$ must commute if they have the same set of eigenvectors. In this case, they are both diagonalized by the same unitary transformation $\mathbf{U}$:

$$\mathbf{U}^\dagger\mathbf{A}\mathbf{U} = \left[a_i \delta_{i,j}\right] \tag{2-152}$$

$$\mathbf{U}^\dagger\mathbf{B}\mathbf{U} = \left[b_i \delta_{i,j}\right]$$

Two diagonal matrices always commute so that

$$(\mathbf{U}^\dagger\mathbf{A}\mathbf{U})(\mathbf{U}^\dagger\mathbf{B}\mathbf{U}) = (\mathbf{U}^\dagger\mathbf{B}\mathbf{U})(\mathbf{U}^\dagger\mathbf{A}\mathbf{U}) \tag{2-153}$$

or

$$\mathbf{U}^{\dagger}\mathbf{ABU}=\mathbf{U}^{\dagger}\mathbf{BAU} \tag{2-154}$$

It follows, therefore, that **A** and **B** commute. It can also be shown that there exists a unitary transformation that diagonalizes both **A** and **B** if they commute.

Initially, the numerical procedures for obtaining the eigenvalues and eigenfunctions of a Hermitian matrix **H** were based either on iterative procedures or on the diagonalization of **H**. The latter procedure was called the Jacobi method. Today, the computer programs for deriving the eigenvalues and eigenfunctions are based on a procedure that is called the Givens-Householder tridiagonalization method. This method is described by Householder, by Ralston and Wilf, and by Wilkinson in books that are listed at the end of this chapter. We do not describe the tridiagonalization method in detail, but we should mention that this method turns out to be much faster than any of the other methods.

## 2-9 Heisenberg's Matrix Mechanics

Now that we have finished our discussion of linear algebra we continue our story on the development of quantum mechanics. As we have seen at the end of the previous chapter, the old quantum theory supplied methods to describe the motions of various systems, but it lacked a sound theoretical foundation and it also lacked consistency.

The first step toward the formulation of a more comprehensive theory was the work of Heisenberg in 1925. Werner Karl Heisenberg was born in 1901, and during the years around 1925 he was associated with Max Born in Göttingen, with Arnold Sommerfeld in München, and with Bohr and Kramers in Copenhagen. During his stay in Copenhagen he worked with Kramers on the theory of interactions between radiation and atoms, the result of which is the Kramers-Heisenberg dispersion formula. It can be seen that certain ideas that were developed for the derivation of the dispersion theory were used later by Heisenberg in the formulation of matrix mechanics.

In a more general sense Heisenberg felt that the quantum mechanical description of atomic systems should be based on physical observable quantities only. Consequently, he felt that the classical orbits and momenta of the electrons within the atom should not be used in a theoretical description because they cannot be observed. Instead, the theory should be based on the experimental information that can be derived from the atomic spectra. Each spectral line is determined by its frequency $\nu$ and also by its intensity, but the intensity is determined by a quantity that appeared in the Kramers-Heisenberg dispersion formula and which is known today as the transition moment. A spectral transition between two stationary states $n$ and $m$ was thus described by the frequency $\nu(n,m)$ and the dipole amplitude $x(n,m)$.

Heisenberg presented his theory in 1925 in a rather complicated paper where he showed how a physical observable or a time-dependent quantity could be expressed in terms of the set of frequencies $\nu(n, m)$ and dipole amplitudes $x(n, m)$. In this way a physical observable could be related to the set of quantities $x(n, m)$. Furthermore, Heisenberg derived the multiplication properties of the sets of dipole amplitudes and he showed that they must satisfy the condition

$$x(n, m) = x^*(m, n) \tag{2-155}$$

We know that the set $x(n, m)$ forms a matrix and, according to Eq. (2-37), a Hermitian matrix in particular. However, Heisenberg, as most physicists in 1925, was not familiar with matrix theory and he did not realize that the various relations that he derived in his paper were well known in linear algebra.

Max Born was one of the few physicists who was somewhat familiar with linear algebra and he recognized the key feature of Heisenberg's work, namely the fact that a physical observable was represented by a Hermitian matrix. Born was a professor at Göttingen and his mathematics colleague was the famous David Hilbert, who had an extensive knowledge of all areas of mathematics. By a fortunate coincidence the well-known book *Methods of Mathematical Physics* by Richard Courant and David Hilbert appeared in 1924. It contained chapters on linear algebra, integral equations, and eigenvalue problems, exactly those parts of mathematics that were relevant to the further development of quantum mechanics.

Ernest Pascual Jordan helped write the chapter on linear algebra in the book and one of the side effects of the publication was a fruitful collaboration between Born and Jordan and, later, among Born, Jordan, and Heisenberg.

First Born and Jordan expressed Heisenberg's ideas in terms of matrices, and next they added some interesting ideas of their own. They considered the coordinate $q$ and the momentum $p$ in a one-dimentional system. Both of them can be represented by an infinite matrix, and the specific form of the matrices depends on the nature of the system. In addition, they considered the Hamiltonian $H(p, q)$ of the system. Since $H$ is a function of $p$ and $q$, it can be represented as a matrix also, the form of the $H$ matrix depending on the form of the $p$ and $q$ matrices.

The specific form of the various matrices depends on the type of system under consideration, but Born and Jordan showed that for all systems the $p$ and $q$ matrices satisfy the relation

$$\mathbf{pq} - \mathbf{qp} = \left(\frac{h}{2\pi i}\right)\mathbf{I} \tag{2-156}$$

where $\mathbf{I}$ is the identity matrix. The elements of $\mathbf{I}$ are given by

$$\begin{aligned} \delta_{i,j} &= 1 \qquad \text{if } i = j \\ &= 0 \qquad \text{if } i \neq j \end{aligned} \tag{2-157}$$

In a subsequent paper by Born, Heisenberg and Jordan it was shown how the stationary states of a given system could be determined. One had to find two matrices **p** and **q** that satisfy the condition (2-157) and that have the property that the Hamiltonian matrix **H**, derived from **p** and **q**, is a diagonal matrix. This procedure offers a logical and consistent method for deriving the stationary states of quantum mechanical systems, but the required mathematics is rather involved. The harmonic oscillator can be solved by applying matrix mechanics and the method is particularly suitable for dealing with angular momentum problems. However, the Schrödinger differential equation was derived a year later, and this equation offers a much more convenient method for dealing with quantum mechanical problems.

In fact, major new theories in quantum mechanics were proposed during 1925 and 1926. The final formulation of quantum mechanics as we know it today was derived during that period. We discuss the work of de Broglie on wave mechanics and the Schrödinger equation in the following chapter. We should mention here first the results of the work by Born, Jordan, and Heisenberg and subsequently the work by Jordan and by the English physicist P. A. M. Dirac which showed that the matrix mechanics formalism and the Schrödinger equation formalism are equivalent. This work is known as transformation theory, and it shows in a very general sense how the various seemingly different mathematical descriptions of quantum mechanics can be transformed into one another.

## 2-10 The Uncertainty Relations

Chronologically, the uncertainty relations were formulated by Heisenberg after the various developments listed in the previous section, namely in 1927. Nevertheless, we wish to discuss them in this chapter because they affect basic philosophical concepts and they are helpful in understanding the fundamental principles of quantum mechanics.

The purpose of any type of mechanics is the prediction of the behavior as a function of time of the system under consideration. In order to make these predictions we must have two kinds of information available to us. First we must know the state of the system at a given time $t_o$. Second we must know the laws of nature that are typical of the mechanics we use. These laws tell us how the state of the system changes from one moment to the next. From these two sets of information we can then predict what the state of the system is at subsequent times $t_1$, $t_2$, $t_3$, and so on, that is, the time evolution of the system.

In classical mechanics the state of a system is described by the set of coordinates $q_1, q_2, \ldots, q_N$ and the conjugate momenta $p_1, p_2, \ldots, p_N$. The state of the system at time $t_o$ is therefore determined by the set of $2N$ quantities $q_1(t_o)$, $p_1(t_o)$, $q_2(t_o)$, $p_2(t_o)$, and so on. As we have seen in Section 1-2, the mathematical expressions of the laws of nature are the equations of motion; by integrating these equations we can predict the values of the quantities $q_i(t)$ and

$p_i(t)$ at a future time $t$ from the values of these same quantities $q_i(t_o)$ and $p_i(t_o)$ at a previous time $t_o$.

In practice our knowledge of the set of quantities $q_i(t_o)$ and $p_i(t_o)$ can only be derived from an experiment and we know that every experiment is subject to a certain margin of error. If we incorporate these experimental errors into our considerations, we should admit that our predictions are subject to comparable error margins and that our theoretical results are not completely accurate.

Classical mechanics assumes an idealized world where experiments can be performed with unlimited accuracy, where we have point particles with infinitely small dimensions, where we have frictionless pistons, weightless rods, and so on. This is of course not the world we live in, but we assume that this idealized world really exists and that we can approach it arbitrarily close by improving our experimental techniques and technical aids. In other words, we may not be able in practice to perform experiments of unlimited accuracy, but we assume that it is possible in principle to approach unlimited accuracy by improving the experimental techniques.

The above assumption seems reasonable for the case of macroscopic systems. We have seen a steady improvement in the accuracy of experimental results and there is no reason to doubt that this improvement will continue also in the future.

Heisenberg recognized that the classical concepts of an exactly determined position, velocity, or momentum were in principle incompatible with the principles of quantum mechanics. Instead, it is necessary to allow for a certain amount of uncertainty $\Delta q$ in the determination of a coordinate $q$ and an uncertainty $\Delta p$ in the determination of the momentum. Heisenberg proposed that the product of the uncertainties $\Delta q_i$ and $\Delta p_i$, belonging to a coordinate $q_i$ and its conjugate momentum $p_i$, must always be larger than the quantity $h/2\pi$:

$$\Delta q_i \cdot \Delta p_i > \frac{h}{2\pi} \tag{2-158}$$

It should be noted that a new symbol, $\hbar$, is now used to denote $h/2\pi$:

$$\hbar = \frac{h}{2\pi} \tag{2-159}$$

Heisenberg argued that the condition (2-158) is consistent with the commutator relation (2-156) for the $p$ and $q$ matrices. He also presented the derivation that we describe in Chapter 3, Section 3-8. This derivation represents the motion of a free particle in one dimension according to the most favorable initial conditions. Heisenberg showed that in this situation the product of $\Delta x$ and $\Delta p$ at the time $t=0$ is given by

$$\Delta x \cdot \Delta p = \frac{\hbar}{2} \tag{2-160}$$

At subsequent times the product of the uncertainties becomes larger.

Since the uncertainty relation (2-158) is connected with the commutator relation (2-156), it refers only to a coordinate and its conjugate momentum. It should be noted that no such relation exists between the uncertainty $\Delta q_i$ in the coordinate $q_i$ and the uncertainty $\Delta p_j$ in a different nonconjugate momentum $p_j$. However, Heisenberg proposed one additional uncertainty relation, namely

$$\Delta E \cdot \Delta t > \hbar \tag{2-161}$$

Here, $\Delta E$ is the uncertainty in the energy of the system and $\Delta t$ is the time interval during which the energy is determined.

It should be noted again that the uncertainty relations (2-158) and (2-161) were proposed by Heisenberg from basic considerations and by taking into account the commutator relations of matrix mechanics. The uncertainty relations state that a coordinate $q_i$ and its conjugate momentum $p_i$ cannot be measured with arbitrary accuracy. The possible errors $\Delta q_i$ and $\Delta p_i$ in the experimental results must satisfy Eq. (2-158). Many scientists considered it a challenge to plan or to describe experiments where the uncertainty relations were violated. This led to some highly sophisticated arguments between challengers and defenders of the uncertainty relations.

The uncertainty relations are quite important in a philosophical sense. They show us that the predictions of quantum mechanics must be of statistical nature as opposed to the exact predictions of classical mechanics. As we shall see, the statistical rules that are associated with quantum mechanics are quite precise and they can lead to quite accurate predictions about the average behavior of a large number of particles. For example, we cannot predict the exact position and momentum of a single particle, but we can predict the properties of a molecular beam because they are determined by the averages over a large number of particles. In everyday life we know that birth and death of one individual are unpredictable events, but we can talk about the birth rate and the mortality tables of the people in a certain country, the latter two quantities capable of being determined quite accurately. Today the uncertainty relations are looked upon as a law of nature just as the conservation of energy and the laws of thermodynamics. As we shall see, it will help us understand various aspects and results of quantum mechanics that will be discussed in this book.

## Problems

**1** Prove that

$$\begin{vmatrix} 0 & 1 & 1 & 1 \\ 1 & 0 & z^2 & y^2 \\ 1 & z^2 & 0 & x^2 \\ 1 & y^2 & x^2 & 0 \end{vmatrix} = (x^2 - y^2 - z^2)^2 - 4y^2z^2$$

**2** Calculate the determinant

$$\begin{vmatrix} x & 1 & 0 & 0 & 0 & 1 \\ 1 & x & 1 & 0 & 0 & 0 \\ 0 & 1 & x & 1 & 0 & 0 \\ 0 & 0 & 1 & x & 1 & 0 \\ 0 & 0 & 0 & 1 & x & 1 \\ 1 & 0 & 0 & 0 & 1 & x \end{vmatrix}$$

**3** Determine the matrix $\mathbf{U}^{\dagger}\mathbf{A}\mathbf{U}$ if

$$\mathbf{A} = \begin{vmatrix} a_{1,1} & a_{1,2} & a_{1,3} \\ a_{2,1} & a_{2,2} & a_{2,3} \\ a_{3,1} & a_{3,2} & a_{3,3} \end{vmatrix} \qquad \mathbf{U} = \begin{vmatrix} \cos\phi & 0 & \sin\phi \\ 0 & 1 & 0 \\ -\sin\phi & 0 & \cos\phi \end{vmatrix}$$

**4** Determine the values of $\lambda$ for which the following set of equations has a solution, and determine each solution:

$$\lambda x + y + u = 0$$

$$\lambda y + z = 0$$

$$x - u = 0$$

$$y - z = 0$$

**5** Find the values of $\lambda$ for which the following set of equations has solutions, and find each solution:

$$\lambda x + y + u = 0$$

$$x + \lambda y + z = 0$$

$$y + \lambda z + u = 0$$

$$x + z + \lambda u = 0$$

**6** Determine the inverse of the matrix

$$\begin{bmatrix} (\cos\alpha\cos\gamma - \sin\alpha\cos\beta\sin\gamma) & (\sin\alpha\cos\gamma + \cos\alpha\cos\beta\sin\gamma) & \sin\beta\sin\gamma \\ (\cos\alpha\sin\gamma + \sin\alpha\cos\beta\cos\gamma) & (-\sin\alpha\sin\gamma + \cos\alpha\cos\beta\cos\gamma) & -\sin\beta\cos\gamma \\ \sin\alpha\sin\beta & \cos\alpha\sin\beta & \cos\beta \end{bmatrix}$$

and show that it is unitary.

**7** Derive the eigenvalues and eigenvectors of the matrix

$$\begin{bmatrix} a & b & 0 & 0 & 0 & b \\ b & a & b & 0 & 0 & 0 \\ 0 & b & a & b & 0 & 0 \\ 0 & 0 & b & a & b & 0 \\ 0 & 0 & 0 & b & a & b \\ b & 0 & 0 & 0 & b & a \end{bmatrix}$$

and the matrix

$$\begin{bmatrix} a & b & 0 & 0 \\ b & a & b & 0 \\ 0 & b & a & b \\ 0 & 0 & b & a \end{bmatrix}$$

**8** Solve the equations

$$3x-2y=7$$

$$2y+2z+u=5$$

$$x-2y-3z-2u=-1$$

$$y+z+u=6$$

**9** Take the square of the matrix

$$M=\begin{bmatrix} a & b & 0 & 0 & 0 \\ b & a & b & 0 & 0 \\ 0 & b & a & b & 0 \\ 0 & 0 & b & a & b \\ 0 & 0 & 0 & b & a \end{bmatrix}$$

and determine the eigenvalues and eigenvectors of $M$ and $M^2$.

**10** The matrices $(x)_{n,m}$ and $(p)_{n,m}$ for the coordinate $x$ and the momentum $p$ of the one-dimensional harmonic oscillator are given by

$$(x)_{n,n+1}=\left[\frac{\alpha(n+1)}{2}\right]^{1/2}$$

$$(x)_{n,n-1}=\left[\alpha\frac{n}{2}\right]^{1/2}$$

$$(x)_{n,m}=0 \qquad \text{if } m\neq n\pm 1$$

$$(p)_{n,n+1}=-i\hbar\left[\frac{(n+1)}{2\alpha}\right]^{1/2}$$

$$(p)_{n,n-1}=i\hbar\left[\frac{n}{2\alpha}\right]^{1/2}$$

$$(p)_{n,m}=0 \qquad \text{if } m\neq n\pm 1$$

Evaluate the following matrices:

(a) $(xp)_{n,m}$

(b) $(px)_{n,m}$

(c) $(x^2)_{n,m}$

(d) $r(p^2)_{n,m}$

Finally, determine the matrices

$$(xp)_{n,m}-(px)_{n,m}$$

and

$$\mathcal{H}=\left(\frac{\alpha\omega}{2\hbar}\right)(p^2)_{n,m}+\left(\frac{\hbar\omega}{2\alpha}\right)(x^2)_{n,m}$$

Show that the forms of these latter two matrices are consistent with Heisenberg's matrix mechanics.

## Recommended Reading

Again, we recommend the book by Jammer for a more complete review of the historical aspects. In addition, we list the book by Courant and Hilbert and a few books on matrices and determinants. The last three books deal with the eigenvalue problem.

Max Jammer, *The Conceptual Development of Quantum Mechanics*, McGraw-Hill Book Company, New York, 1966.

R. Courant and D. Hilbert, *Methods of Mathematical Physics*, Vol. I, Interscience Publishers, Inc., New York, 1953, first published by Julius Springer, Berlin, 1925.

R. A. Frazer, W. J. Duncan, and A. R. Collar, *Elementary Matrices*, Cambridge University Press, London, 1957.

W. L. Ferrar, *Algebra*, Oxford University Press, London, 1957.

Thomas Muir, *The Theory of Determinants in the Historical Order of Development*, Dover Publications, Inc., New York, 1960.

A. S. Householder, *The Theory of Matrices in Numerical Analysis*, Blaisdell, New York, 1964.

J. H. Wilkinson, *The Algebraic Eigenvalue Problem*, Clarendon Press, Oxford, 1965.

A. Ralston and H. S. Wilf, *Mathematical Methods for Digital Computers*, Vol. II, John Wiley & Sons, Inc., New York, 1967.

CHAPTER THREE

# Wave Mechanics and the Schrödinger Equation

## 3-1 Introduction

In the previous chapters, we gave a historical survey of the discovery of quantum mechanics. In the first chapter, we described the initial quantization discovery by Planck and the old quantum theory by Bohr. In the second chapter, we discussed matrix mechanics and the uncertainty relations by Heisenberg. In the present chapter, we complete our historical survey with the treatment of de Broglie's work on wave-particle dualism and of the Schrödinger equation. Subsequent development such as the Dirac equation and other theories dealing with relativistic effects, second quantization, and so on fall outside the range of this book. The discovery of the electron spin and the exclusion principle will be presented subsequently when we consider many-electron systems.

In the next section, we sketch the course of events that led to the discovery and the interpretation of the Schrödinger equation. In the subsequent sections we do not follow the historical course of events because we feel that there is a more effective way of presenting the material. It is possible to rationalize the Schrödinger equation from the mathematical description of classical wave mechanics and from de Broglie's postulate on wave-particle dualism. The interpretation of the physical meaning of the wave function can be based on the Heisenberg uncertainty relation. We follow this approach in presenting the material, starting with Section 3-3, even though this approach differs from the true historical course of events. The latter is described in Section 3-2, just to set the record straight.

## 3-2 Historical Sketch

At the end of Section 1-4, we mentioned that many physicists began to consider a radiation field as an assembly of energy quanta or photons as they became known later. It is tempting to look upon the photons as particles. By doing so, Einstein derived a satisfactory description of the photo-electric effect. The result of these various theories led to a dualistic description of electromagnetic radiation. In some respects it could be considered particles and in other respects it should be described as a wave phenomenon. We call this the wave-particle dualism, and many scientists had difficulty in accepting or comprehending this dual theoretical description.

Louis, Duc de Broglie, was born in 1892. He became interested in the problems associated with wave-particle dualism while he was a student at the Sorbonne pursuing a Doctor of Science degree. An elder brother, Maurice de Broglie, was an experimental physicist interested in X-rays, and it seems that he directed Louis' attention to the problems associated with the wave-particle theories of X-ray radiation.

According to Louis de Broglie the various different viewpoints could be united by assuming that the motion of a particle, for example, a photon, is associated with a wave. In order to be consistent the same assumption should be valid also for the motion of different particles such as electrons, protons, and so on. By making use of Fermat's and Maupertuis's principles, de Broglie deduced that the wave length $\lambda$ of the wave and the momentum $p$ of the particle must satisfy the relation

$$\lambda = \frac{h}{p} \tag{3-1}$$

where $h$ is Planck's constant. Louis de Broglie presented his various ideas in the form of a doctoral thesis to the Faculty of Sciences toward the end of 1924. He successfully defended the thesis and he was awarded the D.Sc. degree.

The experimental verification of these ideas seemed fairly obvious. If the motion of an electron is associated with a wave, then these waves should exhibit interference just as optical waves and they should lead to diffraction phenomena just as X-rays. By conducting scattering experiments with electron beams on crystals, it should be possible to observe diffraction patterns if the motion of the electrons is associated with a set of waves.

Such scattering experiments had already been performed at Bell Telephone Laboratories by C. G. Davisson and C. H. Kunsman in 1921. It was shown later by Franck and Elsasser that these experimental results could be explained by treating them as diffraction effects, analogous to X-ray diffraction patterns. Naturally, de Broglie's work triggered off an increasing interest in these types of scattering experiments. We mention the work by Davisson and Germer and the work by G. P. Thomson which showed quite clearly that a beam of electrons could produce diffraction patterns that are in every respect analogous with X-ray patterns.

There is little doubt that Schrödinger was familiar with de Broglie's work when he derived his differential equation, but it is not clear to what extent he was influenced by it. Erwin Schrödinger was born in Vienna in 1875 and he was a professor at the University of Zürich from 1921 to 1927. At that time, the Dutchman Peter Debye was a professor at the Technische Hochschule in Zürich, and he invited Schrödinger to come and give a seminar about de Broglie's theories. Schrödinger felt that these theories could be easier understood if they were presented in the form of a differential equation. In this way, the energies of the stationary states could be obtained as the eigenvalues of a differential equation with boundary conditions. As we all know, Schrödinger implemented his plans and the resulting differential equation, which bears his name, is the fundamental equation for nonrelativistic quantum mechanics.

It should be noted that the Schrödinger equation is not relativistically invariant and that it does not include relativistic effects. Schrödinger tried at first to derive a differential equation that was compatible with relativity theory, but at the time that was not possible because the existence of the electron spin had not yet been discovered. The relativistic equation was derived a few years later by Dirac, but it is more complicated and we do not discuss it in our book.

According to Schrödinger the problem of finding the stationary states of a system was equivalent to finding the eigenvalues of a differential equation with boundary conditions. Schrödinger solved his differential equation for some simple systems, the harmonic oscillator, the rigid rotor, and the hydrogen atom. The results were in perfect agreement with the experiments. In a final paper Schrödinger derived perturbation theory and he used it to calculate the perturbation of the hydrogen atom by an electric field, that is, the Stark effect.

Schrödinger was more interested in the eigenvalues of his equation than in the corresponding eigenfunctions. It was left to Max Born to offer the correct physical interpretation of the meaning of the eigenfunctions. He proposed that the wave function $\psi$ of a particle determines its probability density $P$ through the relation

$$P(x, y, z; t) = \psi(x, y, z; t) \cdot \psi^*(x, y, z; t) \tag{3-2}$$

It was shown later by various investigators that matrix mechanics and wave mechanics are formally equivalent. This can be understood from the general theory of differential equations. It was well known that a differential equation with boundary conditions (known in one dimension as the Sturm-Liouville problem) can be transformed to an integral equation. The integral equation can then be solved by treating it as a matrix of infinite order. We should mention also that Heisenberg illustrated the validity of the uncertainty relations by considering the time evolution of a specific wave packet. We will discuss this derivation later in this chapter as an application of the properties of wave packets.

In this chapter, we first discuss the mathematical description of classical wave mechanics together with Fourier analysis and some related mathematical topics. Then we introduce de Broglie's postulates and we incorporate them into the mathematical formalism of wave mechanics. We discuss some properties of wave packets. We then derive the Schrödinger equation for a free particle from wave mechanics and, finally, we generalize the equation to particles moving in a potential field and to bound particles.

## 3-3 Waves in One Dimension

The observation of one-dimensional waves is one of the few experiments that can be performed in a classroom without any special preparation. When we hold the cord of a window shade and move the wrist up and down, the cord exhibits the pattern of a running wave very clearly. The mathematical description of this running wave is an expression for the deviation from equilibrium of each point on the cord as a function of position and time. In deriving this mathematical description we make use of our observations rather than follow the correct but more complicated method of solving the differential equation of motion for the cord.

Let us imagine that we take a picture of the cord at a time $t_o$; the result looks like Fig. 3-1. This is a periodic curve with wavelength $\lambda$, which means that we get the same deviation $\psi$ if we move a distance $\lambda$ to the right or left:

$$\psi(x \pm \lambda) = \psi(x) \tag{3-3}$$

The best known periodic function is the sine or cosine function, and if we make a careful inspection of our picture, as reproduced in Fig. 3-1, we find that this is indeed a sine or cosine function. The most general function of this kind can be written as

$$\psi(x) = A \sin[2\pi(\alpha x + \beta)] \tag{3-4a}$$

or

$$\psi(x) = A \cos[2\pi(\alpha' x + \beta')] \tag{3-4b}$$

since

$$\cos\alpha = \sin\left(\frac{\pi}{2} - \alpha\right) = \sin\left(\alpha + \frac{\pi}{2}\right) \tag{3-5}$$

These two expressions are entirely equivalent because the cosine function

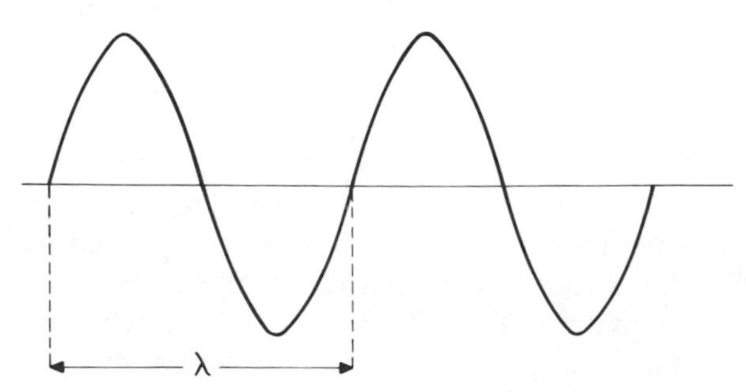


**Fig. 3-1** Sketch of a running wave in one dimension.

(3-4b) can be obtained from Eq. (3-4a) by a judicious selection of the parameters, and vice versa. Without any loss of generality we can consider only Eq. (3-4a).

It is easily seen that the maximum value of the function $\psi(x)$ of Eq. (3-4a) is equal to the parameter $A$. This parameter is therefore equal to the maximum deviation of any point on the cord from its equilibrium and is known as the amplitude of the wave. One of the other parameters, namely $\alpha$, is related to the quantity $\lambda$ of Fig. 3-1 which is known as the wavelength. It follows from Eq. (3-3) that

$$\alpha\lambda = 1 \tag{3-6}$$

Hence Eq. (3-4a) becomes

$$\psi(x) = A\sin\left[2\pi\left(\frac{x}{\lambda} + \beta\right)\right] \tag{3-7}$$

The parameter $\beta$ in this expression is known as the phase constant. A change in $\beta$ causes the curve of Fig. 3-1 to shift to the left or to the right, but it does not change the shape of the curve itself.

Now that we have analyzed the shape of the cord at a particular time $t_o$, we proceed to study the time dependence of the pattern. It follows from our observation that this time dependence is a simple one; we see that the whole pattern moves with a constant velocity $v$. Let us assume that the pattern moves in the direction of the positive $x$ axis (to the right in Fig. 3-1; in this case $v$ is positive).

We have just mentioned that the horizontal motion of the pattern of Fig. 3-1 is determined only by changes in the phase constant $\beta$, and since the time dependence of the wave motion consists solely of such horizontal motion, it follows that the time dependence of the wave function $\psi(x,t)$ must be concentrated in $\beta$:

$$\psi(x,t) = A\sin\left[2\pi\left(\frac{x}{\lambda} + \beta(t)\right)\right] \tag{3-8}$$

The wave function $\psi$ represents the deviation from equilibrium of each point on the cord as a function of position and time. Let us now consider the deviation $\psi(x_o, t_o)$ of a point $x_o$ at a certain time $t_o$. If we wait a certain amount of time $\Delta t$, the deviation at $x_o$ changes to the deviation that the point $x_o - v\Delta t$ had at the time $t_o$, since the whole pattern moved a distance $v\Delta t$ to the right during the time interval $\Delta t$. Hence we have the condition

$$\psi(x_o, t_o + \Delta t) = \psi(x_o - v\Delta t, t_o) \tag{3-9}$$

or

$$\psi(x_o, t_o) = \psi(x_o - v\Delta t, t_o - \Delta t) \tag{3-10}$$

This condition is satisfied only if $\psi$ is a function of $(x-vt)$; consequently, it follows from Eq. (3-8) that $\psi$ must be

$$\psi(x,t)=A\sin\left[\frac{2\pi}{\lambda}(x-vt)+\gamma\right] \tag{3-11}$$

where $\gamma$ is again a phase constant, but now it is time independent.

We would have arrived at essentially the same result if we had started from Eq. (3-4b), containing the cosine instead of the sine function of Eq. (3-4a) because of Eq. (3-5).

At a given time $t_o$ the position of the cord as a function of the position $x$ is described by the function

$$\psi(x,t_o)=A\sin\left[\frac{2\pi}{\lambda}(x-vt_o)+\gamma\right] \tag{3-12}$$

The position of the cord changes as a function of time, but we observe that there is a time $t_o+T$ when the position of the cord is exactly the same as it was at time $t_o$. This time $T$, which is known as the period of the oscillation motion, is determined by the condition

$$\psi(x,t+T)=\psi(x,t) \tag{3-13}$$

which leads to the expression

$$T=\frac{\lambda}{v} \tag{3-14}$$

The inverse of the period is known as the frequency $\nu$ of the wave and is given by

$$\nu=\frac{1}{T}=\frac{v}{\lambda} \tag{3-15}$$

Another quantity that is often used in the description of waves is the wave number $\sigma$, which is the inverse of the wavelength,

$$\sigma=\frac{1}{\lambda} \tag{3-16}$$

The function $\psi(x,t)$ of Eq. (3-12) can also be expressed in terms of $\sigma$ and $\nu$:

$$\psi(x,t)=A\sin[2\pi(\sigma x-\nu t)+\gamma] \tag{3-17}$$

## 3-4 Complete Sets

A mathematical discussion of the properties of waves is based mostly on the branch of applied mathematics known as Fourier analysis. We outline the main features of Fourier analysis without giving rigorous mathematical proofs of the various theorems that are quoted. We also use this opportunity to discuss a few aspects of approximations in the mean, since this will prove helpful when we treat eigenfunction expansions at a later stage in this book.

We consider two functions $f$ and $g$ of a real variable $x$, defined in the interval $a \leq x \leq b$. The functions may take on complex values. By definition the symbol $\langle f | g \rangle$ stands for

$$\langle f | g \rangle = \int_a^b f^*(x) g(x)\, dx \tag{3-18}$$

When the condition

$$\langle f | g \rangle = 0 \tag{3-19}$$

is satisfied, we call $f$ and $g$ orthogonal to each other. When

$$\langle f | f \rangle = 1 \tag{3-20}$$

we say that $f$ is normalized to unity or just normalized. A set of normalized functions $\phi_1, \phi_2, \phi_3, \ldots$, all of which are orthogonal to one another, is called an orthonormal set. We can write this as

$$\langle \phi_n | \phi_m \rangle = \delta_{n,m} \tag{3-21}$$

We introduced the symbol $\delta_{n,m}$ in Eq. (2-157); it is known as the Kronecker $\delta$ symbol and it is equal to unity when $n = m$ and equal to zero when $n \neq m$. The orthonormal set of functions $\phi_n$ can consist either of a finite or of an infinite number of functions. A simple example of an orthonormal set is the functions

$$\phi_n(x) = \frac{1}{\sqrt{2\pi}} e^{inx} \qquad n = 0, \pm 1, \pm 2, \ldots, \text{ and so on} \tag{3-22}$$

in the interval $0 \leq x \leq 2\pi$, since the integral

$$\int_0^{2\pi} \phi_n^*(x) \phi_m(x)\, dx = \frac{1}{2\pi} \int_0^{2\pi} e^{i(m-n)x}\, dx \tag{3-23}$$

is zero for $n \neq m$ and unity for $n = m$.

A complete set of functions $u_n(x)$ in the interval $a \leq x \leq b$ is defined by the property that there exists no other function $\chi$ in this interval such that $\chi$ is orthogonal to all functions $u_n$. An example of a complete set is the functions (3-22) for all integer values of $n$.

The various complete sets of functions that are considered in this book are all orthonormal. In addition, they usually satisfy certain boundary conditions, that is, linear relationships between the values of the functions $u_n(x)$ and their derivatives $u_n'(x)$ at the points $x = a$ and $x = b$. For example, in the case of the functions (3-22) we have

$$\phi_n(0)=\phi_n(2\pi)$$
$$\phi_n'(0)=\phi_n'(2\pi) \tag{3-24}$$

for all values of $n$. It can be shown that a function $f(x)$ that satisfies the same boundary conditions as the functions $u_n(x)$ of a complete set can be expanded as

$$f(x)=\sum_{n=1}^{\infty} c_n u_n(x) \tag{3-25}$$

in the interval $a \leq x \leq b$.

If we multiply Eq. (3-25) on the left by $u_m^*(x)$ and integrate, we obtain

$$c_m=\langle u_m|f\rangle \tag{3-26}$$

It follows that Eq. (3-25) is equivalent to

$$f(x)=\sum_{n=1}^{\infty} \langle u_n|f\rangle u_n(x) \tag{3-27}$$

Let us now show that Eq. (3-27) is always valid if the $u_n(x)$ form a complete set. If Eq. (3-27) were not valid, we would write

$$\chi(x)=f(x)-\sum_{n=1}^{\infty} \langle u_n|f\rangle u_n(x) \tag{3-28}$$

where $\chi(x)$ is not identically zero in the interval $a \leq x \leq b$. Multiplication on the left by $u_m^*(x)$ and subsequent integration of Eq. (3-28) give

$$\langle u_m|\chi\rangle=\langle u_m|f\rangle-\langle u_m|f\rangle=0 \tag{3-29}$$

for all $m$. This result is inconsistent with the definition of a complete set of functions, and we conclude, therefore, that $\chi(x)$ should be identically zero for $a \leq x \leq b$ or that Eq. (3-27) is valid.

It follows easily from Eq. (3-25) and from the orthonormality of the functions $u_n$ that

$$\langle f|f\rangle=\sum_{n=1}^{\infty}\sum_{m=1}^{\infty} c_n^* c_m \langle u_n|u_m\rangle$$
$$=\sum_{n=1}^{\infty}\sum_{m=1}^{\infty} c_n^* c_m \delta_{n,m}$$
$$=\sum_{n=1}^{\infty} c_n c_n^* \tag{3-30}$$

If we truncate the infinite sum then we have

$$\sum_{n=1}^{N} c_n c_n^* \leq \sum_{n=1}^{\infty} c_n c_n^* = \langle f | f \rangle \tag{3-31}$$

The mathematical criterion for the convergence of eigenfunction expansions is related to the convergence of the sequence $s_N$:

$$s_N = \sum_{n=1}^{N} c_n c_n^* \tag{3-32}$$

If for a given expansion the quantities $s_N$ have an upper limit for all values of $N$ and if

$$\lim_{N \to \infty} s_N = \langle f | f \rangle \tag{3-33}$$

we may say that the expansion of the function $f$ in terms of the complete set of functions $u_n$ is permissible and converges.

## 3-5 Fourier Analysis

Let us now apply these general considerations to Fourier series. A Fourier series is an expansion of a function $f(x)$ in terms of the orthonormal set of functions

$$\phi_n(x) = \frac{1}{\sqrt{2\pi}} e^{inx} \qquad n = 0, \pm 1, \pm 2, \ldots, \text{ and so on} \tag{3-34}$$

The expansion is written as

$$f(x) = \sum_{n=-\infty}^{\infty} a_n e^{inx} \tag{3-35}$$

with

$$a_n = \frac{1}{2\pi} \int_{-\pi}^{\pi} e^{-int} f(t)\, dt \tag{3-36}$$

The expansion of Eq. (3-35) is possible, and the Fourier series is convergent if $f$ is bounded and continuous in the interval $-\pi \leq x \leq \pi$. Actually $f$ does not even have to be continuous and bounded for the expansion to be possible; it is sufficient if $f$ satisfies the less restrictive set of Dirichlet's conditions, which require that $f$ have only a finite number of finite discontinuities and a finite number of minima or maxima in the interval $-\pi \leq x \leq \pi$. At any rate, most of the functions that we encounter can be expanded as Fourier series in the interval $-\pi \leq x \leq \pi$.

When the function $f(x)$ is periodic, that is, when it satisfies the condition

$$f(x+2\pi)=f(x) \tag{3-37}$$

the Fourier expansion is also valid outside the interval $-\pi < x < \pi$.

These considerations are easily extended to functions of more than one variable, which may be expanded as multiple Fourier series. For example, the function $F(x, y)$ is expanded as

$$F(x, y)=\sum_{m=-\infty}^{\infty}\sum_{n=-\infty}^{\infty} a_{m,n}e^{i(mx+ny)} \tag{3-38}$$

$$a_{m,n}=\frac{1}{4\pi^2}\int_{-\pi}^{\pi}\int_{-\pi}^{\pi} F(x, y)e^{-i(mx+ny)}\,dx\,dy$$

Similar expansions are possible for the functions $F(x, y, z)$, and so on.

In the following sections we make extensive use of the Fourier integral theorem, which can be made plausible from the Fourier series expansion. We consider a function $f(y)$, which satisfies Dirichlet's conditions in the interval $-\pi \leq y \leq \pi$, so that it can be expanded as

$$f(y)=\sum_{n=-\infty}^{\infty} a_n e^{iny} \tag{3-39}$$

with the coefficients given by

$$a_n=\frac{1}{2\pi}\int_{-\pi}^{\pi} f(s)e^{-ins}\,ds \tag{3-40}$$

We transform from the variables $y$ and $s$ to $x$ and $t$ by means of

$$\begin{aligned} ly &= \pi x \\ ls &= \pi t \end{aligned} \tag{3-41}$$

so that Eqs. (3-39) and (3-40) become

$$f(x)=\sum_{n=-\infty}^{\infty} a_n \exp\frac{in\pi x}{l} \tag{3-42}$$

$$a_n=\frac{1}{2l}\int_{-l}^{l}\exp\frac{-in\pi t}{l}f(t)\,dt$$

Substitution of the second Eq. (3-42) into the first one gives

$$f(x)=\frac{1}{2l}\sum_{n=-\infty}^{\infty}\int_{-l}^{l}f(t)\exp\frac{in\pi(x-t)}{l}dt \tag{3-43}$$

This expression is valid for the interval $-l\leq x\leq l$, and it becomes so for all values of $x$ and $t$ if we let $l$ tend to infinity. We introduce the small quantity $\delta=\pi/l$, which tends to zero if $l$ tends to infinity. In terms of $\delta$ Eq. (3-43) becomes

$$f(x)=\frac{1}{2\pi}\sum_{n=-\infty}^{\infty}\delta\int_{-l}^{l}f(t)e^{in\delta(x-t)}\,dt \tag{3-44}$$

When $l$ tends to infinity, $\delta$ tends to zero, and the summation of Eq. (3-44) can be replaced by an integration over a variable we call $u$. We find then that

$$f(x)=\frac{1}{2\pi}\int_{-\infty}^{\infty}du\int_{-\infty}^{\infty}f(t)e^{-iu(t-x)}\,dt \tag{3-45}$$

which is known as the Fourier integral theorem.

There are two alternate ways of expressing Eq. (3-45). In the first one we rewrite the equation as two separate steps:

$$f(x)=(2\pi)^{-1/2}\int_{-\infty}^{\infty}F(u)e^{iux}\,du \tag{3-46}$$

$$F(u)=(2\pi)^{-1/2}\int_{-\infty}^{\infty}f(t)e^{-iut}\,dt$$

The function $F(u)$ here is called the Fourier transform of $f(t)$. We see that if we take the Fourier transform of a function twice, we get the original function back.

The second way of expressing the Fourier integral theorem makes use of the Dirac $\delta$ function. This function is discussed in more detail in Chapter 8, Section 8-5. Here, we just mention that the function $\delta(x-x_o)$ is very small for all values of the variable $x$ except when $x$ is close to $x_o$, where the $\delta$ function becomes very large. It is defined by the property

$$f(x_o)=\int f(x)\delta(x-x_o)\,dx \tag{3-47}$$

and, consequently,

$$\int\delta(x-x_o)\,dx=1 \tag{3-48}$$

provided $x_o$ is contained within the limits of integration. Equation (3-48) follows from Eq. (3-47) if we take the function $f$ unity. The specific form of the $\delta$ function is not uniquely defined by these conditions, and there are many different ways in which it can be represented. However, it follows from a comparison of Eqs. (3-45) and (3-47) that one of the representations is

$$\delta(x-t)=\frac{1}{2\pi}\int_{-\infty}^{\infty} e^{iu(x-t)}\,du \tag{3-49}$$

Finally we wish to point out that the Fourier integral theorem is also valid for functions of more than one variable. For example, if the Fourier transform $F(\lambda,\mu,\nu)$ of the function $f(u,v,w)$ is given by

$$F(\lambda,\mu,\nu)=\frac{1}{\sqrt{(2\pi)^3}}\int_{-\infty}^{\infty}\int_{-\infty}^{\infty}\int_{-\infty}^{\infty} f(u,v,w)e^{-i(\lambda u+\mu v+\nu w)}\,du\,dv\,dw \tag{3-50}$$

then the function $f(x,y,z)$ is determined by

$$f(x,y,z)=\frac{1}{\sqrt{(2\pi)^3}}\int_{-\infty}^{\infty}\int_{-\infty}^{\infty}\int_{-\infty}^{\infty} F(\lambda,\mu,\nu)e^{i(\lambda x+\mu y+\nu z)}\,d\lambda\,d\mu\,d\nu \tag{3-51}$$

These two equations can also be combined to

$$f(x,y,z)=\frac{1}{(2\pi)^3}\int_{-\infty}^{\infty}\int_{-\infty}^{\infty}\int_{-\infty}^{\infty}\int_{-\infty}^{\infty}\int_{-\infty}^{\infty}\int_{-\infty}^{\infty} f(u,v,w)$$

$$\times e^{-i[\lambda(u-x)+\mu(v-y)+\nu(w-z)]}\,du\,dv\,dw\,d\lambda\,d\mu\,dy \tag{3-52}$$

## 3-6 Waves in Three Dimensions

Since actual physical phenomena occur in three dimensions rather than in one, we wish to extend our one-dimensional mathematical discussion of Section 3-3 to the three-dimensional case. An example of what we wish to consider is the propagation of sound in a gas, which is due to variations in the local density $\rho(x,y,z;t)$. The difference

$$\delta\rho(x,y,z;t)=\rho(x,y,z;t)-\bar{\rho} \tag{3-53}$$

where $\bar{\rho}$ is the average density, behaves like a three-dimensional wave and is connected with the behavior and the propagation of sound. The mathematical representation of $\delta\rho$, that is, a three-dimensional wave, is much more complicated than the one-dimensional case. Therefore, we try to simplify the general case by expressing it as a superposition of simple plane waves. This can be achieved by making use of the Fourier integral theorem of Section 3-5, but it requires the introduction of two new concepts, the use of complex wave functions and the superposition of monochromatic waves. We illustrate these ideas first for the one-dimensional case.

In Section 3-3 it was shown that the wave function $\psi(x,t)$, which represents the deviation of the cord from its equilibrium position, can be written as

$$\psi(x,t)=A\sin\left[2\pi(\sigma x-\nu t)+\gamma\right] \tag{3-54}$$

or

$$\psi(x,t)=A\cos\left[2\pi(\sigma x-\nu t)+\gamma\right] \tag{3-55}$$

Let us now consider the expression

$$\begin{aligned}\phi(x,t)&=Ae^{i\gamma}e^{2\pi i(\sigma x-\nu t)}\\&=A\{\cos[2\pi(\sigma x-\nu t)+\gamma]+i\sin[2\pi(\sigma x-\nu t)+\gamma]\}\end{aligned} \tag{3-56}$$

It is casily verified that Eqs. (3-54) and (3-55) are the real and imaginary parts, respectively, of the function $\phi(x,t)$ and that the wave motion is completely determined by the function $\phi(x,t)$. We can therefore take the complex function $\phi(x,t)$ as the mathematical representation of the wave motion, although in the examples that were used the observable quantities were all real and not complex. Equation (3-56) can be simplified even further if we allow $A$ to be complex; in this case we can include the phase factor $e^{i\gamma}$ in the complex amplitude and write

$$\phi(x,t)=Ae^{2\pi i(\sigma x-\nu t)} \tag{3-57}$$

as the wave function.

Let us now discuss the superposition of waves in one dimension. For this purpose we again consider the motion of the cord of Section 3-3, but now we suppose that instead of one person there are two people moving the cord up and down at different points. The effect of the action of the first person is described by the wave function

$$\phi_1(x,t)=A_1\exp\left[2\pi i(\sigma_1 x-\nu_1 t)\right] \tag{3-58}$$

and the second person's actions are represented by

$$\phi_2(x,t)=A_2\exp\left[2\pi i(\sigma_2 x-\nu_2 t)\right] \tag{3-59}$$

We now assume that the motion of the cord due to the combined actions of the two people is described by the function

$$\phi(x,t)=\phi_1(x,t)+\phi_2(x,t)$$
$$=A_1\exp[2\pi i(\sigma_1 x-\nu_1 t)]+A_2\exp[2\pi i(\sigma_2 x-\nu_2 t)] \tag{3-60}$$

This is called the superposition of the two separate wave motions of Eqs. (3-58) and (3-59). We write Eq. (3-60) in a different form for the case where

$$A_1=A_2=A \tag{3-61}$$

by substituting

$$\sigma_1=\sigma+\sigma' \qquad \nu_1=\nu+\nu' \tag{3-62}$$
$$\sigma_2=\sigma-\sigma' \qquad \nu_2=\nu-\nu'$$

into it. We obtain

$$\phi(x,t)=Ae^{2\pi i(\sigma x-\nu t)}[e^{2\pi i(\sigma' x-\nu' t)}+e^{-2\pi i(\sigma' x-\nu' t)}] \tag{3-63}$$

or

$$\phi(x,t)=2A\cos[2\pi(\sigma' x-\nu' t)]e^{2\pi i(\sigma x-\nu t)} \tag{3-64}$$

When $\sigma'$ is much smaller than $\sigma$, the position of the cord at a time $t_o$ is described by the situation in Fig. 3-2. There is a very rapid oscillation described by $\sigma$, and the amplitude of this wave varies much more slowly; this variation is determined by $\sigma'$. We see that the situation of Fig. 3-2 can be obtained as the sum of two single waves, and we speak here of a superposition of the single waves, leading to the situation of Fig. 3-2.

In the same way we can discuss the superposition of three or more waves or even of an infinite number of single waves. The most convenient way to study

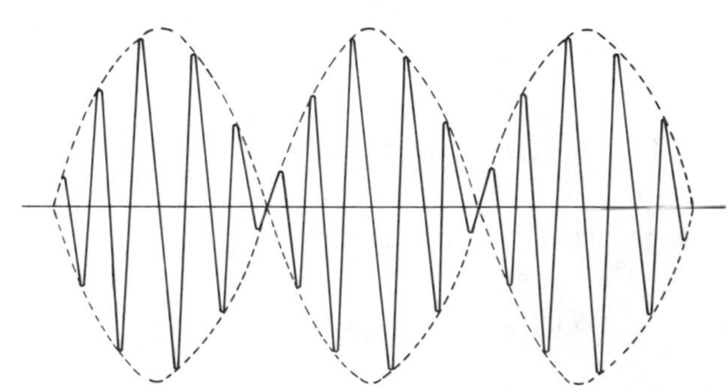

**Fig. 3-2** Superposition of two waves with similar frequencies [compare Eq. (3-64)].

this situation makes use of the Fourier integral theorem of Eq. (3-46). Let us consider an arbitrary motion of the cord, where the deviations are described by a function $D(x, t)$. The deviation $D(x,0)$ at the time $t=0$ can be written as

$$D(x,0)=\frac{1}{\sqrt{2\pi}}\int_{-\infty}^{\infty} F(u)e^{iux}\,du \qquad (3\text{-}65)$$

with

$$F(u)=\frac{1}{\sqrt{2\pi}}\int_{-\infty}^{\infty} D(s,0)e^{-ius}\,ds \qquad (3\text{-}66)$$

We write this in a more suitable form by replacing $u$ by $2\pi\sigma$ and by including the factor $\sqrt{2\pi}$ in the function $F$:

$$D(x,0)=\int_{-\infty}^{\infty} F(\sigma)e^{2\pi i\sigma x}\,d\sigma \qquad (3\text{-}67)$$

$$F(\sigma)=\int_{-\infty}^{\infty} D(x,0)e^{-2\pi i\sigma x}\,dx$$

At the time $t=0$ we have now expressed $D(x,0)$ as a superposition of an infinite number of single waves. The transformation function $F(\sigma)$, which represents the amplitudes of the single waves, is given by the second Eq. (3-67).

It follows that any arbitrary configuration of the cord at a specific time can be expressed as a superposition of single waves. What we would like to do is to predict the time dependence of $D(x, t)$ from our knowledge of the initial configuration $D(x,0)$, and we wonder how and when this can be achieved. We know that the time dependence of one individual wave is given by

$$e^{2\pi i(\sigma x-\nu t)} \qquad (3\text{-}68)$$

and we also know that $\nu$ is related to $\sigma$:

$$\nu=\nu(\sigma) \qquad (3\text{-}69)$$

The dependence of $\nu$ and $\sigma$ depends on the properties of the system. For the system of Section 3-3 this dependence was of the form

$$\nu=v\sigma \qquad (3\text{-}70)$$

but it should be realized that we cannot automatically assume such a simple function with a constant $v$ for waves in general. For example, it might be possible that $v$ depends on $\sigma$ or, if we consider different systems, that the function $\nu(\sigma)$ of Eq. (3-69) is completely different from Eq. (3-70). In general, we should allow for all possibilities for the functions $\nu(\sigma)$, depending on the kind of system under consideration. This function $\nu(\sigma)$ is known as the dispersion relation for the waves in the system that we consider, and it is this function that determines the time dependence of $D(x,t)$. This is easily seen by substituting Eq. (3-68) instead of the time-independent exponential into the first Eq. (3-67), which gives

$$D(x,t)=\int_{-\infty}^{\infty} F(\sigma)\exp 2\pi i[\sigma x-\nu(\sigma)t]\,d\sigma \tag{3-71}$$

with

$$F(\sigma)=\int_{-\infty}^{\infty} D(x,0)e^{-2\pi i\sigma x}\,dx \tag{3-72}$$

From the configuration $D(x,0)$ at $t=0$ we can determine the amplitude function $F(\sigma)$ through Eq. (3-72), and if we know the specific form of the dispersion relation $\nu(\sigma)$, we can predict the time dependence of $D(x,t)$ by means of Eq. (3-71). The motion of the system is thus described as a superposition of single waves, and it is completely determined by the dispersion relation $\nu(\sigma)$ and by the initial configuration $D(x,0)$.

The advantage of this mathematical description is that it can be extended immediately to the three dimensional case that we mentioned at the beginning of this section. For the sake of simplicity we use the symbol $\psi(x,y,z;t)$ or $\psi(\mathbf{r};t)$ instead of $\delta\rho$, and we assume that $\psi(\mathbf{r};0)$ is known. We then have

$$\psi(\mathbf{r};0)=\int_{-\infty}^{\infty} A(\sigma_x,\sigma_y,\sigma_z)e^{2\pi i(\sigma_x x+\sigma_y y+\sigma_z z)}\,d\sigma_x\,d\sigma_y\,d\sigma_z \tag{3-73}$$

$$A(\boldsymbol{\sigma})=\int_{-\infty}^{\infty} \psi(\mathbf{r};0)e^{-2\pi i(\sigma_x x+\sigma_y y+\sigma_z z)}\,dx\,dy\,dz$$

The difference from the one-dimensional case is that now we have three variables $\sigma_x$, $\sigma_y$, and $\sigma_z$, which form a vector $\boldsymbol{\sigma}=(\sigma_x,\sigma_y,\sigma_z)$, instead of the one variable $\sigma$ in one dimension. Let us now imagine that we know the three-dimensional dispersion relation

$$\nu=\nu(\sigma_x,\sigma_y,\sigma_z)=\nu(\boldsymbol{\sigma}) \tag{3-74}$$

The time dependence of

$$\psi(\mathbf{r};t)=\int_{-\infty}^{\infty} A(\boldsymbol{\sigma})\exp\{2\pi i[(\boldsymbol{\sigma}\cdot\mathbf{r})-\nu(\boldsymbol{\sigma})\cdot t]\}\,d\boldsymbol{\sigma} \tag{3-75}$$

is completely analogous to the one-dimensional case. Here the function $A(\boldsymbol{\sigma})$ is derived from $\psi(\mathbf{r};0)$ by means of the second Eq. (3-73).

This simple argument, consisting only of Eqs. (3-73) to (3-75), contains the complete mathematical description of waves in three dimensions. We now consider the physical aspects of the situation. It has been shown how in the one-dimensional case we have, in Eq. (3-71), decomposed a complicated wave pattern in terms of single waves, each of which was described by an exponential function. In Eq. (3-75) we have done the same in three dimensions; that is, we have written $\psi$ as a superposition of single waves that are described by the exponentials

$$g = \exp\{2\pi i[(\boldsymbol{\sigma} \cdot \mathbf{r}) - \nu(\boldsymbol{\sigma}) \cdot t]\} \tag{3-76}$$

A wave described by the simple function $g$ is called a plane wave. In order to appreciate this name we should consider the surfaces of equal phases; these are the surfaces over which $g$ is constant. It is easily verified from Eq. (3-76) that these surfaces are planes satisfying the equation

$$\sigma_x x + \sigma_y y + \sigma_z z = \text{const.} \tag{3-77}$$

All the planes are parallel and they are perpendicular to the vector $\boldsymbol{\sigma}$. This may be understood if we consider the plane of Eq. (3-77), which passes through the origin so that the constant is zero. Let us consider a vector between the origin and a point $r_o$ in this plane. The point $(x_o, y_o, z_o)$ is in the plane when it satisfies the condition

$$\sigma_x x_o + \sigma_y y_o + \sigma_z z_o = 0 \tag{3-78}$$

and this is also the condition that $\mathbf{r}_o$ is perpendicular to $\boldsymbol{\sigma}$.

All planes of equal phases are therefore perpendicular to $\boldsymbol{\sigma}$, and when they move as a function of time, they do so in the direction of propagation $\boldsymbol{\sigma}$. Therefore $\boldsymbol{\sigma}$ gives the direction of propagation of the waves. The velocity of propagation in the $\boldsymbol{\sigma}$ direction is given by

$$v = \frac{\nu}{\sigma} \tag{3-79}$$

and

$$\sigma = \left(\sigma_x^2 + \sigma_y^2 + \sigma_z^2\right)^{1/2} = \frac{1}{\lambda} \tag{3-80}$$

just as in the one-dimensional case. We call $\boldsymbol{\sigma}$ the wave vector of the plane wave; its direction determines the propagation of the wave, and its magnitude is the inverse of the wavelength.

Naturally we could have decomposed $\psi$ in a different way, but since plane waves are so easily visualized, we prefer to write $\psi$ as a superposition of plane waves. This is why Fourier analysis, which describes such superpositions, is so important in wave mechanics.

## 3-7 de Broglie's Quantum Postulate for Free Particles

We already described in Section 3-2 how Louis de Broglie suggested in 1924 that the motion of a free particle is associated with a three-dimensional wave. We mentioned some of the experimental results that confirmed the validity of de Broglie's hypothesis, in particular the work by Davisson and Germer on the diffraction of an electron beam by nickel.

According to de Broglie a free particle moving with a constant momentum $\mathbf{p}$ and energy $E$ is associated with a plane wave, determined by a wave vector $\boldsymbol{\sigma}$ and frequency $\nu$. The connection between these quantities is given by the relations

$$\mathbf{p} = h\boldsymbol{\sigma}$$

$$E = h\nu \tag{3-81}$$

Here $h$ is Planck's constant. We have seen that for an understanding of the time dependence of such a wave it is essential to know the dispersion relation. This is easily derived since we know from classical mechanics that

$$E = \frac{p^2}{2m} \tag{3-82}$$

If we substitute Eqs. (3-81) we find that

$$\nu = \frac{h}{2m}\left(\sigma_x^2 + \sigma_y^2 + \sigma_z^2\right) \tag{3-83}$$

is the desired dispersion formula. This expression is not relativistically invariant, but throughout this book we limit ourselves to nonrelativistic quantum mechanics.

The mathematical description of quantum mechanical systems by means of Schrödinger's differential equation was proposed in the beginning of 1926. It is not known to what extent Schrödinger was influenced by de Broglie's ideas, but it is quite clear that there was no direct connection between de Broglie's results and Schrödinger's equation.

It is possible to rationalize the Schrödinger equation starting with the mathematical description of the de Broglie waves. We give this derivation in the present chapter because we feel that it is instructive and it is useful to know the mathematical description of wave packets. However, it should be noted that this "derivation" of the Schrödinger equation did not occur in the historical course of events in 1926.

de Broglie showed that the wave hypothesis coupled with Eq. (3-1) leads to Ehrenfest's quantum conditions of Eq. (1-139) for the hydrogen atom. Let us consider an electron in a circular orbit around a hydrogen nucleus and moving with a constant momentum $p$. The motion of the electron is coupled to a wave whose length $\lambda$ is given by

$$\lambda=\frac{h}{p} \tag{3-84}$$

The radius of the circle is $r$, and in one complete orbit the electron covers a distance $2\pi r$. It can be understood that the electronic motion is stationary only when this complete orbit is an integer times the wavelength, otherwise the motion pattern would be different in successive orbits. We have, therefore, the condition

$$2\pi r=n\lambda \tag{3-85}$$

or, substituting Eq. (3-84),

$$pr=\frac{nh}{2\pi} \tag{3-86}$$

This is identical with the Ehrenfest condition of Eq. (1-139) (or with the more general Sommerfeld quantization condition).

We now return to the mathematical description of wave motion. We have seen that a single plane wave can be represented by a wave function

$$g(\mathbf{r};t)=\exp\{2\pi i[(\boldsymbol{\sigma}\cdot\mathbf{r})-\nu t]\} \tag{3-87}$$

If we substitute the expressions (3-81) for $\boldsymbol{\sigma}$ and $\nu$, we have the motion of a particle that is described by one plane wave. In general, we have to represent the motion of a particle by a superposition of plane waves; in this way we obtain wave mechanical descriptions that bear at least some resemblance to the ideas of classical mechanics. The most general wave motion can be represented as

$$\psi(\mathbf{r};t)=\iiint A(\boldsymbol{\sigma})\exp\{2\pi i[(\boldsymbol{\sigma}\cdot\mathbf{r})-\nu(\boldsymbol{\sigma})\cdot t]\}\,d\boldsymbol{\sigma} \tag{3-88}$$

with the amplitude function $A(\boldsymbol{\sigma})$ determined by

$$A(\boldsymbol{\sigma})=\iiint\psi(\mathbf{r};0)\exp[-2\pi i(\boldsymbol{\sigma}\cdot\mathbf{r})]\,d\mathbf{r} \tag{3-89}$$

and the function $\nu(\boldsymbol{\sigma})$ given by Eq. (3-83). Since the dispersion relation is known, the behavior of the probability pattern of the particle depends entirely on the initial choice of $\psi(\mathbf{r};0)$. We can approach the classical concept of a localized particle moving in a well-defined orbit by choosing $\psi(\mathbf{r};0)$ in such a way that it has a sharp maximum around a point $\mathbf{r}_o$. This assumption leads to a quantum mechanical description in which, at each time, the position and momentum of the particle can be determined within certain boundaries that are compatible with the Heisenberg uncertainty relations. Such a superposition of waves that corresponds to approximate localization of the particle is called a wave packet. We give a more detailed mathematical description of the properties of wave packets in the following section.

As we mentioned in Section 3-2, Schrödinger pointed out the importance of the function

$$\rho(x, y, z; t) = \psi(x, y, z; t)\psi^*(x, y, z; t) \tag{3-90}$$

but it was Born who related the function $\rho$ to the probability density of the particle. According to Born's interpretation the particle is localized, but we cannot make exact predictions about its position in space. Instead, we can only make statistical predictions by means of a probability density function. By definition, if we enclose a small volume $d\tau$ around a point $\mathbf{r}_o$, the probability of capturing the particle at a time $t_o$ is given by

$$P = \rho(\mathbf{r}_o; t)\, d\tau \tag{3-91}$$

According to this model, any predictions with regard to the positions of the particle are of a statistical nature, but the statistics are exactly defined.

It may be helpful to give a simple example. If we flip a coin, the probabilities for either heads or tails up are exactly equal. Consequently, the probabilities are exactly given. On the other hand, we cannot make any prediction about the result of one coin flip. We can make statistical predictions about a large number of coin flips, for example, if we flip 10,000 coins we may expect heads and tails to appear in about the same number, 5000 each. This means that probability predictions are fairly precise if we apply them to a large number of events. In this way, a probabilistic prediction is full of uncertainties even though the probability distribution at the basis of our prediction is an exact and precise quantity.

Let us now return to our quantum mechanics interpretation. Since the particle must be somewhere in space, it follows that the density function $\rho$ must satisfy the condition

$$\int \rho(x, y, z; t)\, dx\, dy\, dz = 1 \tag{3-92}$$

this condition is usually achieved by multiplying $\psi$ with a suitable normalization constant.

We should make certain that this normalization constant is time independent, otherwise the procedure above would not be consistent. In order to prove this time independence, we note that the function $A(\boldsymbol{\sigma})$ is time independent, so that we can normalize it to unity by multiplying it with a constant:

$$\int A(\boldsymbol{\sigma})A^*(\boldsymbol{\sigma})\, d\boldsymbol{\sigma} = 1 \tag{3-93}$$

The integral (3-92) can now be expressed according to Eq. (3-88):

$$\int \psi(\mathbf{r};t)\psi^*(\mathbf{r};t)\,d\mathbf{r} = \int\int A(\boldsymbol{\sigma}_1)A^*(\boldsymbol{\sigma}_2)\exp\{-2\pi it[\nu(\sigma_1)-\nu(\sigma_2)]\}\,d\boldsymbol{\sigma}_1\,d\boldsymbol{\sigma}_2$$

$$\times \int \exp\{2\pi i[(\boldsymbol{\sigma}_1-\boldsymbol{\sigma}_2)\cdot\mathbf{r}]\}\,d\mathbf{r} \tag{3-94}$$

The last integral is just the representation (3-49) of the $\delta$ function, so that we can write

$$\int \psi(\mathbf{r};t)\psi^*(\mathbf{r};t)\,dr = \int\int A(\boldsymbol{\sigma}_1)A^*(\boldsymbol{\sigma}_2)e^{-2\pi it[\nu(\sigma_1)-\nu(\sigma_2)]}\delta(\boldsymbol{\sigma}_1-\boldsymbol{\sigma}_2)\,d\boldsymbol{\sigma}_1\,d\boldsymbol{\sigma}_2$$

$$= \int A(\boldsymbol{\sigma})A^*(\boldsymbol{\sigma})\,d\boldsymbol{\sigma} \tag{3-95}$$

where we have introduced the three-dimensional $\delta$ function

$$\delta(\boldsymbol{\sigma}) = \delta(\sigma_x)\delta(\sigma_y)\delta(\sigma_z) \tag{3-96}$$

Since the last integral of Eq. (3-95) is time independent, the normalization integral for $\psi$ is time independent also.

Let us now consider the questions of the most probable position and momentum of the particle. One way of answering the first question would be to determine the maximum of the function $\rho(\mathbf{r};t)$ as a function of $\mathbf{r}$. However, this procedure is rather tedious, and besides, it is not consistent with other aspects of quantum mechanics. Instead, we define the expectation value $\bar{\mathbf{r}}(t)$ of $\mathbf{r}$ by means of

$$\bar{\mathbf{r}}(t) = \int \psi^*(\mathbf{r};t)\mathbf{r}\psi(\mathbf{r};t)\,d\mathbf{r} \tag{3-97}$$

and we consider it the average position of the particle at the time $t$. We can write Eq. (3-97) in a different form if we make use of the Dirac notation:

$$\bar{\mathbf{r}}(t) = \langle\psi(\mathbf{r};t)|\mathbf{r}|\psi(\mathbf{r};t)\rangle = \int \psi^*(\mathbf{r};t)\mathbf{r}\psi(\mathbf{r};t)\,d\mathbf{r} \tag{3-98}$$

Obviously, this is a shorthand notation for the three equations

$$\begin{aligned}\bar{x}(t) &= \langle\psi(\mathbf{r};t)|x|\psi(\mathbf{r};t)\rangle\\ \bar{y}(t) &= \langle\psi(\mathbf{r};t)|y|\psi(\mathbf{r};t)\rangle\\ \bar{z}(t) &= \langle\psi(\mathbf{r};t)|z|\psi(\mathbf{r};t)\rangle\end{aligned} \tag{3-99}$$

where all integrations are performed over the three dimensions $x$, $y$, and $z$.

The expectation value $\mathbf{p}$ of the momentum is derived by means of the de Broglie postulates (3-81) from the expectation value $\bar{\boldsymbol{\sigma}}$ of the wave vector, which is defined as

$$\bar{\boldsymbol{\sigma}}=\langle A(\boldsymbol{\sigma})|\boldsymbol{\sigma}|A(\boldsymbol{\sigma})\rangle=\int A^*(\boldsymbol{\sigma})\boldsymbol{\sigma}A(\boldsymbol{\sigma})\,d\boldsymbol{\sigma} \tag{3-100}$$

This illustrates that the connection between $\boldsymbol{\sigma}$ and $A(\boldsymbol{\sigma})$ is similar to that between $\mathbf{r}$ and $\psi(\mathbf{r};t)$. In fact, the probability distribution of $\boldsymbol{\sigma}$ depends on $A(\boldsymbol{\sigma})$ in exactly the same way as the $\mathbf{r}$ probability density depends on $\psi(\mathbf{r};t)$. The components $\sigma_x$, $\sigma_y$, and $\sigma_z$ define a three-dimensional space that is known as the momentum space. If we define a volume element $d\boldsymbol{\sigma}$ in momentum space around the point $\boldsymbol{\sigma}_o$, then the probability of finding $\boldsymbol{\sigma}$ in this volume element is given by

$$A(\boldsymbol{\sigma}_o)A^*(\boldsymbol{\sigma}_o)\,d\boldsymbol{\sigma} \tag{3-101}$$

in complete analogy to Eq. (3-91). The only difference between the probability patterns of $\mathbf{r}$ and $\boldsymbol{\sigma}$ for a free particle is that the first is time dependent and the second is not. This is related to one of the principles of classical mechanics that says that a free particle that is not subject to exterior forces has a momentum that is constant in time.

The definitions (3-98) and (3-100) are special cases of more general definitions which state that the expectation values $\bar{f}$ and $\bar{g}$ of a function $f(\mathbf{r})$ of the coordinates and of a function $g(\boldsymbol{\sigma})$ of the wave vector are given by

$$\bar{f}=\langle\psi(\mathbf{r};t)|f(\mathbf{r})|\psi(\mathbf{r};t)\rangle \tag{3-102}$$

and

$$\bar{g}=\langle A(\boldsymbol{\sigma})|g(\boldsymbol{\sigma})|A(\boldsymbol{\sigma})\rangle \tag{3-103}$$

respectively. According to these definitions the average square deviation $(\Delta x)^2$ of the coordinate $x$ with respect to the expectation value $\bar{x}$ is given by

$$(\Delta x)^2=\langle\psi(\mathbf{r};t)|(x-\bar{x})^2|\psi(\mathbf{r};t)\rangle \tag{3-104}$$

The average square deviation $(\Delta\sigma_x)^2$ of $\sigma_x$ with respect to $\bar{\sigma}_x$ is similarly defined as

$$(\Delta\sigma_x)^2=\langle A(\boldsymbol{\sigma})|(\sigma_x-\bar{\sigma}_x)^2|A(\boldsymbol{\sigma})\rangle \tag{3-105}$$

It is easily verified that Eqs. (3-104) and (3-105) are equivalent to

$$(\Delta x)^2=\langle\psi(\mathbf{r};t)|x^2-\bar{x}^2|\psi(\mathbf{r};t)\rangle \tag{3-106}$$

and

$$(\Delta\sigma_x)^2=\langle A(\boldsymbol{\sigma})|\sigma_x^2-\bar{\sigma}_x^2|A(\boldsymbol{\sigma})\rangle \tag{3-107}$$

We use these definitions for a more detailed discussion of the properties of wave packets in the next section.

## 3-8 Properties of Wave Packets and the Uncertainty Principle

The quantum mechanical analogue of the classical motion of a localized particle is a moving wave packet. Here the wave function has been chosen in such a way that the function $\psi\psi^*$ is very small everywhere except in the vicinity of the point $\mathbf{r}_o$, where it has a sharp maximum. The point $\mathbf{r}_o$ moves with a momentum that is approximately equal to $\mathbf{p}_o$; consequently, the wave vector of the wave packet should be approximately equal to $\boldsymbol{\sigma}_o$ and $h\boldsymbol{\sigma}_o = \mathbf{p}_o$. We first study a specific example of these wave packets; this will enable us to understand and discuss some of their general properties.

The specific example that we choose is identical with the wave packet that Heisenberg analyzed in his famous paper on the uncertainty relations. It is a one-dimensional wave packet whose wave function $\psi(x;t)$ at $t=0$ is given by

$$\psi(x;o) = (2\tau)^{1/4} \exp(-\pi\tau x^2 + 2\pi i\sigma_o x) \tag{3-108}$$

This function is normalized to unity.

In order to prove the normalization we report the following integration result

$$I = \int_{-\infty}^{\infty} \exp(-x^2)\,dx = \sqrt{\pi} \tag{3-109}$$

This can be derived by using polar coordinates,

$$I^2 = \int_{-\infty}^{\infty}\int_{-\infty}^{\infty} \exp(-x^2-y^2)\,dx\,dy = \int_0^{\infty} r\exp(-r^2)\,dr \int_0^{2\pi} d\phi = \pi \tag{3-110}$$

It follows from Eqs. (1-55), (1-57a), and (1-58) that

$$\int_{-\infty}^{\infty} \exp(-\alpha x^2)\,dx = \left(\frac{\pi}{\alpha}\right)^{1/2}$$

$$\int_{-\infty}^{\infty} x^2 \exp(-\alpha x^2)\,dx = \frac{1}{2\alpha}\left(\frac{\pi}{\alpha}\right)^{1/2} \tag{3-111}$$

The reader can easily verify that the function (3-108) is normalized to unity.

From Eq. (3-111) it can be derived also that

$$x_0 = \int_{-\infty}^{\infty} \psi^*(x;0)\,x\,\psi(x;o)\,dx = 0$$

$$(\Delta x)^2 = \int_{-\infty}^{\infty} \psi^*(x;o)\,x^2\,\psi(x;o)\,dx = \left(\frac{1}{4\pi\tau}\right) \tag{3-112}$$

We saw in the previous section that the time dependence of the wave function is determined by first evaluating the amplitude function $A(\sigma)$. According to Eq. (3-73), this function is given by

$$A(\sigma)=\int_{-\infty}^{\infty}\psi(x;0)e^{-2\pi i\sigma x}\,dx \tag{3-113}$$

which in the present case becomes

$$A(\sigma)=(2\tau)^{1/4}\int_{-\infty}^{\infty}e^{-\pi\tau x^2-2\pi i(\sigma-\sigma_o)x}\,dx \tag{3-114}$$

This integral can be derived from the result

$$\begin{aligned}\int_{-\infty}^{\infty}\exp(-ps^2-2iups)\,ds&=\exp(-pu^2)\int_{-\infty}^{\infty}\exp[-p(s^2+2ius+i^2u^2)]\,ds\\&=\exp(-pu^2)\int_{-\infty}^{\infty}\exp[-p(s+iu)^2]\,d(s+iu)\\&=\exp(-pu^2)\left(\frac{\pi}{p}\right)^{1/2}\end{aligned} \tag{3-115}$$

By combining Eqs. (3-114) and (3-115) we obtain

$$A(\sigma)=\left(\frac{2}{\tau}\right)^{1/4}\exp\left[-\frac{\pi(\sigma-\sigma_o)^2}{\tau}\right] \tag{3-116}$$

This function is also normalized to unity.

Let us now calculate the expectation values of $\sigma$ and $(\Delta\sigma)^2$. It is easily verified that

$$\bar{\sigma}=\left(\frac{2}{\tau}\right)^{1/2}\int\sigma\exp\left[-\frac{2\pi(\sigma-\sigma_o)^2}{\tau}\right]d\sigma=\sigma_o \tag{3-117}$$

and

$$(\Delta\sigma)^2=\left(\frac{2}{\tau}\right)^{1/2}\int(\sigma-\sigma_o)^2\exp[-2\pi(\sigma-\sigma_o)^2\tau]\,d\sigma=\frac{\tau}{4\pi} \tag{3-118}$$

Let us now compare the results for $\Delta x$ and $\Delta\sigma$ from Eqs. (3-112) and (3-118):

$$\Delta x=\frac{1}{\sqrt{4\pi\tau}}\qquad\Delta\sigma=\frac{\sqrt{\tau}}{\sqrt{4\pi}} \tag{3-119}$$

We can make $\Delta x$ smaller by increasing $\tau$, and in this way we can reduce the uncertainty in the position of the particle; but this will increase the uncertainty $\Delta\sigma$. If, on the other hand, we decrease $\tau$ in order to make $\Delta\sigma$ smaller, the uncertainty $\Delta x$ becomes larger. We see that the product

$$\Delta x \cdot \Delta\sigma = \frac{1}{4\pi} \tag{3-120}$$

is independent of $\tau$. According to de Broglie's postulates, we have

$$\Delta x \cdot \Delta p = \frac{\hbar}{2} \tag{3-121}$$

This result illustrates the validity of Heisenberg's uncertainty relations.

We now proceed to determine $\psi(x,t)$. It follows from Eqs. (3-88) and (3-116) that

$$\psi(x;t) = \left(\frac{2}{\tau}\right)^{1/4} \int_{-\infty}^{\infty} \exp\left[-\frac{\pi(\sigma-\sigma_o)^2}{\tau} + 2\pi i\sigma x - \frac{\pi i h\sigma^2}{m}t\right] d\sigma \tag{3-122}$$

since the dispersion relation $\nu(\sigma)$ is

$$\nu = \frac{h\sigma^2}{2m} \tag{3-123}$$

In order to evaluate this integral, we first change to a new integration variable $s = \sigma - \sigma_o$, which gives

$$\psi(x;t) = \left(\frac{2}{\tau}\right)^{1/4} \exp\left[2\pi i\left(\sigma_o x - \frac{h\sigma_o^2 t}{2m}\right)\right]$$
$$\int_{-\infty}^{\infty} \exp\left[-\left(\frac{\pi}{\tau} + \frac{\pi i h t}{m}\right)s^2\right] \exp\left[2\pi i\left(x - \frac{h\sigma_o t}{m}\right)s\right] ds \tag{3-124}$$

This integral can again be calculated by using Eq. (3-115). The result is

$$\psi(x,t) = (2\tau)^{1/4}\left(1 + \frac{ih\tau t}{m}\right)^{-1/2} \exp\left[-\pi\tau\left(1 + \frac{ih\tau t}{m}\right)^{-1}\left(x - \frac{h\sigma_o t}{m}\right)^2\right]$$
$$\times \exp\left[2\pi i\left(\sigma_o x - \frac{h\sigma_o^2 t}{2m}\right)\right] \tag{3-125}$$

It is also convenient to write the probability density function $\rho(x,t)$, which is

$$\rho(x,t) = (2\tau')^{1/2} \exp\left[-2\pi\tau'(x - v_o t)^2\right] \tag{3-126}$$

where we have introduced the new parameters

$$\tau'=\tau\left(1+\frac{h^2\tau^2t^2}{m^2}\right)^{-1} \qquad v_o=\frac{h\sigma_o}{m} \tag{3-127}$$

Equation (3-125) gives the complete description of the motion of the wave packet defined by Eq. (3-108). It follows from Eq. (3-125) that the expectation value of $x$ as a function of time is given by

$$x(t)=v_o t \tag{3-128}$$

which is in agreement with classical mechanics. The expectation value $(\Delta x)_t^2$ of $(\Delta x)^2$ as a function of $t$ is given by

$$(\Delta x)_t^2=\frac{1}{4\pi\tau'}=\frac{1}{4\pi\tau}\left(1+\frac{h^2\tau^2t^2}{m^2}\right) \tag{3-129}$$

It follows that $\Delta x$ increases with time, and since $\Delta\sigma$ is time independent, the product $\Delta x\cdot\Delta\sigma$ also increases as a function of time. Apparently Eq. (3-108) is a very good choice for the wave function at $t=0$, since at this time the product of the uncertainties $\Delta x$ and $\Delta\sigma$ is as small as we can ever hope to achieve. However, as time goes on, the uncertainty $\Delta x$ becomes larger and larger, and eventually there comes a time when $x$ is so large that we no longer know where the particle is supposed to be. This means that in quantum mechanics we can only follow the motion of the wave packet for a limited time, after which the accumulated uncertainties become so large that we cannot make any reliable predictions.

This argument can be extended to three dimensions without any difficulties. If we represent the three-dimensional wave function at $t=0$ as

$$\psi(\mathbf{r};0)=(8\tau_1\tau_2\tau_3)^{1/4}\exp\left[-\pi\left(\tau_1x^2+\tau_2y^2+\tau_3z^2\right)\right]\exp\left[2\pi i(\boldsymbol{\sigma}_0\cdot\mathbf{r})\right] \tag{3-130}$$

it follows from Eq. (3-106) that

$$A(\boldsymbol{\sigma})=\left(\frac{8}{\tau_1\tau_2\tau_3}\right)^{1/4}\exp\left[-\pi\left(\frac{(\sigma_x-\sigma_{o,x})^2}{\tau_1}+\frac{(\sigma_y-\sigma_{o,y})^2}{\tau_2}+\frac{(\sigma_z-\sigma_{o,z})^2}{\tau_3}\right)\right] \tag{3-131}$$

By analogy to Eq. (3-126) we find

$$\rho(r;t)=(8\tau_1'\tau_2'\tau_3')^{1/2}\exp\left\{-2\pi\left[\tau_1'(x-v_{o,x}t)^2+\tau_2'(y-v_{o,y}t)^2+\tau_3'(z-v_{o,z}t)^2\right]\right\} \tag{3-132}$$

with

$$\tau_i' = \tau_i \left(1 + \frac{h^2 \tau_i^2 t^2}{m^2}\right)^{-1} \quad (3\text{-}133)$$

$$m\mathbf{v}_o = h\boldsymbol{\sigma}_o$$

Obviously all our conclusions for one-dimensional motion are equally valid for each dimension in three-dimensional motion.

We saw in Section 3-3 that the propagation of a single wave is described by a velocity **v**, which is given by

$$v = \frac{\nu}{\sigma} \quad (3\text{-}134)$$

This is called the phase velocity of the individual waves. A wave packet, constructed as a superposition of waves as described above, has a second velocity **u** that is called the group velocity. This is the velocity at which the maximum of the function $\psi\psi^*$ moves along, and it is defined as the time derivative of the expectation value $\mathbf{r}(t)$. We show that **u** has the same direction as the expectation value of $\boldsymbol{\sigma}$ and that its magnitude is given by

$$u = \frac{d\nu}{d\sigma} \quad (3\text{-}135)$$

In order to prove this, let us first investigate how the point $\mathbf{r}(t)$ is characterized from a physical point of view. At $\mathbf{r}(t)$ the function $\psi(\mathbf{r}, t)$ has a maximum value, owing to the fact that the various waves from which the wave packet is constructed reinforce each other. They can do this only if they have approximately the same phase, and therefore $\mathbf{r}(t)$ is determined by the condition

$$x_o\, d\sigma_x + y_o\, d\sigma_y + z_o\, d\sigma_z - t_o\, d\nu = 0 \quad (3\text{-}136)$$

At a small time interval $dt$ later, the point $\mathbf{r}(t)$ has moved to $\mathbf{r}_o + d\mathbf{r}$, and we have

$$(x_o + dx)\, d\sigma_x + (y_o + dy)\, d\sigma_y + (z_o + dz)\, d\sigma_z - (t_o + dt)\, d\nu = 0 \quad (3\text{-}137)$$

If we subtract Eq. (3-136) from Eq. (3-137) and divide by $dt$, we obtain

$$\frac{dx}{dt} d\sigma_x + \frac{dy}{dt} d\sigma_y + \frac{dz}{dt} d\sigma_z - d\nu = 0 \quad (3\text{-}138)$$

From the definition of the group velocity **u** we know that the time derivatives in Eq. (3-138) are just the components of the group velocity, so that

$$u_x\, d\sigma_x + u_y\, d\sigma_y + u_z\, d\sigma_z - d\nu = 0 \quad (3\text{-}139)$$

The frequency $\nu$ is a function of $\boldsymbol{\sigma}$, and we have

$$d\nu = \frac{\partial \nu}{\partial \sigma_x} d\sigma_x + \frac{\partial \nu}{\partial \sigma_y} d\sigma_y + \frac{\partial \nu}{\partial \sigma_z} d\sigma_z \tag{3-140}$$

Substitution into Eq. (3-139) gives

$$\left(u_x - \frac{\partial \nu}{\partial \sigma_x}\right) d\sigma_x + \left(u_y - \frac{\partial \nu}{\partial \sigma_y}\right) d\sigma_y + \left(u_z - \frac{\partial \nu}{\partial \sigma_z}\right) d\sigma_z = 0 \tag{3-141}$$

Since this equation should be valid for all possible $d\sigma_\alpha$, we find that

$$u_x = \frac{\partial \nu}{\partial \sigma_x} \qquad u_y = \frac{\partial \nu}{\partial \sigma_y} \qquad u_z = \frac{\partial \nu}{\partial \sigma_z} \tag{3-142}$$

In most cases $\nu$ depends only on the modulus $\sigma$ of $\boldsymbol{\sigma}$. Then we find

$$\frac{\partial \nu}{\partial \sigma_\alpha} = \frac{d\nu}{d\sigma} \frac{\sigma_\alpha}{\sigma} \tag{3-143}$$

or

$$\mathbf{u} = \frac{1}{\sigma} \frac{d\nu}{d\sigma} \boldsymbol{\sigma} \tag{3-144}$$

by substituting Eq. (3-143) into Eq. (3-142). This means that $\mathbf{u}$ has the same direction as $\boldsymbol{\sigma}$ and that its magnitude is given by Eq. (3-135).

For a free particle, where the dispersion relation is given by Eq. (3-123), the group velocity is

$$u = \frac{p}{m} \tag{3-145}$$

We see that the motion of a quantum mechanical wave packet is in every respect equivalent to the classical motion of a localized particle.

## 3-9 The Schrödinger Equation for a Free Particle

In Section 3-7 we saw that the wave functions for a free particle can be written as

$$\psi(\mathbf{r}; t) = \int A(\boldsymbol{\sigma}) e^{2\pi i(\boldsymbol{\sigma}\cdot\mathbf{r} - \nu t)} d\boldsymbol{\sigma} \tag{3-146}$$

which is the most general superposition of plane waves of the form

$$g(\mathbf{r}; t) = e^{2\pi i(\boldsymbol{\sigma}\cdot\mathbf{r} - \nu t)} \tag{3-147}$$

Each of these plane waves has to satisfy the condition that $\nu$ and $\boldsymbol{\sigma}$ are related by means of the dispersion formula

$$\nu = \frac{h\sigma^2}{2m} \tag{3-148}$$

We show that each function $g(\mathbf{r}; t)$ of Eq. (3-147) that satisfies the condition of Eq. (3-148) is the solution of a specific differential equation. To this end we observe that

$$\begin{aligned}
\frac{1}{2\pi i}\frac{\partial g}{\partial x} &= \sigma_x g \\
\frac{1}{2\pi i}\frac{\partial g}{\partial y} &= \sigma_y g \\
\frac{1}{2\pi i}\frac{\partial g}{\partial z} &= \sigma_z g \\
-\frac{1}{2\pi i}\frac{\partial g}{\partial t} &= \nu g
\end{aligned} \tag{3-149}$$

and also that

$$\begin{aligned}
-\frac{1}{4\pi^2}\frac{\partial^2 g}{\partial x^2} &= \sigma_x^2 g \\
-\frac{1}{4\pi^2}\frac{\partial^2 g}{\partial y^2} &= \sigma_y^2 g \\
-\frac{1}{4\pi^2}\frac{\partial^2 g}{\partial z^2} &= \sigma_z^2 g
\end{aligned} \tag{3-150}$$

If $g$ satisfies the condition of Eq. (3-148), we can write

$$\left[\frac{h}{2m}\left(\sigma_x^2 + \sigma_y^2 + \sigma_z^2\right) - \nu\right] g(\mathbf{r}; t) = 0 \tag{3-151}$$

and if we make the substitutions of Eqs. (3-149) and (3-150), this becomes

$$-\frac{h}{8\pi^2 m}\left(\frac{\partial^2}{\partial x^2} + \frac{\partial^2}{\partial y^2} + \frac{\partial^2}{\partial z^2}\right) g + \frac{1}{2\pi i}\frac{\partial g}{\partial t} = 0 \tag{3-152}$$

It follows that every plane wave of the form of Eq. (3-147), whose frequency $\nu$ and wave vector $\boldsymbol{\sigma}$ are related by the dispersion relation (3-148), satisfies the differential equation (3-152).

Since we can choose arbitrary values for $\sigma_x$, $\sigma_y$, and $\sigma_z$, Eq. (3-152) has an infinite number of solutions, and as long as $\nu$ is given by Eq. (3-148), $g$ is the solution of Eq. (3-152). The most general solution of Eq. (3-152) is a linear combination of all the specific solutions of Eq. (3-152). We have already observed that the specific solutions of the differential equation are the functions $g(\mathbf{r}; t)$ of Eq. (3-147), where we have a free choice for $\sigma_x$, $\sigma_y$, and $\sigma_z$. The most general linear combination of these specific solutions is the function (3-146), and therefore $\psi(\mathbf{r}; t)$ is the general solution of Eq. (3-152). Any special solution of Eq. (3-152) can be obtained by making a specific choice for the function $A(\boldsymbol{\sigma})$ occurring in Eq. (3-146). Equation (3-152) can be written in a slightly different way:

$$-\frac{\hbar^2}{2m}\left(\frac{\partial^2}{\partial x^2}+\frac{\partial^2}{\partial y^2}+\frac{\partial^2}{\partial z^2}\right)\psi(\mathbf{r}; t)=-\frac{\hbar}{i}\frac{\partial}{\partial t}\psi(\mathbf{r}; t) \qquad (3\text{-}153)$$

In this form it is known as the Schrödinger equation for a free particle.

It may be helpful to recall the procedure that led to the formulation of the Schrödinger equation in anticipation of our intention to generalize it for nonfree particles. We started by equating the energy $E$ of the particle with the Hamiltonian as a function of the coordinates and momenta:

$$E=H(\mathbf{r}; p) \qquad (3\text{-}154)$$

For a free particle the Hamiltonian does not depend on $\mathbf{r}$, but generally, for bound particles, we have to allow for the possibility that $H$ contains the coordinate $\mathbf{r}$. Let us now combine Eq. (3-154) with de Broglie's postulates. This leads to

$$H(\mathbf{r}; h\boldsymbol{\sigma})-h\nu=0 \qquad (3\text{-}155)$$

From this result we derive that

$$\left[H(\mathbf{r}; h\boldsymbol{\sigma})-h\nu\right]g(\mathbf{r}; t)=0 \qquad (3\text{-}156)$$

and we found that we can replace the multiplications by $\boldsymbol{\sigma}$ with differentiations with respect to $x$, $y$, and $z$ when $g$ satisfies the dispersion relations. The differential equation (3-153) is therefore derived from

$$H\left(\mathbf{r}; \frac{\hbar}{i}\nabla\right)\psi(\mathbf{r}; t)=-\frac{\hbar}{i}\frac{\partial\psi}{\partial t} \qquad (3\text{-}157)$$

The left side of the Schrödinger equation (3-153) is obtained from the Hamiltonian by replacing the vector $\boldsymbol{\sigma}$ by the three-component operator $-i\hbar\nabla$ with $\nabla$ defined by Eq. (1-12).

The above Schrödinger equation, either in the form of Eq. (3-153) or Eq. (3-157) represents the motion of a free particle moving in a potential field that is zero at every point. In the following section we generalize the description to bound particles moving in nonzero potential fields.

## 3-10 The Schrödinger Equation for a Bound Particle

In Section 3-9 we discussed the quantum mechanical description of a free particle, whose energy was given by the expression

$$E = \frac{p^2}{2m} \tag{3-158}$$

This expression describes a particle that moves in a potential field that is everywhere equal to zero. Let us now consider a particle that moves in a potential field that is everywhere equal to a constant $U$. The energy expression is then

$$E = \frac{p^2}{2m} + U \tag{3-159}$$

The difference between Eqs. (3-158) and (3-159) is caused by a shift of the energy zero point from zero to $-U$, and obviously this should not cause any change in the physical behavior of the particle. We wish to investigate whether this change in energy zero point causes any changes in the wave functions.

Let the wave function $\psi(\mathbf{r};0)$ be known. The amplitude function $A(\sigma)$ is then given by

$$A(\boldsymbol{\sigma}) = \int \psi(\mathbf{r};0) e^{-2\pi i(\boldsymbol{\sigma}\cdot\mathbf{r})}\, d\mathbf{r} \tag{3-160}$$

Obviously, $A(\sigma)$ does not depend on the dispersion relation, and therefore it is independent of the energy zero point. On the other hand, the function $\psi(\mathbf{r}; t)$ is given by

$$\psi(\mathbf{r}; t) = \int A(\boldsymbol{\sigma}) e^{2\pi i(\boldsymbol{\sigma}\cdot\mathbf{r})} e^{-2\pi i h\sigma^2 t/2m}\, d\boldsymbol{\sigma} \tag{3-161}$$

The function $\psi'(\mathbf{r}; t)$ that we derive from the energy relation (3-159) is

$$\psi'(r; t) = \int A(\boldsymbol{\sigma}) e^{2\pi i(\boldsymbol{\sigma}\cdot\mathbf{r})} e^{-2\pi i Ut/h} e^{-2\pi i h\sigma^2 t/2m}\, d\boldsymbol{\sigma} \tag{3-162}$$

A comparison of Eqs. (3-161) and (3-162) shows that

$$\psi'(\mathbf{r}; t) = \psi(\mathbf{r}; t) e^{-2\pi i Ut/h} \tag{3-163}$$

Although the two functions $\psi'$ and $\psi$ are different, the corresponding probability density functions $\rho$ are identical, and it can be shown that the two sets of expectation values that are derived from $\psi$ and $\psi'$ are identical in every respect. We see that a change in energy zero point changes the wave function by a phase factor of unit modulus but that it does not affect any of our predictions of physical observables.

The functions $\psi$ and $\psi'$, however, satisfy different Schrödinger equations. From Eq. (3-163) we see that $\psi$ is a solution of the Schrödinger equation

$$-\frac{\hbar^2}{2m}\left(\frac{\partial^2}{\partial x^2}+\frac{\partial^2}{\partial y^2}+\frac{\partial^2}{\partial z^2}\right)\psi=-\frac{\hbar}{i}\frac{\partial\psi}{\partial t} \tag{3-164}$$

whereas $\psi'$ is a solution of

$$-\frac{\hbar^2}{2m}\left(\frac{\partial^2}{\partial x^2}+\frac{\partial^2}{\partial y^2}+\frac{\partial^2}{\partial z^2}\right)\psi'+U\psi'=-\frac{\hbar}{i}\frac{\partial\psi'}{\partial t} \tag{3-165}$$

as follows from Eq. (3-159).

What happens now when the potential field is zero in one part of space, for example, $x<0$, and the potential is equal to the constant $U$ in the rest of space, $x\geq 0$? It seems logical to expect that the wave function is a solution of Eq. (3-164) for $x<0$ and that it is a solution of Eq. (3-165) for $x\geq 0$. In addition, we expect the two solutions to match at the plane $x=0$, so that we impose the condition that both the wave function and its gradient are continuous in every point of the plane $x=0$.

We can extend this argument to the situation where there are three, four, or more regions of different potential, and finally we can consider the case of an infinite number of regions with different potentials. In particular, we construct a system where space is divided into an infinite number of cubes with dimensions $dx\,dy\,dz$ and where each cube has a different potential. If we number the cubes $1,2,3,\ldots,$ and so on, and if we denote the potential in the $n$th cube by $U_n$, then the wave function is determined by the equation

$$-\frac{\hbar^2}{2m}\left(\frac{\partial^2}{\partial x^2}+\frac{\partial^2}{\partial y^2}+\frac{\partial^2}{\partial z^2}\right)\psi_n+U_n\psi_n=-\frac{\hbar}{i}\frac{\partial\psi_n}{\partial t} \tag{3-166}$$

and by the condition that the $\psi_n$ match at the boundaries of the cubes. Since $U_n$ is constant throughout the $n$th cube, we can introduce the notation

$$U_n=U(\mathbf{r}_n) \tag{3-167}$$

where $\mathbf{r}_n$ is a point inside the $n$th cube.

Let us finally consider the limiting case where the dimensions of each cube tend to zero. The differential equation for the wave function then becomes

$$-\frac{\hbar^2}{2m}\left(\frac{\partial^2}{\partial x^2}+\frac{\partial^2}{\partial y^2}+\frac{\partial^2}{\partial z^2}\right)\psi(\mathbf{r};t)+U(\mathbf{r})\psi(\mathbf{r};t)=-\frac{\hbar}{i}\frac{\partial}{\partial t}\psi(\mathbf{r};t) \quad (3\text{-}168)$$

This is the Schrödinger equation for a particle that moves in a potential field $U(\mathbf{r})$.

We realize fully that this argument does not constitute a rigorous proof for the validity of the Schrödinger equation. On the other hand, we hope that it helps make the Schrödinger equation acceptable to the reader, and we feel that it gives us an idea of how the Schrödinger equation may be related to de Broglie's postulates. The ultimate justification of the Schrödinger equation lies in the fact that all its solutions lead to excellent agreement with observations.

Equation (3-168), known as the time-dependent Schrödinger equation, was published in 1926 by Schrödinger. It has an infinite number of specific solutions and a general solution that is a linear combination of all specific solutions. In many cases we are interested in the solutions that describe the stationary states of the system, that is, the situations that have specific, well-defined energies. Such a stationary state with an energy $E$ is described by a wave function with a specific frequency $\nu$, since $E=h\nu$. We can write such a wave function $\psi(\mathbf{r};t)$ as

$$\psi(\mathbf{r};t)=\phi(\mathbf{r})e^{-iEt/\hbar} \quad (3\text{-}169)$$

where $\phi(\mathbf{r})$ is now time independent. Substitution of the function of Eq. (3-169) into Eq. (3-168) leads to the following differential equation for $\phi$:

$$-\frac{\hbar^2}{2m}\left(\frac{\partial^2}{\partial x^2}+\frac{\partial^2}{\partial y^2}+\frac{\partial^2}{\partial z^2}\right)\phi(\mathbf{r})+U(\mathbf{r})\phi(\mathbf{r})=E\phi(\mathbf{r}) \quad (3\text{-}170)$$

This is known as the time-independent Schrödinger equation for the stationary states of the system. We can introduce a new symbol $\Delta$ for the partial differentiations, namely

$$\Delta=\nabla^2=\frac{\partial^2}{\partial x^2}+\frac{\partial^2}{\partial y^2}+\frac{\partial^2}{\partial z^2} \quad (3\text{-}171)$$

which is known as the Laplace operator. The Schrödinger equation can then be written as

$$\left[-\frac{\hbar^2}{2m}\Delta+U(\mathbf{r})\right]\phi(\mathbf{r})=E\phi(\mathbf{r}) \quad (3\text{-}172)$$

Each solution of Eq. (3-172) contains $E$ as a parameter, and since $E$ is a continuous variable, there are an infinite number of solutions. However, many of these solutions are not permissible because of physical requirements. We should realize that a suitable wave function should not only be a solution of the Schrödinger equation, but it should also describe the behavior of the particle in a realistic way. The latter requirement poses some restrictions on the wave function. Our first condition is that the wave function be continuous and that its gradient be continuous everywhere in space. The second is that it be single-valued everywhere. The third is that it be normalizable, that is, that the integral $\langle \phi | \phi \rangle$ be finite.

It is said that every rule has its exceptions, and we can think of a number of situations where a physically acceptable wave function violates some of the conditions above. However, we think it best to disregard these exceptions for the time being and to postpone their discussion until we have fully appreciated the meaning of the conditions above.

The solutions of the Schrödinger equation (3-172) can be written in the form $\phi(\mathbf{r}; E)$, since they contain the continuous variable $E$ as a parameter. In most situations we find that many of these solutions do not satisfy the three conditions that were stated above. Only if we substitute certain specific values $E_1, E_2, E_3, \ldots,$ and so on for the parameter $E$ do the corresponding functions $\phi(\mathbf{r}; E_1), \phi(\mathbf{r}; E_2), \ldots,$ and so on satisfy the necessary conditions. We call these values $E_1, E_2, \ldots,$ and so on the energy eigenvalues of the Schrödinger equation and the functions $\phi_1(\mathbf{r}) = \phi(\mathbf{r}; E_1), \phi_2(\mathbf{r}) = \phi(\mathbf{r}; E_2), \ldots,$ and so on the corresponding eigenfunctions.

In the following chapter we discuss a number of simple examples to show how the Schrödinger equation is solved and how its eigenvalues and eigenfunctions are obtained. We also show how the Schrödinger equation can be expressed in terms of operator language and we derive some properties of the eigenvalues and eigenfunctions of operators.

## Problems

**1** We consider a one-dimensional wave packet and we denote the expectation value of the coordinate $x$ at time $t$ by $\bar{x}(t)$ and the expectation value of the wave number $\sigma$ by $\bar{\sigma}$. Prove that for a quantum mechanical wave packet we always have

$$\bar{x}(t) = \bar{x}(0) + \left(\frac{ht}{m}\right)\bar{\sigma}$$

**2** If $A(\sigma)$ is real and the origin is chosen so that at $t=0$ the expectation value of $x$, $\bar{x}(0)$ is zero, show that at later times

$$(\Delta x)_t^2 = (\Delta x)_{t=0}^2 + \left(\frac{h^2 t^2}{m^2}\right)(\Delta\sigma)^2$$

**3** Prove by using Fourier analysis that in the interval $-\pi<x<\pi$ we have

$$\frac{x}{2}=\frac{\sin x}{1}-\frac{\sin 2x}{2}+\frac{\sin 3x}{3}-\frac{\sin 4x}{4}+\cdots=\sum_{n=1}^{\infty}\frac{(-1)^{n-1}\sin nx}{n}$$

**4** The function $f(x)$ is defined in the interval $-\pi<x<\pi$ as

$$f(x)=0 \quad (x<0), \qquad f(x)=\left(\frac{A}{2}\right) \quad (x=0), \qquad f(x)=A \quad (x>0)$$

Expand $f(x)$ in a Fourier series.

**5** Evaluate the expectation value of $r^2=x^2+y^2+z^2$ with respect to the three-dimensional wave function $\psi(\mathbf{r};0)$ of Eq. (3-130).

**6** Prove that the integral

$$I=\int f(\mathbf{r}_1)g(\mathbf{r}_2)h(\mathbf{r}_1-\mathbf{r}_2)\,d\mathbf{r}_1\,d\mathbf{r}_2$$

is equal to

$$I=(2\pi)^{-3}\int F(-\boldsymbol{\sigma})G(\boldsymbol{\sigma})H(\boldsymbol{\sigma})\,d\boldsymbol{\sigma}$$

with

$$F(\boldsymbol{\sigma})=\int f(\mathbf{t})e^{-i\boldsymbol{\sigma}\cdot\mathbf{t}}\,d\mathbf{t}$$

$$G(\boldsymbol{\sigma})=\int g(\mathbf{t})e^{-i\boldsymbol{\sigma}\cdot\mathbf{t}}\,d\mathbf{t}$$

$$H(\boldsymbol{\sigma})=\int h(\mathbf{t})e^{-i\boldsymbol{\sigma}\cdot\mathbf{t}}\,d\mathbf{t}$$

## Recommended Reading

The various historical sections of this chapter are discussed in detail in the text by Jammer, and the subject of wave packets and wave propagation is treated in Kramers' textbook on quantum mechanics. A standard textbook on Fourier analysis is *Fourier Transforms* by Sneddon.

Max Jammer, *The Conceptual Development of Quantum Mechanics*, McGraw-Hill Book Company, New York, 1966.

H. A. Kramers, *Quantum Mechanics*, North-Holland Publishing Company, Amsterdam, 1958.

Ian N. Sneddon, *Fourier Transforms*, McGraw-Hill Book Company, New York, 1951.

CHAPTER FOUR

# Some Properties and Simple Applications of the Schrödinger Equation

## 4-1 Introduction

In this chapter we discuss some properties and some simple applications of the Schrödinger equation. The various applications that we illustrate all refer to one-dimensional problems where the potential function is presented by a step function. As we shall see, these problems are not necessarily simple, but the differential equations that must be solved are fairly straightforward. They can all be written in the forms

$$\frac{d^2y}{dx^2} = -\lambda^2 y \tag{4-1}$$

or

$$\frac{d^2y}{dx^2} = \mu^2 y \tag{4-2}$$

where $\lambda$ and $\mu$ are real positive parameters.

We present a more comprehensive treatment of the theory of differential equations in Chapter 5. We explain there that every second-order differential equation has two linearly independent solutions and that the general solution can be written as a linear combination of these two solutions. It is easily verified that the two independent solutions of the first Eq. (4-1) can be written either as

$$y_1 = \sin\lambda x \qquad y_2 = \cos\lambda x \tag{4-3}$$

or as

$$y_1' = e^{i\lambda x} \qquad y_2' = e^{-i\lambda x} \tag{4-4}$$

The general solution of Eq. (4-1) can therefore be written as

$$y = A_1 \sin \lambda x + A_2 \cos \lambda x = B_1 e^{i\lambda x} + B_2 e^{-i\lambda x} \tag{4-5}$$

The two representations of Eq. (4-5) are entirely equivalent since $A_1$, $A_2$, $B_1$, and $B_2$ are arbitrary parameters. In our applications we choose one or the other, whichever of the two seems to be more suitable.

It can also be verified that the two solutions of Eq. (4-2) are

$$y_1 = e^{\mu x} \qquad y_2 = e^{-\mu x} \tag{4-6}$$

The general solution of Eq. (4-2) is given as

$$y = C_1 e^{\mu x} + C_2 e^{-\mu x} \tag{4-7}$$

Various applications that involve more complicated differential equations will be discussed in Chapter 5, where we also discuss the general theory of differential equations. In the present chapter we discuss the particle in a box, the particle in a rectangular potential well, the particle tunneling through a square potential barrier, and the particle moving in a periodic step-function-type potential field.

We also discuss various properties of operators and their eigenvalues and eigenfunctions since the Schrödinger equation may be considered as an eigenvalue problem. In Section 4-2 we define operators and in Section 4-3 we derive some properties of their eigenfunctions and eigenvalues. It should be noted that parts of Section 4-2, in particular the integral representation of operators, was developed by Max Born together with Norbert Wiener during Born's visit to the Massachusetts Institute of Technology in 1925. The integral representation is the basis for the formal mathematical description of operator properties.

In the subsequent sections of this chapter we alternate the various applications with general discussions of eigenvalue problems. We feel that this alternation promotes a better understanding of both topics.

## 4-2 Operators

A procedure that transforms a given function $f$ into a different function $g$ can be represented as

$$g = \Omega f \tag{4-8}$$

where $\Omega$ is called an operator.

In quantum mechanics we encounter mostly multiplicative and differential operators. For example, multiplication of $f$ by a function $h$, which we write as

$$g = hf \tag{4-9}$$

can also be written in the operator representation of Eq. (4-8) if we take the operator $\Omega$ equal to $h$. If $g$ is obtained as the derivative of $f$,

$$g=\frac{\partial f}{\partial x} \tag{4-10}$$

then we have to take

$$\Omega=\frac{\partial}{\partial x} \tag{4-11}$$

We have already introduced the Laplace operator

$$\Delta=\frac{\partial^2}{\partial x^2}+\frac{\partial^2}{\partial y^2}+\frac{\partial^2}{\partial z^2} \tag{4-12}$$

Obviously, we have

$$\Delta f=\frac{\partial^2 f}{\partial x^2}+\frac{\partial^2 f}{\partial y^2}+\frac{\partial^2 f}{\partial z^2} \tag{4-13}$$

These examples are all linear operators, which are defined by the property

$$\Omega(af_1+bf_2)=a\Omega f_1+b\Omega f_2 \tag{4-14}$$

Not all operators are linear. If we take the function $g$ of Eq. (4-8) as the square of $f$, we can write this in operator language as

$$g=(\mathrm{sq})_{\mathrm{op}}f \tag{4-15}$$

In general,

$$(\mathrm{sq})_{\mathrm{op}}(f_1+f_2)\neq(\mathrm{sq})_{\mathrm{op}}f_1+(\mathrm{sq})_{\mathrm{op}}f_2 \tag{4-16}$$

and it follows from the definition (4-14) that the operator sq is nonlinear. Throughout this book we limit our considerations to linear operators. We assume, therefore, that all operators that we encounter satisfy Eq. (4-14).

The sum $\Omega$ of two operators $\Omega_1$ and $\Omega_2$ is defined by

$$\Omega f=\Omega_1 f+\Omega_2 f \tag{4-17}$$

Similarly, we dcfine the product

$$\Lambda=\Omega_1\Omega_2 \tag{4-18}$$

by

$$\Lambda f=\Omega_1\Omega_2 f=\Omega_1(\Omega_2 f) \tag{4-19}$$

It is easily verified that

$$\Omega_1 + \Omega_2 = \Omega_2 + \Omega_1 \tag{4-20}$$

is always valid, but often

$$\Omega_1\Omega_2 \neq \Omega_2\Omega_1 \tag{4-21}$$

If we take

$$\Omega_1 = x \qquad \Omega_2 = \frac{\partial}{\partial x} \tag{4-22}$$

then

$$\Omega_1\Omega_2 f = x\frac{\partial f}{\partial x} \tag{4-23}$$

and

$$\Omega_2\Omega_1 f = \frac{\partial}{\partial x}(xf) = x\frac{\partial f}{\partial x} + f \tag{4-24}$$

This is an example of the inequality (4-21). On the other hand, if we take

$$\Omega_1 = \frac{\partial}{\partial x} \qquad \Omega_2 = \frac{\partial}{\partial y} \tag{4-25}$$

then

$$\frac{\partial}{\partial x}\left(\frac{\partial f}{\partial y}\right) = \frac{\partial}{\partial y}\left(\frac{\partial f}{\partial x}\right) \tag{4-26}$$

so that

$$\Omega_1\Omega_2 = \Omega_2\Omega_1 \tag{4-27}$$

We say that two operators $\Omega_1$ and $\Omega_2$ commute when they satisfy Eq. (4-27). The commutator $[\Omega_1, \Omega_2]$ of the two operators is defined as

$$[\Omega_1, \Omega_2] = \Omega_1\Omega_2 - \Omega_2\Omega_1 \tag{4-28}$$

Obviously the commutator is equal to zero when the two operators commute.
The complex conjugate $\Omega^*$ of an operator $\Omega$ is defined by

$$(\Omega f)^* = \Omega^* f^* \tag{4-29}$$

An operator is defined to be Hermitian if for any two functions $f$ and $g$ it satisfies the condition

$$\int f\Omega g\,dq = \int g\Omega^* f\,dq \tag{4-30}$$

Here the symbol $q$ represents all integration variables. It will be seen that Hermitian operators play a very important role in quantum mechanics.

It is convenient to introduce a new notation for the integrals of Eq. (4-30). We define

$$\langle f|\Omega|g\rangle = \int f^*\Omega g\,dq \tag{4-31}$$

The advantage of this notation is that we do not have to write out all the integration variables. In our new notation we find that $\Omega$ is Hermitian if

$$\langle f|\Omega|g\rangle = \langle g^*|\Omega^*|f^*\rangle = \langle g|\Omega|f\rangle^* \tag{4-32}$$

It is interesting to let an operator $\Omega$ work on the functions $\phi_n$ of a complete set:

$$g_n = \Omega\phi_n \tag{4-33}$$

The result can now be expanded in terms of the complete set

$$g_n = \sum_m a_{n,m}\phi_m \tag{4-34}$$

so that we have

$$\Omega\phi_n = \sum_m a_{n,m}\phi_m \tag{4-35}$$

In Chapter 3, where we discussed the properties of complete sets of functions, we mentioned that usually the functions are orthonormal:

$$\langle\phi_n|\phi_m\rangle = \delta_{n,m} \tag{4-36}$$

We assume that this is the case here. If we multiply Eq. (4-35) on the left by $\phi_k^*$ and integrate, we obtain

$$a_{n,k} = \langle\phi_k|\Omega|\phi_n\rangle = \Omega_{k,n} \tag{4-37}$$

The effect of $\Omega$ on any of the functions $\phi_n$ can thus be described as

$$\Omega\phi_n = \sum_k \Omega_{k,n}\phi_k \tag{4-38}$$

In this way we have represented the operator as a matrix of infinite order.

If we take a different complete set of functions $\psi_n$ as a starting point,we obtain

$$\Omega\psi_n = \sum_k \Omega'_{k,n}\psi_k$$

$$\Omega'_{k,n} = \langle \psi_k | \Omega | \psi_n \rangle \tag{4-39}$$

This gives a different matrix of infinite order as the representation of the operator. In general, there is an infinite number of ways in which we can describe the operator by way of a matrix. All these matrices can have different elements, but certain essential characteristics of the operator are contained in every matrix, and certain matrix representations are often useful for deriving some properties of the operator.

Some of the procedures in Section 3-5, where we discussed Fourier transforms, can also be expressed in terms of operators. For instance, in Eq. (3-46) we introduced the Fourier transform $F(u)$ of a function $f(t)$ by means of

$$F(u) = (2\pi)^{-1/2} \int_{-\infty}^{\infty} e^{-iut} f(t)\, dt \tag{4-40}$$

Since this describes a specific procedure for deriving a function $F$ from a function $f$, we can also represent this procedure as

$$F = \Lambda f = \int_{-\infty}^{\infty} \Lambda(x;x') f(x')\, dx' \tag{4-41}$$

where

$$\Lambda(x;x') = (2\pi)^{-1/2} \exp(-ixx') \tag{4-42}$$

is the Kernel of the integral operator $\Lambda$. This is an example of an integral representation of an operator. It can be shown that many operators may be described in this fashion.

The unit operator $\Omega_E$ describes the multiplication of a function by unity and is defined by

$$f = \Omega_E f \tag{4-43}$$

The integral representation of this operator can be written as

$$f(x) = \int_{-\infty}^{\infty} \Omega_E(x, x') f(x')\, dx' \tag{4-44}$$

and if we compare this with Eq. (3-47),

$$f(x) = \int_{-\infty}^{\infty} \delta(x - x') f(x')\, dx' \tag{4-45}$$

we conclude that

$$\Omega_E(x, x') = \delta(x - x') \tag{4-46}$$

In operator language, therefore, we may call the Dirac $\delta$ function the integral representation of the unit operator. From the $\delta$ function we can also derive the integral representation of the differential operator

$$D = \frac{\partial}{\partial x} \tag{4-47}$$

This integral representation is defined by

$$\frac{\partial f}{\partial x} = \int_{-\infty}^{\infty} D(x, x') f(x')\, dx' \tag{4-48}$$

From Eq. 4-45 we see that

$$\frac{\partial f(x)}{\partial x} = \int_{-\infty}^{\infty} \delta(x - x') \frac{\partial f(x')}{\partial x'}\, dx' \tag{4-49}$$

Integration by parts gives

$$\begin{aligned} \frac{\partial f(x)}{\partial x} &= \delta(x - x') f(x') \Big|_{-\infty}^{\infty} - \int_{-\infty}^{\infty} \left[ \frac{\partial}{\partial x'} \delta(x - x') \right] f(x')\, dx' \\ &= \int_{-\infty}^{\infty} \left[ \frac{\partial}{\partial x} \delta(x - x') \right] f(x')\, dx' = \int_{-\infty}^{\infty} \delta'(x - x') f(x')\, dx' \end{aligned} \tag{4-50}$$

By comparing Eqs. (4-48) and (4-50) we find that

$$\delta'(x - x') = D(x, x') \tag{4-51}$$

so that the derivative of the $\delta$ function is the integral representation of the operator $D$.

## 4-3 Eigenvalues and Eigenfunctions of Operators

We have seen that a given operator $\Lambda$, when working on an arbitrary function $f$, usually produces a completely different function $g$. However, there usually exists a set of functions $f_n$ for which the result of the operation is proportional to the original function:

$$\Lambda f_n = \lambda_n f_n \tag{4-52}$$

We call $\lambda_n$ an eigenvalue of the operator $\Lambda$ and $f_n$ its corresponding eigenfunction. An eigenvalue is called nondegenerate when it has only one eigenfunction; otherwise we call it degenerate. When there are $s$ different, linearly independent eigenfunctions $f_{k,s}$, all belonging to the same eigenvalue $\lambda_k$, then we call $\lambda_k$ an $s$-fold degenerate eigenvalue.

It is easily seen that the Schrödinger equation

$$-\frac{\hbar^2}{2m}\left(\frac{\partial^2}{\partial x^2}+\frac{\partial^2}{\partial y^2}+\frac{\partial^2}{\partial z^2}\right)\psi+V(x,y,z)\psi=E\psi \tag{4-53}$$

can also be written as an eigenvalue problem. If we take the Hamiltonian operator $H$ as

$$H=-\frac{\hbar^2}{2m}\left(\frac{\partial^2}{\partial x^2}+\frac{\partial^2}{\partial y^2}+\frac{\partial^2}{\partial z^2}\right)+V(x,y,z) \tag{4-54}$$

then we can write the time-independent Schrödinger equation as

$$H\psi=E\psi \tag{4-55}$$

The solutions are given by

$$H\psi_n=E_n\psi_n \tag{4-56}$$

and we see that $E_n$ and $\psi_n$ are eigenvalues and eigenfunctions, respectively, of the operator $H$.

From a mathematician's point of view the one-dimensional Schrödinger equation can be regarded as a special case of a Sturm-Liouville equation. This is a second-order differential equation over an interval $a\leq x\leq b$, where the solutions are restricted by the condition that they have certain prescribed values at the boundaries $x=a$ and $x=b$. In the case of the particle in a box, the boundary conditions are that the solution is zero at the two boundaries. It is useful to know that all these different problems are related to one another, since it enables us to describe the same system from different mathematical approaches.

In quantum mechanics Hermitian operators play an important role, since in the formal presentation of the subject a basic assumption is that every observable can be represented by a Hermitian operator. We briefly discuss some properties of the eigenvalues and eigenfunctions of Hermitian operators.

It is easily shown that the eigenvalues $h_n$ of a Hermitian operator $H$ are all real. We write

$$H\psi_n=h_n\psi_n \tag{4-57}$$

Multiplication on the left by $\psi_n^*$ and subsequent integration give

$$\langle\psi_n|H|\psi_n\rangle=h_n\langle\psi_n|\psi_n\rangle \tag{4-58}$$

The complex conjugate of this equation is

$$\langle \psi_n | H | \psi_n \rangle^* = h_n^* \langle \psi_n | \psi_n \rangle^* = h_n^* \langle \psi_n | \psi_n \rangle \tag{4-59}$$

Since $H$ is Hermitian, we can also write this as

$$\langle \psi_n | H | \psi_n \rangle = h_n^* \langle \psi_n | \psi_n \rangle \tag{4-60}$$

It now follows from Eqs. (4-58) and (4-60) that

$$h_n = h_n^* \tag{4-61}$$

which means that the eigenvalue is real.

Two eigenfunctions $\psi_n$ and $\psi_m$ that belong to two different eigenvalues $h_n$ and $h_m$, respectively, of a Hermitian operator $H$ are orthogonal. We prove this by first writing the two equations

$$\begin{aligned} H\psi_n &= h_n \psi_n \\ (H\psi_m)^* &= h_m \psi_m^* \end{aligned} \tag{4-62}$$

Multiplication of these equations on the left by $\psi_m^*$ and by $\psi_n$, respectively, and subsequent integration give

$$\begin{aligned} \langle \psi_m | H | \psi_n \rangle &= h_n \langle \psi_m | \psi_n \rangle \\ \langle \psi_n | H | \psi_m \rangle^* &= h_m \langle \psi_n | \psi_m \rangle^* \end{aligned} \tag{4-63}$$

Since $H$ is Hermitian, we can also write the second Eq. (4-63) as

$$\langle \psi_m | H | \psi_n \rangle = h_m \langle \psi_m | \psi_n \rangle \tag{4-64}$$

If we subtract this from the first Eq. (4-63), we obtain

$$(h_n - h_m) \langle \psi_m | \psi_n \rangle = 0 \tag{4-65}$$

and it follows that $\psi_n$ and $\psi_m$ are orthogonal if $h_n$ and $h_m$ are different.

If the eigenvalue $h_n$ is degenerate, then its eigenfunctions are not necessarily orthogonal. However, by taking suitable linear combinations we can obtain a new set of orthogonal functions. Let us imagine that $h_n$ is $s$-fold degenerate and that originally its eigenfunctions are $\psi_{n,1}, \psi_{n,2}, \ldots, \psi_{n,s}$. As a first step we now construct a new set of eigenfunctions:

$$\begin{aligned} \psi'_{n,1} &= \psi_{n,1} \\ \psi'_{n,2} &= \langle \psi_{n,1} | \psi_{n,1} \rangle \psi_{n,2} - \langle \psi_{n,1} | \psi_{n,2} \rangle \psi_{n,1} \\ &\cdots \\ \psi'_{n,s} &= \langle \psi_{n,1} | \psi_{n,1} \rangle \psi_{n,s} - \langle \psi_{n,1} | \psi_{n,s} \rangle \psi_{n,1} \end{aligned} \tag{4-66}$$

The functions $\psi'_{n,2}, \psi'_{n,3}, \ldots, \psi'_{n,s}$ and each linear combination of them are now orthogonal to $\psi'_{n,1}$. We now repeat this procedure with the functions $\psi'_{n,2}$, $\psi'_{n,3}, \ldots, \psi'_{n,s}$, and we construct a new set of functions $\psi''_{n,3}, \psi''_{n,4}, \ldots, \psi''_{n,s}$ that are orthogonal to $\psi''_{n,2}$. Repeated applications of these transformations lead to a new orthogonal set of eigenfunctions $\phi_{n,1}, \phi_{n,2}, \ldots, \phi_{n,s}$ of the eigenvalue $h_n$. Thus we see that we can select a completely orthogonal set of eigenfunctions of the operator $H$. If, finally, we normalize every eigenfunction to unity, we obtain an orthonormal set of functions.

The reader may have detected some similarities between these derivations and the discussion in Chapter 2 of the eigenvalues and eigenvectors of a Hermitian matrix. This is not surprising, since the two problems are analogous to each other and we can transform the eigenvalue problem of an operator into an eigenvalue problem of a matrix.

Let us take an arbitrary complete set of functions $\phi_n$ as the basis for our considerations, and let us consider the eigenvalue problem

$$\Omega\psi = \lambda\psi \tag{4-67}$$

We denote the eigenvalues by $\lambda_n$ and the corresponding eigenfunctions by $\psi_n$. Every function $\psi$ can be expanded in terms of our complete set:

$$\psi = \sum_k c_k \phi_k \tag{4-68}$$

and if we substitute this expansion into Eq. (4-67), the eigenvalue problem becomes

$$\sum_k c_k \Omega \phi_k = \lambda \sum_k c_k \phi_k \tag{4-69}$$

The problem is now to determine the coefficients $c_k$ in such a way that the function $\psi$ is an eigenfunction of $\Omega$. We multiply Eq. (4-69) on the left by $\phi_m$ and integrate. This gives

$$\sum \Omega_{m,k} c_k = \lambda c_m$$

$$\Omega_{m,k} = \langle \phi_m | \Omega | \phi_k \rangle \tag{4-70}$$

For the sake of simplicity we assume that $\Omega$ is a Hermitian operator. Then we have

$$\Omega_{m,k} = \langle \phi_m | \Omega | \phi_k \rangle = \langle \phi_k | \Omega | \phi_m \rangle^* = \Omega^*_{k,m} \tag{4-71}$$

and we see that the matrix $\Omega_{m,k}$ is also Hermitian. It follows that a Hermitian operator is represented by a Hermitian matrix.

A comparison of Eqs. (4-67) and (4-70) shows that we have transformed the eigenvalue problem (4-67) of a Hermitian operator into the eigenvalue problem

(4-70) of an infinite Hermitian matrix, and the two problems are therefore equivalent. Since $\Omega_{m,k}$ is Hermitian, there exists a unitary transformation, defined by a matrix $[u_{k,n}]$, that diagonalizes the matrix $\Omega$:

$$\sum_m \sum_k u_{l,m}^{\dagger} \Omega_{m,k} u_{k,n} = \lambda_n \delta_{l,n} \tag{4-72}$$

Let us now multiply these equations by $u_{i,l}$ and sum over $l$:

$$\sum_l \sum_m \sum_k u_{i,l} u_{l,m}^{\dagger} \Omega_{m,k} u_{k,n} = \lambda_n \sum_l u_{i,l} \delta_{l,n} \tag{4-73}$$

Since $U$ is unitary, we have

$$\sum_l u_{i,l} u_{m,l}^{*} = \sum_l u_{i,l} u_{l,m}^{\dagger} = \delta_{i,m} \tag{4-74}$$

so that we can rewrite Eq. (4-73) as

$$\sum_m \sum_k \delta_{i,m} \Omega_{m,k} u_{k,n} = \lambda_n u_{i,n} \tag{4-75}$$

or

$$\sum_k \Omega_{i,k} u_{k,n} = \lambda_n u_{i,n} \tag{4-76}$$

By comparing this expression with Eq. (4-70) we can deduce that $\lambda_n$ is an eigenvalue of $\Omega$ and that the corresponding eigenvector is

$$\mathbf{u}_n = (u_{1,n}, u_{2,n}, u_{3,n}, \ldots) \tag{4-77}$$

It follows, therefore, that $\lambda_n$ is also an eigenvalue of the operator $\Omega$ and that the corresponding eigenfunction is

$$\psi_n = \sum_k u_{k,n} \phi_k \tag{4-78}$$

In general, different operators have different eigenvalues and eigenfunctions, but commuting Hermitian operators have the same eigenfunctions. In order to prove this statement, we consider two commuting Hermitian operators $\Omega$ and $\Lambda$:

$$\Omega\Lambda = \Lambda\Omega \tag{4-79}$$

Their eigenvalues are $\omega_n$ and $\lambda_n$, and the corresponding eigenfunctions are $w_n$ and $u_n$:

$$\Lambda u_n = \lambda_n u_n$$

$$\Omega w_n = \omega_n w_n \tag{4-80}$$

First we show that for a nondegenerate eigenvalue $\lambda_n$ of $\Lambda$ the corresponding eigenfunction $u_n$ is also an eigenfunction of $\Omega$. We have

$$\Lambda(\Omega u_n)=\Omega\Lambda u_n=\Omega\lambda_n u_n=\lambda_n(\Omega u_n) \tag{4-81}$$

Hence $(\Omega u_n)$ is also an eigenfunction of $\Lambda$ belonging to $\lambda_n$. Since $\lambda_n$ is nondegenerate, there is only one eigenfunction, so that $(\Omega u_n)$ and $u_n$ have to be proportional:

$$\Omega u_n=\sigma_n u_n \tag{4-82}$$

Consequently, $u_n$ is an eigenfunction of both $\Lambda$ and $\Omega$.

If $\lambda_n$ is degenerate, the situation becomes more complicated. We assume that the degeneracy is $s$-fold and that $u_{n,1}, u_{n,2}, u_{n,3}, \ldots, u_{n,s}$ are a set of linearly independent eigenfunctions belong to $\lambda_n$. Again, we can write

$$\Lambda(\Omega u_{n,i})=\Omega\Lambda u_{n,i}=\Omega\lambda_n u_{n,i}=\lambda_n(\Omega u_{n,i}) \tag{4-83}$$

for an arbitrary eigenfunction $u_{n,i}$. It follows again that $(\Omega u_{n,i})$ is an eigenfunction of $\Lambda$, belonging to $\lambda_n$, but this time we can conclude only that $(\Omega u_{n,i})$ is a linear combination of the $u_{n,j}$:

$$\Omega u_{n,i}=\sum_j c_{i,j} u_{n,j} \tag{4-84}$$

We assume that the $u_{n,i}$ are orthonormal, so that multiplication on the left by $u_{n,k}$ and subsequent integration give

$$c_{i,k}=\Omega_{k,i}=\langle u_{n,k}|\Omega|u_{n,i}\rangle \tag{4-85}$$

It follows, therefore, that

$$\Omega u_{n,i}=\sum_k \Omega_{k,i} u_{n,k} \tag{4-86}$$

where $[\Omega_{k,i}]$ is a Hermitian matrix of order $s$.

Apparently, the functions $u_{n,i}$ are not necessarily eigenfunctions of $\Omega$. However, let us attempt to find a new set of functions,

$$v=\sum_i b_i u_{n,i} \tag{4-87}$$

such that they are eigenfunctions of $\Omega$. The functions $v$ must then satisfy the condition

$$\Omega v=\sum_i b_i \Omega u_{n,i}=\sigma v=\sigma\sum_i b_i u_{n,i} \tag{4-88}$$

Multiplication on the left by $u_{n,j}$ and subsequent integration give

$$\sum_i \Omega_{j,i} b_i = \sigma b_j \tag{4-89}$$

This is an eigenvalue problem of a finite Hermitian matrix of order $s$, and it can always be solved. Let its eigenvalues be $\sigma_k$ and the corresponding eigenvectors $b_{k,j}$. Then the $\sigma_k$ are also eigenvalues of $\Omega$, and the corresponding eigenfunctions are

$$v_k = \sum_j b_{k,j} u_{n,j} \tag{4-90}$$

The functions $v_k$ are also eigenfunctions of the operator $\Lambda$, belonging to the eigenvalue $\lambda_n$, and they are therefore eigenfunctions of both $\Omega$ and $\Lambda$. It may be concluded that two commuting Hermitian operators always have a set of common eigenfunctions.

## 4.4 The Quantum Mechanical Description of a Particle in a Box

Now that we have discussed the general theory of operators and eigenvalue problems, we proceed finally to a practical application of the Schrödinger equation. In the present section we consider the simplest quantum mechanical problem, namely the particle in a one-dimensional box. A slightly more complex problem, the particle in a rectangular potential, will be discussed in Section 4-5.

A particle in a one-dimensional box moves in a potential field $V(x)$ which is zero for $0 \leq x \leq a$ and which is infinite elsewhere. This potential is graphically

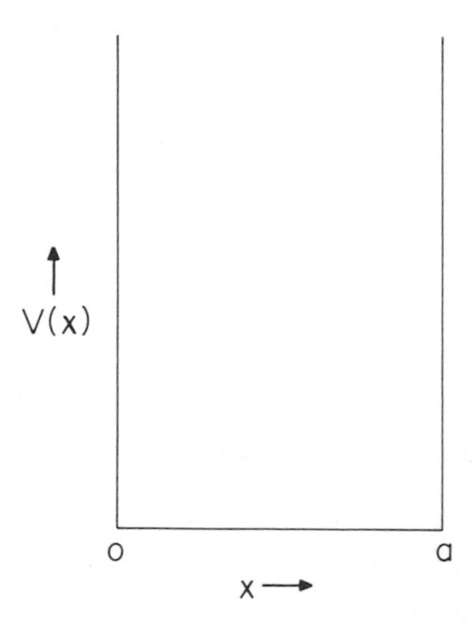


**Fig. 4-1** The potential function of a particle in a one-dimensional box.

represented in Fig. 4-1 and its algebraic description is

$$\begin{aligned} V(x)&=\infty \qquad x<0 \qquad &\text{(II)} \\ V(x)&=0 \qquad 0\leq x\leq a \qquad &\text{(I)} \\ V(x)&=\infty \qquad x>a \qquad &\text{(III)} \end{aligned} \tag{4-91}$$

The one-dimensional Schrödinger equation is

$$-\frac{\hbar^2}{2m}\frac{\partial^2\phi}{\partial x^2}+V(x)\phi=E\phi \tag{4-92}$$

In regions II and III we have to substitute $\infty$ for $V(x)$, and it is easily seen that here the Schrödinger equation can be satisfied only if we take $\phi(x)=0$. In the middle region, region I, the Schrödinger equation is

$$-\frac{\hbar^2}{2m}\frac{\partial^2\phi}{\partial x^2}=E\phi \tag{4-93}$$

We can treat this as an eigenvalue problem with boundary conditions. Within region I, $0\leq x\leq a$, the eigenfunction is determined by Eq. (4-93). Outside region I, $x<0$ and $x>a$, the eigenfunction is zero. We want the wave function to be continuous everywhere, and this means that we impose the conditions

$$\phi(0)=\phi(a)=0 \tag{4-94}$$

for the solutions of the differential equation (4-93).

In order to solve the differential equation we note that the energy $E$ is positive so that we can write Eq. (4-93) as

$$\frac{d^2\phi}{dx^2}=-k^2\phi \qquad k^2=\frac{2mE}{\hbar^2} \tag{4-95}$$

We make use of the results of Section 4-1, in particular of Eq. (4-5), and we find that the solution of Eq. (4-95) can be written as

$$\phi(x)=A\sin kx+B\cos kx \tag{4-96}$$

with $A$ and $B$ being undetermined parameters.

We now make use of the boundary conditions (4-94) in order to determine the various parameters. First we note that

$$\phi(0)=B=0 \tag{4-97}$$

so that the solution reduces to

$$\phi(x)=A\sin kx \tag{4-98}$$

Next we use the continuity condition at the point $x=a$ or the second boundary condition (4-94),

$$\phi(a)=A\sin ka=0 \tag{4-99}$$

The parameter $A$ must be different from zero, otherwise the wave function would be zero everywhere and could not be normalized. Consequently, we have

$$\sin ka=0 \tag{4-100}$$

whose solutions are

$$ka=n\pi \qquad n=\pm 1, \pm 2, \pm 3, \ldots, \text{ and so on} \tag{4-101}$$

By substituting this result back into Eq. (4-98) we find that the allowed wave functions are

$$\phi_n(x)=A_n\sin\left(\frac{n\pi x}{a}\right) \tag{4-102}$$

The remaining parameter $A_n$ is determined from the normalization condition

$$\int_0^a \phi_n^*(x)\phi(x)\,dx=A_n^*A_n\int_0^a \sin^2\left(\frac{n\pi x}{a}\right)dx=1 \tag{4-103}$$

Since the last integral is given by

$$\int_0^a \sin^2\left(\frac{n\pi x}{a}\right)dx=\frac{a}{2} \tag{4-104}$$

the equation for $A_n$ is

$$A_nA_n^* = \frac{2}{a} \tag{4-105}$$

The solution is

$$A_n=\left(\frac{2}{a}\right)^{1/2}e^{i\gamma} \tag{4-106}$$

since Eq. (4-105) determines only the absolute value of $A$ and not its argument. The wave functions are obtained by substituting this value $A_n$ into Eq. (4-102):

$$\phi_n(x)=e^{i\gamma}\left(\frac{2}{a}\right)^{1/2}\sin\left(\frac{n\pi x}{a}\right) \tag{4-107}$$

where $\gamma$ can take any value. Since the physical significance of the wave function does not depend on $\gamma$, it may be seen that $\phi_n$ and $(-\phi_n)$ are the same wave functions and therefore $\phi_n$ and $\phi_{-n}$ can be considered identical. Our final conclusions are thus that the eigenvalues of the particle in a box are given by

$$E_n = \frac{n^2h^2}{8ma^2} \tag{4-108}$$

according to Eqs. (4-95) and (4-101), and the corresponding eigenfunctions are

$$\phi_n(x) = e^{i\gamma}\left(\frac{2}{a}\right)^{1/2} \sin\frac{n\pi x}{a} \tag{4-109}$$

with

$$n = 1, 2, 3, 4, \ldots, \text{ and so on} \tag{4-110}$$

It follows that only certain discrete values of the energy lead to permissible wave functions; this situation is typical for bound states.

The observant reader may have noticed that we have been careful to avoid any reference to the continuity of the derivative of the wave function at the points $x=0$ and $x=a$. Actually these derivatives are not continuous at these points, as is easily verified. However, there is an explanation for this. The potential function $U(x)$ is not only discontinuous at $x=0$ and $x=a$, but it even has an infinite discontinuity, since the potential jumps from zero to infinity. It can be shown that in such a case the derivative of the wave function does not have to be continuous. Thus we see that in our first example we already encounter an exception to the rules that were outlined in Chapter 3.

## 4-5 The Particle in a Rectangular Potential Well

In this section we consider a problem that is similar to the particle in a box but which is slightly more complicated, namely the particle in a finite potential well. We have sketched the potential function $V(x)$ in Fig. 4-2. The algebraic

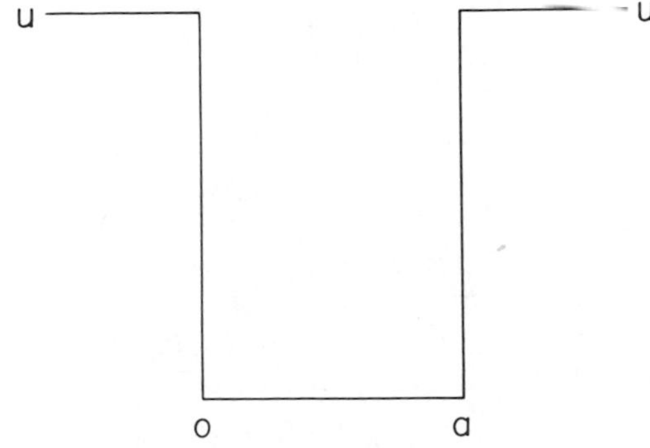


**Fig. 4-2** The potential function of a particle in a one-dimensional rectangular potential well.

expression is

$$V(x)=U \qquad x<0$$
$$V(x)=0 \qquad 0\leq x\leq a$$
$$V(x)=U \qquad x>a \tag{4-111}$$

In the previous Section 4-4 we considered the situation where $U$ is infinite. Now we wish to study the situation where $U$ is finite. The one-dimensional Schrödinger equation can be written in the form

$$\frac{d^2\psi}{dx^2}=-\frac{2m}{\hbar^2}[E-V(x)]\psi \tag{4-112}$$

and again we derive its solutions for the different regions I, II, and III, where $x<0$, $0\leq x\leq a$, and $x>a$, respectively. In region II we have $V(x)=0$, and if we substitute

$$\frac{2mE}{\hbar^2}=\alpha^2 \tag{4-113}$$

then the Schrödinger equation in this region becomes

$$\frac{d^2\psi_{\text{II}}}{dx^2}=-\alpha^2\psi_{\text{II}} \tag{4-114}$$

We discussed this equation in Section 4-1 and found that the general solution is

$$\psi_{\text{II}}(x)=A_{\text{II}}e^{i\alpha x}+B_{\text{II}}e^{-i\alpha x} \tag{4-115}$$

In regions I and III the Schrödinger equation is

$$\frac{d^2\psi}{dx^2}=-\frac{2m}{\hbar^2}(E-U)\psi \tag{4-116}$$

It is useful to distinguish between the two cases $E<U$ and $E>U$. In the first case we substitute

$$\frac{2m}{\hbar^2}(E-U)=-\beta^2 \tag{4-117}$$

and we obtain the equation

$$\frac{d^2\psi}{dx^2}=\beta^2\psi \tag{4-118}$$

The solutions are

$$\begin{aligned} \psi_{\mathrm{I}}(x) &= A_{\mathrm{I}} e^{\beta x} + B_{\mathrm{I}} e^{-\beta x} \qquad x<0 \\ \psi_{\mathrm{III}}(x) &= A_{\mathrm{III}} e^{\beta x} + B_{\mathrm{III}} e^{-\beta x} \qquad x>a \end{aligned} \tag{4-119}$$

Let us now consider the behavior of the wave function when $x$ tends to plus or minus infinity. When $x$ tends to plus infinity, we have to take $\psi_{\mathrm{III}}(x)$ as the wave function. It is easily seen that the first term of $\psi_{\mathrm{III}}(x)$, namely $e^{\beta x}$, tends to infinity when $x$ tends to plus infinity. Since we cannot allow this, the constant $A_{\mathrm{III}}$ has to be zero. According to the same argument, we find that $B_{\mathrm{I}}$ has to be zero, since $e^{-\beta x}$ tends to infinity when $x$ tends to minus infinity. We find, therefore, that for $E<U$ the wave function is

$$\psi(x) = \begin{cases} A_{\mathrm{I}} e^{\beta x} & x<0 \\ A_{\mathrm{II}} e^{i\alpha x} + B_{\mathrm{II}} e^{-i\alpha x} & 0 \le x \le a \\ B_{\mathrm{III}} e^{-\beta x} & x>a \end{cases} \tag{4-120}$$

The derivatives are

$$\frac{d\psi}{dx} = \begin{cases} \beta A_{\mathrm{I}} e^{\beta x} & x<0 \\ i\alpha\left( A_{\mathrm{II}} e^{i\alpha x} - B_{\mathrm{II}} e^{-i\alpha x}\right) & 0 \le x \le a \\ -\beta B_{\mathrm{III}} e^{-\beta x} & x>a \end{cases} \tag{4-121}$$

We now consider the behavior of $\psi(x)$ and its derivative at the point $x=0$. The condition that $\psi$ is continuous at $x=0$ leads to the equation

$$A_{\mathrm{I}} = A_{\mathrm{II}} + B_{\mathrm{II}} \tag{4-122}$$

and the condition that the derivative is continuous at $x=0$ gives

$$\beta A_{\mathrm{I}} = i\alpha( A_{\mathrm{II}} - B_{\mathrm{II}}) \tag{4-123}$$

The condition that $\psi$ and its derivative are continuous at the point $x-a$ yields the equations

$$A_{\mathrm{II}} e^{i\alpha a} + B_{\mathrm{II}} e^{-i\alpha a} = B_{\mathrm{III}} e^{-\beta a} \tag{4-124}$$

and

$$i\alpha\left( A_{\mathrm{II}} e^{i\alpha a} - B_{\mathrm{II}} e^{-i\alpha a}\right) = -\beta B_{\mathrm{III}} e^{-\beta a} \tag{4-125}$$

From Eqs. (4-122) and (4-123) we find

$$A_{\mathrm{II}} = \frac{1}{2}\left(1 - \frac{i\beta}{\alpha}\right) A_{\mathrm{I}} \qquad B_{\mathrm{II}} = \frac{1}{2}\left(1 + \frac{i\beta}{\alpha}\right) A_{\mathrm{I}} \tag{4-126}$$

Substitution of this result in Eqs. (4-124) and (4-125) gives

$$(\alpha\cos\alpha a+\beta\sin\alpha a)A_{\mathrm{I}}-\alpha e^{-\beta a}B_{\mathrm{III}}=0$$

$$(\beta\cos\alpha a-\alpha\sin\alpha a)A_{\mathrm{I}}+\beta e^{-\beta a}B_{\mathrm{III}}=0 \tag{4-127}$$

Elimination of $B_{\mathrm{III}}$ yields

$$\left[2\alpha\beta\cos\alpha a+(\beta^2-\alpha^2)\sin\alpha a\right]A_{\mathrm{I}}=0 \tag{4-128}$$

Hence the expression

$$2\alpha\beta\cos\alpha a+(\beta^2-\alpha^2)\sin\alpha a=0 \tag{4-129}$$

or

$$tg\,\alpha a=-\frac{2\alpha\beta}{\beta^2-\alpha^2} \tag{4-130}$$

is the equation for the eigenvalues.

We solve the equation by graphical methods. It is convenient to transform Eq. (4-130) into

$$\sin\alpha a=-\frac{2\alpha\beta}{\beta^2+\alpha^2} \tag{4-131}$$

We also introduce the new parameter $\sigma$ by defining

$$\sigma=\left(\frac{E}{U}\right)^{1/2} \qquad 0\leq\sigma\leq 1 \tag{4-132}$$

It follows then from the definitions (4-113) and (4-117) that Eq. (4-131) can be written in the form

$$\sin 2\pi p\sigma=-\left[1-(2\sigma^2-1)^2\right]^{1/2} \tag{4-133}$$

with the parameter $p$ defined as:

$$p=\frac{a(2mU)^{1/2}}{h} \tag{4-134}$$

The solutions of Eq. (4-133) are derived by plotting the left and right sides of Eq. (4-133) as a function of $\sigma$. The points where the two curves intersect give us the energy eigenvalues. Obviously, we must know the value of the parameter $p$ in order to find the eigenvalues. In Fig. 4-3 we have sketched two different

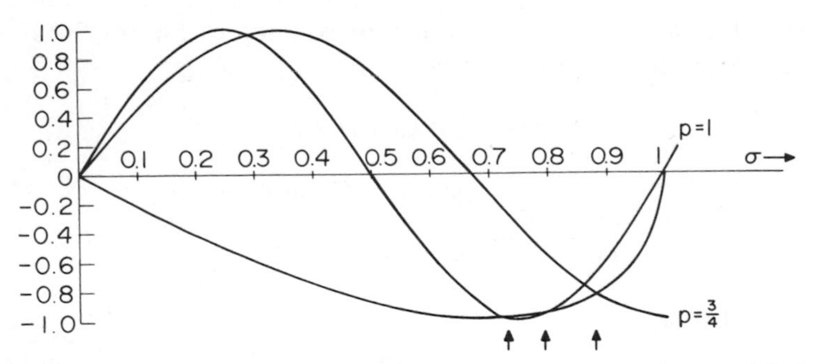


**Fig. 4-3** The graphical method for deriving the eigenvalues of a particle in a finite potential well according to Eq. (4-133).

cases, namely $p=\frac{3}{4}$, and $p=1$. It may be seen that there are no solutions if $p<\frac{1}{2}$, there is one solution if $\frac{1}{2}<p<1$, and so on. There is always a point of intersection for $\sigma=0$, but this does not lead to an acceptable eigenfunction and we should disregard it as an acceptable solution of the problem.

The eigenfunctions can be derived from Eq. (4-127) by substituting Eq. (4-131). It follows that

$$e^{-\beta a}B_{\mathrm{III}}=-A_{\mathrm{I}} \tag{4-135}$$

The complete wave function is derived from Eq. (4-120) by substituting the values of the various coefficients. The result is

$$\begin{aligned} &=Ae^{\beta x} && x<0 \\ \psi(x)&=A\left[\cos\alpha x+\frac{\beta}{\alpha}\sin\alpha x\right] && 0\leq x\leq a \\ &=-Ae^{\beta(a-x)} && x>a \end{aligned} \tag{4-136}$$

Here $A$ is a normalization constant, and we leave it to the reader to derive its value.

Let us now proceed to the situation where $E>U$. In this case the Schrödinger equation for regions I and III becomes

$$\frac{d^2\psi}{dx^2}=-\lambda^2\psi \qquad \lambda^2=\frac{2m}{\hbar^2}(E-U) \tag{4-137}$$

Its general solution is

$$\psi(x)=Ae^{i\lambda x}+Be^{-i\lambda x} \tag{4-138}$$

which can also be written in the form

$$\psi(x)=C\cos\lambda x+D\sin\lambda x \tag{4-139}$$

If we choose the same representation for the solution of Eq. (4-114) in region II, then the complete wave function is

$$\begin{aligned} \psi_{\mathrm{I}}(x) &= C_{\mathrm{I}}\cos\lambda x + D_{\mathrm{I}}\sin\lambda x \\ \psi_{\mathrm{II}}(x) &= C_{\mathrm{II}}\cos\alpha x + D_{\mathrm{II}}\sin\alpha x \\ \psi_{\mathrm{III}}(x) &= C_{\mathrm{III}}\cos\lambda x + D_{\mathrm{III}}\sin\lambda x \end{aligned} \tag{4-140}$$

and the derivative is

$$\begin{aligned} \psi_{\mathrm{I}}'(x) &= \lambda(D_{\mathrm{I}}\cos\lambda x - C_{\mathrm{I}}\sin\lambda x) \\ \psi_{\mathrm{II}}'(x) &= \alpha(D_{\mathrm{II}}\cos\alpha x - C_{\mathrm{II}}\sin\alpha x) \\ \psi_{\mathrm{III}}'(x) &= \lambda(D_{\mathrm{III}}\cos\lambda x - C_{\mathrm{III}}\sin\lambda x) \end{aligned} \tag{4-141}$$

From the continuity of $\psi$ and its derivative at the point $x=0$, we find that

$$C_{\mathrm{II}} = C_{\mathrm{I}} = C \qquad D_{\mathrm{II}} = \frac{\lambda}{\alpha}D_{\mathrm{I}} = \frac{\lambda}{\alpha}D \tag{4-142}$$

The condition of continuity of $\psi$ and $\psi'$ at the point $x=a$ yields the equations

$$\begin{aligned} C_{\mathrm{II}}\cos\alpha a + D_{\mathrm{II}}\sin\alpha a &= C_{\mathrm{III}}\cos\lambda a + D_{\mathrm{III}}\sin\lambda a \\ \alpha(D_{\mathrm{II}}\cos\alpha a - C_{\mathrm{II}}\sin\alpha a) &= \lambda(D_{\mathrm{III}}\cos\lambda a - C_{\mathrm{III}}\sin\lambda a) \end{aligned} \tag{4-143}$$

which lead to the solutions

$$\begin{aligned} C_{\mathrm{III}} &= \left(\cos\lambda a\cos\alpha a + \frac{\alpha}{\lambda}\sin\lambda a\sin\alpha a\right)C + \left(\frac{\lambda}{\alpha}\cos\lambda a\sin\alpha a - \sin\lambda a\cos\alpha a\right)D \\ D_{\mathrm{III}} &= \left(\sin\lambda a\cos\alpha a - \frac{\alpha}{\lambda}\cos\lambda a\sin\alpha a\right)C + \left(\frac{\lambda}{\alpha}\sin\lambda a\sin\alpha a + \cos\lambda a\cos\alpha a\right)D \end{aligned} \tag{4-144}$$

If we combine the results of Eqs. (4-140), (4-142), and (4-144), we find that the allowed wave functions are

$$\begin{aligned} \psi_{\mathrm{I}}(x) &= \cos\lambda x & & x<0 \\ &= \cos\alpha x & & 0\le x\le a \\ &= \left(\cos\lambda a\cos\alpha a + \frac{\alpha}{\lambda}\sin\lambda a\sin\alpha a\right)\cos\lambda x & & \\ &\quad + \left(\sin\lambda a\cos\alpha a - \frac{\alpha}{\lambda}\cos\lambda a\sin\alpha a\right)\sin\lambda x & & x>a \end{aligned} \tag{4-145}$$

and

$$\begin{aligned}
\psi_2(x) &= \sin \lambda x && x<0 \\
&= \frac{\lambda}{\alpha} \sin \alpha x && 0 \leq x \leq a \\
&= \left( \frac{\lambda}{\alpha} \cos \lambda a \sin \alpha a - \sin \lambda a \cos \alpha a \right) \cos \lambda x \\
&\quad + \left( \frac{\lambda}{\alpha} \sin \lambda a \sin \alpha a + \cos \lambda a \cos \alpha a \right) \sin \lambda x && x>a
\end{aligned} \tag{4-146}$$

It follows that our argument does not lead to any conditions for the energy, and we conclude that all values of $E$ that are larger than $U$ are allowed. The total spectrum of allowed energy values consists, therefore, of a finite number of discrete values of $E \leq U$ and of all values of $E > U$. This combination of a discrete spectrum and a continuum occurs frequently in quantum mechanical problems.

## 4-6 Some General Properties of Eigenvalues and Eigenfunctions

In the previous two sections we discussed two different applications of the Schrödinger equation: the particle in a box and the particle in a finite potential well. Particularly in the latter application it could be seen that for a given energy $E$ we should differentiate always between the two cases $U<E$ and $U>E$ in each region. In the first case, where the potential energy is larger than the total energy, the Schrödinger equation has the form of Eq. (4-2) and the solutions have the form of Eq. (4-7), the wave function behaves like an exponential. In the second case, where the potential energy is smaller than the total energy, the Schrödinger equation has the form of Eq. (4-1) and the solutions have the form of Eq. (4-5), the wave function behaves in an oscillatory fashion as a sine or a cosine function.

The above considerations apply to step-function potentials only, but we may derive similar conclusions for more general cases by a qualitative study of the Schrödinger equation. We consider a particle moving in a potential field $V(x)$ which has the general behavior of Fig. 4-4. The function $V(x)$ tends to infinity for large negative $x$ and to an asymptotic value $U_\infty$ for large positive $x$.

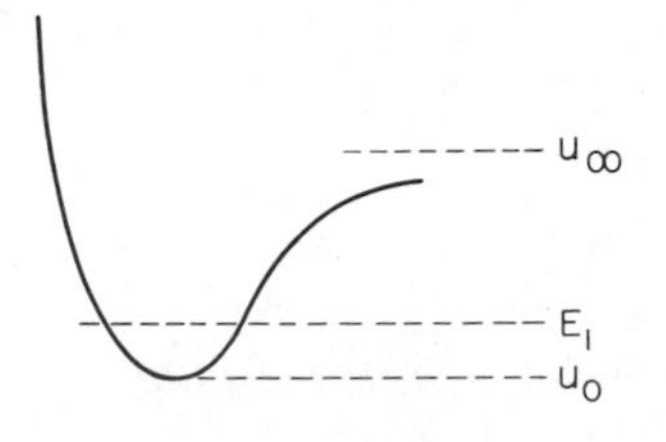


**Fig. 4-4** Potential field that is used for a general discussion of the eigenvalue spectrum.

In between it has a minimum value $U_o$. We write the one-dimensional Schrödinger equation of the particle as

$$\phi'' = \frac{d^2\phi}{dx^2} = \frac{2m(V-E)\phi}{\hbar^2} \tag{4-147}$$

Let us first assume that $E$ has a value that is slightly larger than $U_o$, for example, the value $E_1$ of Fig. 4-4. There exists then a region of the variable $x$ in which $E > V$. In this region the ratio between $\phi''$ and $\phi$ is negative:

$$\frac{\phi''}{\phi} < 0 \tag{4-148}$$

as follows from Eq. (4-147). We know that in this case the function $\phi(x)$ is concave toward the $x$ axis, as we have symbolically indicated in Fig. 4-5*a*. We say that such a function behaves in an oscillatory fashion. We can understand this if we imagine that over a small interval $V$ is almost constant, so that we can write

$$\phi'' = -\lambda^2\phi \tag{4-149}$$

This has the solutions $\sin\lambda x$ and $\cos\lambda x$, which are oscillatory functions.

On the left and on the right there are regions where $E < V(x)$, and here we have

$$\frac{\phi''}{\phi} > 0 \tag{4-150}$$

In these regions $\phi(x)$ is convex toward the $x$ axis, like the functions we have sketched in Fig. 4-5*b*. Now we say that this function behaves in an exponential fashion. We can understand this if we imagine that in a small interval we can write the Schrödinger equation as

$$\phi'' = \mu^2\phi \tag{4-151}$$

which has the solutions $e^{\mu x}$ and $e^{-\mu x}$. We can conclude that $\phi(x)$ behaves in

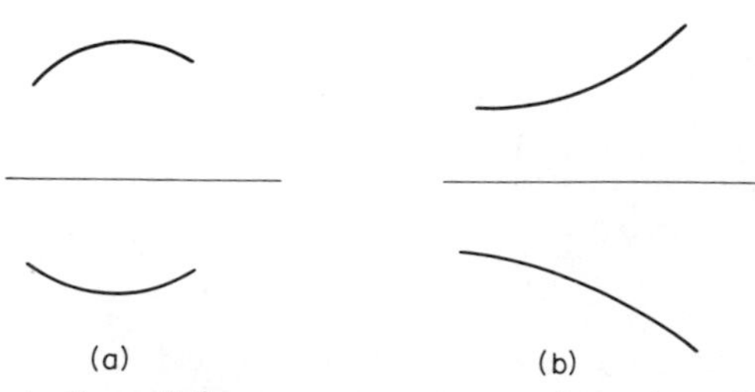


**Fig. 4-5** Illustration of the concave or convex behavior of wave functions.

an oscillatory fashion in the regions where $E > V(x)$ and that $\phi(x)$ behaves in an exponential fashion in the regions where $E < V(x)$.

Let us now consider the asymptotic behavior of the wave function $\phi(x)$ for the potential field that we have sketched in Fig. 4-6. We imagine that we know the value of $\phi(P)$ in a point $P$, together with the derivative of $\phi$, for example, the value and the derivative that we have indicated in Fig. 4-6. Now we attempt to analyze qualitatively the behavior of $\phi(x)$ at the right side of $P$. For a value of $E$ that is only slightly larger than the minimum of $V(x)$, for example, $E_1$, the second derivative of $\phi(x)$ is not large enough to bend $\phi(x)$ very much toward the $x$ axis. In this case $\phi'(x)$ is still positive when we cross over to the area where $\phi(x)$ behaves exponentially, and consequently the asymptotic behavior of $\phi(x)$ for $x \rightarrow \infty$ is exponential in $x$. Since $\phi(x)$ tends to infinity when $x$ tends to infinity, it is not normalizable, and $E_1$ is therefore not an eigenvalue. For a larger energy, for example, $E_2$, the second derivative of $\phi(x)$ is so much larger that $\phi(x)$ has bent around to a negative first derivative of $\phi(x)$ when the wave function crosses into the exponential area. Now $\phi(x)$ tends to minus infinity when $x$ tends to infinity, and since $\phi(x)$ cannot be normalized, we conclude that $E_2$ is not an eigenvalue either.

Since $\phi(x; E)$ is a continuous function of the parameter $E$ and since $\phi(x; E_1)$ tends to plus infinity for $x \rightarrow \infty$ and $\phi(x; E_2)$ tends to minus infinity for $x \rightarrow \infty$, we argue that there is an energy value $E_3$ between $E_1$ and $E_2$ such that $\phi(x; E_3)$ tends to zero for $x \rightarrow \infty$. For the value $E_3$ it is possible that the wave function can be normalized, and we expect that $E_3$ is an eigenvalue of the Schrödinger equation. The next eigenvalue is $E_4$, for which we have indicated the behavior of the wave function in Fig. 4-6. We see that $\phi(x; E_3)$ does not intersect with the $x=0$ axis, whereas $\phi(x; E_4)$ intersects $x=0$ once and approaches the $x=0$ asymptotically from the negative side. In general, the eigenfunction belonging to the lowest eigenvalue has no zero points, or nodes; that belonging to the next eigenvalue has one node; and that belonging to the $n$th eigenvalue has $(n-1)$ nodes.

Let us now return to the potential field of Fig. 4-4. We again consider a point $P$ in the left region of the potential well, and we assume that $\phi(x)$ has a

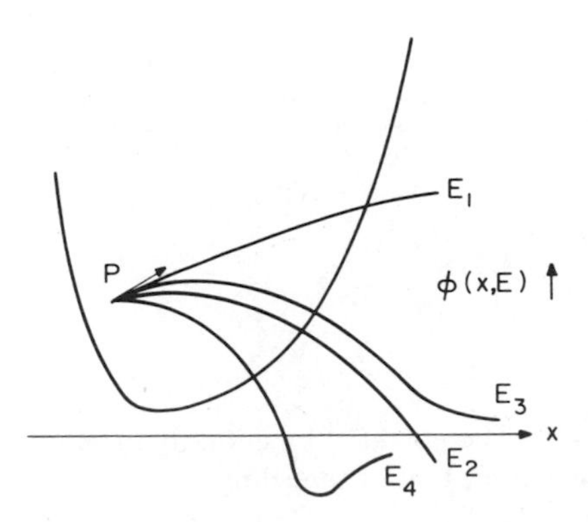


**Fig. 4-6** Behavior of various solutions of the Schrödinger equation of the potential function of Fig. 4-4.

positive derivative at that point. In the same way as we did for the potential field of Fig. 4-6, we can argue that there is a lowest eigenvalue $E_1$ with an eigenfunction $\phi_1$, a second eigenvalue $E_2,\ldots,$ and so on. However, there is an important difference between the two cases: the potential field of Fig. 4-6 has an infinite number of discrete eigenvalues, but the situation in Fig. 4-4 contains only a finite number of discrete eigenvalues. In the latter case the wave functions are oscillatory for $x \to \infty$ as soon as $E > U_\infty$, and they are not normalizable in the ordinary sense of the word. All values of $E > U_\infty$ are equally permissible, and we speak of a continuum of eigenvalues as opposed to the discrete spectrum of Fig. 4-6. We discuss the continuum situation at a later stage. We only mention here that a bound particle usually has a discrete eigenvalue spectrum and that a free particle has a continuous eigenvalue spectrum.

## 4-7 Tunneling Through a Potential Barrier

After the digression in the previous section on the general behavior of eigenvalues and eigenfunctions we now return to our presentation of various applications of the Schrödinger equation. In the present section we consider the encounter between a free particle and a rectangular potential barrier of height $U$ and length $a$. The corresponding potential $V(x)$ is shown in Fig. 4-7, and its algebraic form is

$$V(x) = \begin{cases} 0 & \text{if} \quad x < 0 \\ U & \text{if} \quad 0 \le x \le a \\ 0 & \text{if} \quad x > a \end{cases} \tag{4-152}$$

The classical description of this situation is quite straightforward. If the energy of the free particle is larger than $U$, then its motion is not essentially affected by the barrier. If its energy is smaller than $U$, it is totally reflected. The quantum mechanical theory leads to different predictions. Here it is found that the particle can pass the barrier even when its energy is smaller than $U$. This phenomenon is known as tunneling, and it is a quantum mechanical effect that has found many applications.

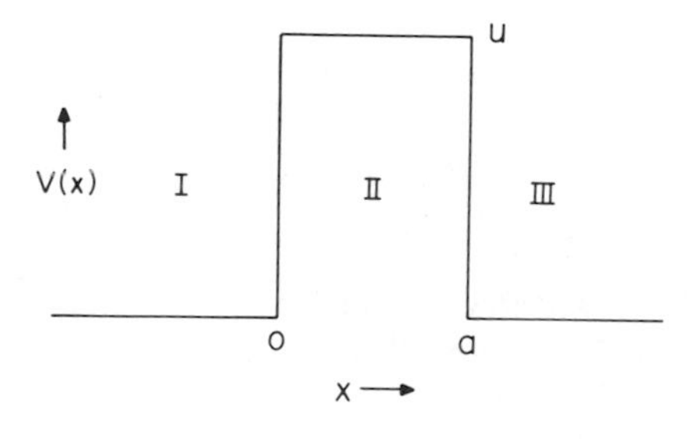


**Fig. 4-7** The potential function that is used for the description of tunneling.

Let us consider the Schrödinger equation for the potential function of Eq. (4-152). In regions I and III we may write the Schrödinger equation as

$$\frac{d^2\phi}{dx^2} = -\alpha^2\phi \qquad \alpha^2 = \frac{2mE}{\hbar^2} \tag{4-153}$$

In region II we write it as

$$\frac{d^2\phi}{dx^2} = \lambda^2\phi \qquad \lambda^2 = \frac{2m(U-E)}{\hbar^2} \tag{4-154}$$

when $E \leq U$ and as

$$\frac{d^2\phi}{dx^2} = -\mu^2\phi \qquad \mu^2 = \frac{2m(E-U)}{\hbar^2} \tag{4-155}$$

when $E > U$.

The solutions of Eq. (4-153) are $\exp(i\alpha x)$ and $\exp(-i\alpha x)$ according to Eq. (4-4). We should realize that these two solutions $\exp(i\alpha x)$ and $\exp(-i\alpha x)$ describe a particle that moves from left to right or from right to left, respectively. We wish to study the effect of the potential barrier on the motion of a particle that encounters the barrier from the left, and therefore we take the wave function as

$$\begin{aligned} \phi_{\mathrm{I}}(x) &= e^{i\alpha x} + Re^{-i\alpha x} \qquad & x<0 \\ \phi_{\mathrm{III}}(x) &= Te^{i\alpha x} \qquad & x>a \end{aligned} \tag{4-156}$$

Here $\phi_{\mathrm{III}}(x)$ is the transmitted wave, and the two terms of $\phi_{\mathrm{I}}(x)$ represent the incident and the reflected waves, respectively.

We first consider the case where $E \leq U$. The behavior of the wave function in region II is then described by Eq. (4-154), and the solutions are

$$\phi_{\mathrm{II}}(x) = Ae^{\lambda x} + Be^{-\lambda x} \tag{4-157}$$

The coefficients $A$, $B$, $R$, and $T$ are related through the continuity conditions for the wave function at the points $x$ and $a$. These conditions for the continuity of the wave function and its derivative at the point $x=0$ lead to the equations

$$\begin{aligned} 1+R &= A+B \\ i\alpha(1-R) &= \lambda(A-B) \end{aligned} \tag{4-158}$$

The same conditions for the point $x=a$ give

$$\begin{aligned} Te^{i\alpha a} &= Ae^{\lambda a} + Be^{-\lambda a} \\ i\alpha Te^{i\alpha a} &= \lambda(Ae^{\lambda a} - Be^{-\lambda a}) \end{aligned} \tag{4-159}$$

We eliminate $A$ and $B$ from Eqs. (4-158) and (4-159) and obtain

$$(\lambda - i\alpha)R - e^{-\lambda a}e^{i\alpha a}(\lambda + i\alpha)T + (\lambda + i\alpha) = 0$$

$$(\lambda + i\alpha)R - e^{\lambda a}e^{i\alpha a}(\lambda - i\alpha)T + (\lambda - i\alpha) = 0 \tag{4-160}$$

The solution of this equation is

$$R = \frac{(\alpha^2 + \lambda^2)\sinh \lambda a}{(\alpha^2 - \lambda^2)\sinh \lambda a + 2i\lambda\alpha \cosh \lambda a}$$

$$e^{i\alpha a}T = \frac{2i\lambda\alpha}{(\alpha^2 - \lambda^2)\sinh \lambda a + 2i\lambda\alpha \cosh \lambda a} \tag{4-161}$$

The transmission coefficient $D$ of the potential barrier is defined as $TT^*$. From Eq. (4-161) we find that

$$D = \frac{4\lambda^2\alpha^2}{(\alpha^2 + \lambda^2)^2 \sinh^2 \lambda a + 4\lambda^2\alpha^2} \tag{4-162}$$

since

$$\cosh^2 x - \sinh^2 x = 1 \tag{4-163}$$

When the product $\lambda a$ is much larger than unity, $D$ behaves asymptotically like

$$D \approx 16\gamma(1-\gamma)e^{-2\lambda a} \qquad \gamma = \frac{E}{U} \tag{4-164}$$

We have already mentioned that according to classical mechanics $D$ is exactly zero, but according to our quantum mechanical argument it has a finite value given by Eq. (4-162). This phenomenon is the tunneling effect mentioned above. Since it is predicted only by quantum mechanical and not by classical arguments, it is called a quantum mechanical effect by the experimentalists.

Let us now study the situation where $E > U$. The wave function in region II is now determined by Eq. (4-155), which has the solutions

$$\phi_{\text{II}}(x) = P\cos \mu x + Q \sin \mu x \tag{4-165}$$

The wave function in regions I and III is again given by Eq. (4-156). The conditions for the continuity of the wave function and its derivative at the points $x=0$ and $x=a$ lead to the equations

$$1 + R = P$$

$$i\alpha(1 - R) = \mu Q \tag{4-166}$$

and

$$Te^{i\alpha a} = P\cos\mu a + Q\sin\mu a$$
$$i\alpha Te^{i\alpha a} = \mu(-P\sin\mu a + Q\cos\mu a) \tag{4-167}$$

respectively. The solution of these equations is

$$R = \frac{(\alpha^2-\mu^2)\sin\mu a}{(\alpha^2+\mu^2)\sin\mu a + 2\alpha\mu i\cos\mu a}$$

$$T = \frac{2\alpha\mu i e^{-i\alpha a}}{(\alpha^2+\mu^2)\sin\mu a + 2\alpha\mu i\cos\mu a}$$

$$P = \frac{2\alpha(\alpha\sin\mu a + i\mu\cos\mu a)}{(\alpha^2+\mu^2)\sin\mu a + 2\alpha\mu i\cos\mu a} \tag{4-168}$$

$$Q = \frac{2\alpha(-\alpha\cos\mu a + i\mu\sin\mu a)}{(\alpha^2+\mu^2)\sin\mu a + 2\alpha\mu i\cos\mu a}$$

The transmission coefficient $TT^*$ is now

$$D = \frac{4\alpha^2\mu^2}{(\alpha^2-\mu^2)^2\sin^2\mu a + 4\alpha^2\mu^2} = \frac{4E(E-U)}{U^2\sin^2\mu a + 4E(E-U)} \tag{4-169}$$

It is easily verified that $D$ approaches unity when $E$ is large compared with $U$, so that for large energies the particle is not affected by the potential barrier. The possibility that the particle is reflected is given by $RR^*$, which is

$$RR^* = \frac{U^2\sin^2\mu a}{U^2\sin^2\mu a + 4E(E-U)} \tag{4-170}$$

We see that

$$TT^* + RR^* = 1 \tag{4-171}$$

which is in agreement with our expectation that the sum of the probabilities for transmission and for reflection should be equal to unity. For large values of $E$ in comparison with $U$ the reflection coefficient approaches zero. Furthermore, it follows from Eq. (4-170) that the reflection coefficient is zero when

$$\mu a = \frac{[2m(E-U)]^{1/2}a}{\hbar} = n\pi \tag{4-172}$$

where $n$ is an integer. In this case there is perfect transmission through the potential barrier.

## 4-8 Periodic Potentials

Our final example of a particle moving in a step-function potential is illustrated in Fig. 4-8. The potential function is periodic and it consists of a set of equidistant potential barriers of height $U$ and width $\delta$. The system was used in the early development of solid-state physics as a simple model for the motion of an electron in a metal and it is known as the Kronig-Penney model. In the model the height $U$ of the potential barriers is assumed to be quite large and the width $\delta$ is supposed to be small. Eventually we consider the limit where $U$ tends to infinity and $\delta$ tends to zero, but for the moment we take $U$ and $\delta$ to be finite.

Our results will show that we deal basically with a free particle and with a continuum eigenvalue spectrum. The interesting feature of the result is the occurrence of energy bands and gaps, and we will see that the Kronig-Penney model, crude as it may be, explains some interesting properties of the electrons moving in a metal.

In order to derive the eigenvalues and eigenfunctions for the system we observe that, in the regions where $V=0$, we can write the Schrödinger equation as

$$\frac{d^2\phi}{dx^2} = -\lambda^2\phi \qquad \lambda^2 = \frac{2mE}{\hbar^2} \tag{4-173}$$

Its solutions are described by Eq. (4-5). In the regions where $V=U$ we may assume that the energy $E$ of the particle is smaller than $U$ since $U$ is assumed to be large. In those regions the Schrödinger equation is

$$\frac{d^2\phi}{dx^2} = \mu^2\phi \qquad \mu^2 = \frac{2m(U-E)}{\hbar^2} \tag{4-174}$$

Its solutions are described by Eq. (4-7).

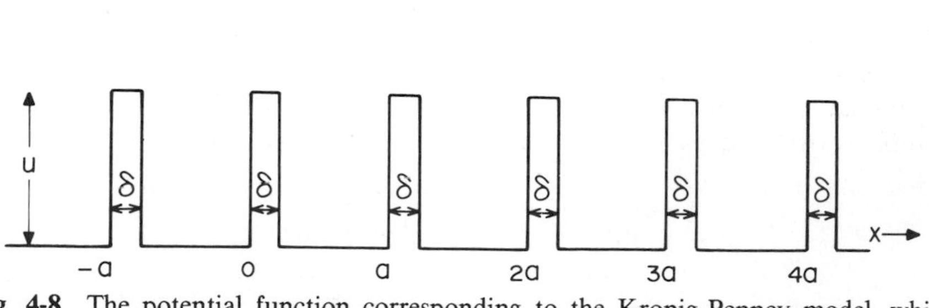


**Fig. 4-8** The potential function corresponding to the Kronig-Penney model, which represents the motion of an electron in a metal.

In every region of space we write the wave function as a linear combination of the possible solutions, and the problem that we have to solve is how to choose the coefficients so that the wave function and its derivative are continuous everywhere.

The most convenient approach to this problem is first to study the behavior of the wave function in the vicinity of the $k$th potential barrier, extending from $ka$ to $ka+\delta$. We have here

$$\begin{aligned}
\phi(x)&=a_k e^{i\lambda x}+b_k e^{-i\lambda x} && (k-1)a+\delta<x<ka\\
\phi(x)&=c_k e^{\mu x}+d_k e^{-\mu x} && ka\leq x\leq ka+\delta\\
\phi(x)&=a_{k+1}e^{i\lambda x}+b_{k+1}e^{-i\lambda x} && ka+\delta<x<(k+1)a
\end{aligned}\tag{4-175}$$

for the wave function and

$$\begin{aligned}
\phi'(x)&=i\lambda(a_k e^{i\lambda x}-b_k e^{-i\lambda x}) && (k-1)a+\delta<x<ka\\
\phi'(x)&=\mu(c_k e^{\mu x}-d_k e^{-\mu x}) && ka\leq x\leq ka+\delta\\
\phi'(x)&=i\lambda(a_{k+1}e^{i\lambda x}-b_{k+1}e^{-i\lambda x}) && ka+\delta<x<(k+1)a
\end{aligned}\tag{4-176}$$

for the derivative. The conditions that the wave function and its derivative are continuous at the points $ka$ and $ka+\delta$ lead to the equations

$$\begin{aligned}
a_k e^{i\lambda ka}+b_k e^{-i\lambda ka}&=c_k e^{\mu ka}+d_k e^{-\mu ka}\\
i\lambda\left(a_k e^{i\lambda ka}-b_k e^{-i\lambda ka}\right)&=\mu\left(c_k e^{\mu ka}-d_k e^{-\mu ka}\right)\\
a_{k+1}e^{i\lambda(ka+\delta)}+b_{k+1}e^{-i\lambda(ka+\delta)}&=c_k e^{\mu(ka+\delta)}+d_k e^{-\mu(ka+\delta)}\\
i\lambda\left[a_{k+1}e^{i\lambda(ka+\delta)}-b_{k+1}e^{-i\lambda(ka+\delta)}\right]&=\mu\left[c_k e^{\mu(ka+\delta)}-d_k e^{-\mu(ka+\delta)}\right]
\end{aligned}\tag{4-177}$$

In order to solve these equations, we substitute

$$\begin{aligned}
a_k&=e^{-i\lambda ka}\rho^k A & c_k&=e^{-\mu ka}\rho^k C\\
b_k&=e^{i\lambda ka}\rho^k B & d_k&=e^{\mu ka}\rho^k D
\end{aligned}\tag{4-178}$$

into Eq. (4-177). We then obtain

$$\begin{aligned}
A+B&=C+D\\
i\lambda(A-B)&=\mu(C-D)\\
\rho[e^{-i\lambda(a-\delta)}A+e^{i\lambda(a-\delta)}B]&=Ce^{\mu\delta}+De^{-\mu\delta}\\
i\lambda\rho[e^{-i\lambda(a-\delta)}A-e^{i\lambda(a-\delta)}B]&=\mu(Ce^{\mu\delta}-De^{-\mu\delta})
\end{aligned}\tag{4-179}$$

These equations are independent of $k$, so that they are the same for each potential barrier. Consequently, if we find a set of values of $\rho$, $A$, $B$, $C$, and $D$ for which Eq. (4-179) is satisfied, then the wave function and its derivative are continuous everywhere.

In order to solve Eq. (4-179), we write its second pair of equations as

$$\rho[(A+B)\cos\gamma - i(A-B)\sin\gamma] = Ce^{\mu\delta} + De^{-\mu\delta}$$

$$i\lambda\rho[(A-B)\cos\gamma - i(A+B)\sin\gamma] = \mu(Ce^{\mu\delta} - De^{-\mu\delta}) \tag{4-180}$$

where we have introduced the abbreviation

$$\gamma = \lambda(a-\delta) \tag{4-181}$$

If we now eliminate $A$ and $B$ by means of the first pair of Eqs. (4-179), then we obtain

$$(\lambda\rho\cos\gamma - \lambda\cosh\mu\delta)(C+D) - (\mu\rho\sin\gamma + \lambda\sinh\mu\delta)(C-D) = 0$$

$$(\lambda\rho\sin\gamma - \mu\sinh\mu\delta)(C+D) + (\mu\rho\cos\gamma - \mu\cosh\mu\delta)(C-D) = 0 \tag{4-182}$$

These equations have solutions only when they are proportional to each other, that is, the following condition is satisfied:

$$\lambda\mu(\rho^2+1) = 2\lambda\mu\rho\cos\gamma\cosh\mu\delta + (\mu^2-\lambda^2)\rho\sin\gamma\sinh\mu\delta \tag{4-183}$$

Let us first consider the possible values for the parameter $\rho$. It is easily verified that for $|\rho|>1$ the wave function tends to infinity when $k$ tends to infinity, so that this is not allowed. If $|\rho|<1$, then the wave function becomes infinite when $k$ tends to minus infinity, so that this is not allowed either. Consequently $|\rho|$ has to be equal to unity, and we write it as

$$\rho = e^{i\sigma a} \tag{4-184}$$

If we substitute this into Eq. (4-183), we obtain

$$\cos\sigma a = \cos\lambda(a-\delta)\cosh\mu\delta + \frac{\mu^2-\lambda^2}{2\lambda\mu}\sin\lambda(a-\delta)\sinh\mu\delta \tag{4-185}$$

as the condition for the energy eigenvalues.

Equation (4-185) can be solved graphically, but we prefer to simplify our model before attempting its solution. We let the height $U$ of the potential barrier become very large, and at the same time we let $\delta$ become very small in such a way that the product $U\delta$ remains finite. This finite product is defined as

$$P = \frac{mU\delta}{\hbar^2} \tag{4-186}$$

In the limiting case where $U \to \infty$, $\delta \to 0$, and $U\delta \to (\hbar^2 P/m)$, Eq. (4-185) becomes

$$\cos \sigma a = \cos \lambda a + \frac{P}{\lambda} \sin \lambda a \tag{4-187}$$

In Fig. 4-9 we plot the right side of this equation as a function of $\lambda a$ for a certain value of $Pa$. An important feature of Eq. (4-187) now is that it has solutions only for those values of $\lambda a$ where the right side of the equation has values between minus unity and plus unity, since the left side of the equation has to stay within those limits. It follows, therefore, from Fig. 4-9 that the lowest eigenvalue occurs when $a\lambda = p$, the projection of the point where the curve intersects with the 1 axis. This lowest eigenvalue corresponds to the value $\sigma = 0$. All energy values between the points $p$ and $q$ are allowed, but at the point $q$, which is the projection of the intersection with the $-1$ axis, we enter a region of forbidden energy values. The region between the points $p'$ and $q'$ leads again to allowed energy values, and so on. The energy eigenvalues depend on $\sigma$, and in Fig. 4-10 we have a typical plot of $\lambda$ versus $\sigma$. We see that there are certain regions that contain a continuum of allowed energy values, and these regions are separated by finite regions that contain no eigenvalues.

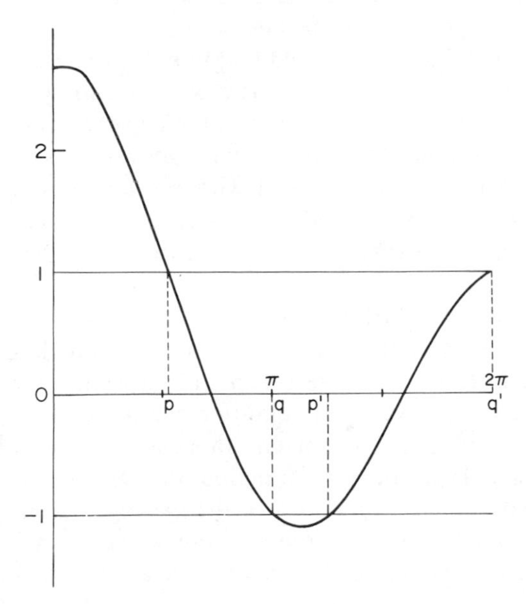


**Fig. 4-9** The graphic solutions of Eq. (4-187), representing the eigenvalue problem of the Kronig-Penney potential of Fig. 4-8.

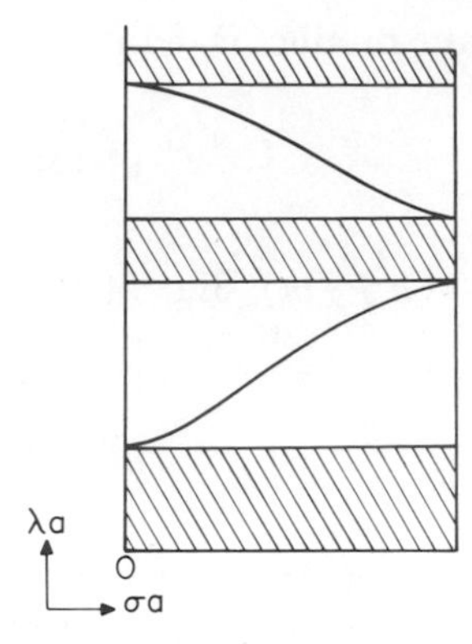


**Fig. 4-10** The energy bands and gaps representing the eigenvalue spectrum of the Kronig-Penney model.

In a situation like this we speak of energy bands, which are the eigenvalue regions, and of energy gaps, which are the remaining energy regions. This band structure plays a very important role in the theory of solids.

## 4-9 Eigenfunction Expansions

The final section of this chapter deals with expansions in terms of the complete set of eigenfunctions of a Hermitian operator. We already discussed expansions in terms of a complete set of functions in Section 3-4, but in the present section we want to consider a more specific case where the complete set of functions is formed by the set of eigenfunctions of a Hermitian operator. An important feature of the latter set of functions is the possible contributions from continuum eigenstates to the expansion. We considered the general form of the eigenvalue spectrum in Section 4-6 and we argued that in many cases the total set of eigenvalues of an operator consists of a discrete part and of a continuum part. It will appear that both parts contribute to the eigenfunction expansions. We will also see that we need to know the eigenvalues and eigenfunctions of a particle in a box in order to deal with the continuum contributions. This may explain why we postponed our discussion of eigenfunction expansions to the present chapter rather than include it in the previous chapter.

In order to consider the expansion of an arbitrary function in terms of the complete set of eigenfunctions of a Hermitian operator we must ask whether such a set of eigenfunctions constitutes a complete set in the sense that we defined in Section 3-4. It is not easy to answer this question for the general case because it is a mathematical problem of current interest. It is well established that in the one-dimensional case the eigenfunctions of a Hermitian operator form a complete set. Also, the eigenfunctions of the hydrogen atom and of the helium atom form complete sets if the continuum states are included. It has not yet been proved that the eigenfunctions of an arbitrary atom form a complete set, but much progress on this problem is being made and it is to be expected that it will be proved in the near future. Naturally, it becomes more difficult to prove the completeness of the sets of eigenfunctions for larger molecules such as benzene, anthracene, and so on. Intuitively we feel that, in general, the eigenfunctions of any Hermitian operator form a complete set as long as we include the continuum states, and we will assume so in the following arguments. This means that an arbitrary function that satisfies the same boundary conditions as the eigenfunctions can be expanded in terms of them and that this expansion is convergent in the mean.

This situation is quite straightforward if the eigenvalues from which we derive the eigenfunctions form a denumerable set, as in the case of the particle in a box (or the harmonic oscillator which will be discussed in the next chapter). If we call the operator $\Lambda$, the eigenvalues $\lambda_n$, and the eigenfunctions $\phi_n$, then we can expand a function $f$ as

$$f = \sum_n c_n \phi_n \tag{4-188}$$

We assume that $\Lambda$ is Hermitian and that the $\phi_n$ have been chosen to be orthonormal. Then it is easily found that

$$c_n = \langle \phi_n | f \rangle \tag{4-189}$$

so that

$$f = \sum_n \langle \phi_n | f \rangle \phi_n \tag{4-190}$$

We have assumed that the set $\phi_n$ is complete, which implies that the expansion (4-190) is allowed and is convergent in the mean.

The problem becomes much more interesting when the eigenvalue spectrum contains both a discrete part and a continuum. Here we must be aware of the difficulties in normalizing the wave functions. We will study this problem for a simple one-dimensional example, namely the eigenvalues of the one-dimensional Hamiltonian $H$ whose potential function $U(x)$ is given by Fig. 4-11. The function $U(x)$ tends to infinity for $x \rightarrow -\infty$ and asymptotically to zero for $x \rightarrow \infty$. It has a minimum value $U_o$ for $x=0$. The operator $H$ has a set of eigenvalues $E_n$ with corresponding eigenfunctions $\phi_n(x)$ for $E \leq 0$, and it also has a set of eigenfunctions $\phi(x; E)$ for $E > 0$. All positive values of $E$ are allowed, and since there is a wave function belonging to each positive value of $E$, we write these functions as $\phi(x; E)$.

We suspect that an arbitrary function $\psi(x)$, which tends to zero for $x \rightarrow -\infty$, can be expanded as

$$\psi(x) = \sum_n \gamma_n \phi_n(x) + \int_0^{\infty} \gamma(E) \phi(x; E)\, dE \tag{4-191}$$

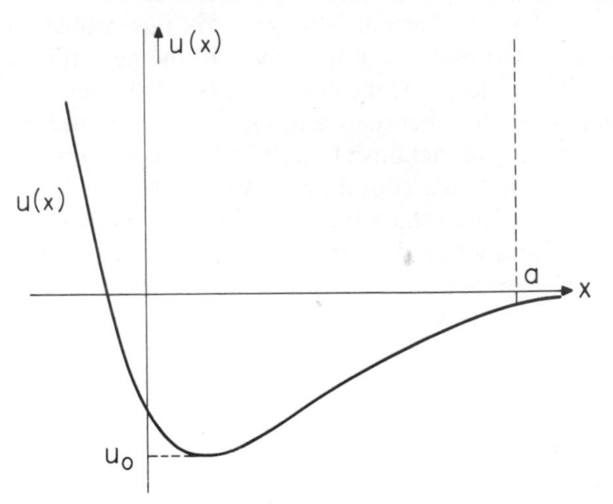


**Fig. 4-11** The potential that is used to discuss the expansion of a function in terms of both the discrete and the continuum parts of a complete set of eigenfunctions.

We wish to investigate whether such an expansion is allowed and how the functions $\phi(x, E)$ ought to be normalized.

In order to answer these questions, we consider the eigenvalues and eigenfunctions of the operator $H'$:

$$H' = T + V(x) \tag{4-192}$$

where we define

$$\begin{aligned} V(x) &= U(x) \qquad x \leq a \\ V(x) &= \infty \qquad x > a \end{aligned} \tag{4-193}$$

We take $a$ to be a very large but finite number. The eigenvalue spectrum of $H'$ is now discrete, and we denote the eigenvalues by $\varepsilon_n$ and the eigenfunctions by $\tilde{\phi}_n(x)$. For positive eigenvalues, $\varepsilon_n > 0$, it will prove to be convenient to write the corresponding eigenfunctions as $\tilde{\phi}(x; \varepsilon_n)$. We will assume that, in the case that $a$ tends to infinity,

$$\begin{aligned} \lim_{a \to \infty} \varepsilon_n &= E_n \\ \lim_{a \to \infty} \tilde{\phi}_n(x) &= \phi_n(x) \\ \lim_{a \to \infty} \tilde{\phi}(x; \varepsilon_n) &= \phi(x; E) \end{aligned} \tag{4-194}$$

Any function $\psi(x)$ that is defined for $x \leq a$ and that is zero for $x = a$ can now be expanded as

$$\psi(x)=\sum_{n} c_n\tilde{\phi}_n(x)+\sum_{k} c_k\tilde{\phi}(x;\varepsilon_k) \tag{4-195}$$

where the first summation is to be performed over all states with negative (or zero) eigenvalues and the second summation over all states with positive eigenvalues. We assume that the functions $\tilde{\phi}_n(x)$ are normalized to unity. The $\tilde{\phi}(x;\varepsilon_k)$ do not have to be normalized, and therefore we introduce the quantities

$$N(a;\varepsilon)=\int_{-\infty}^{a} \tilde{\phi}(x;\varepsilon)\tilde{\phi}^*(x;\varepsilon)\,dx \tag{4-196}$$

We also define

$$c(a;\varepsilon)=\int_{-\infty}^{a} \tilde{\phi}^*(x;\varepsilon)\psi(x)\,dx \tag{4-197}$$

The expansion (4-195) can now be written as

$$\psi(x)=\sum_{n} c_n\tilde{\phi}_n(x)+\sum_{k} \frac{c(a;\varepsilon_k)\tilde{\phi}(x;\varepsilon_k)}{N(a;\varepsilon_k)} \tag{4-198}$$

with

$$c_n=\langle\tilde{\phi}_n(x)|\psi(x)\rangle \tag{4-199}$$

Let us now investigate what happens to Eq. (4-198) when $a$ tends to infinity. The first sum offers no difficulties. We have already assumed that

$$\lim_{a\to\infty}\tilde{\phi}_n(x)=\phi_n(x) \tag{4-200}$$

and therefore we have

$$\lim_{a\to\infty} c_n=\gamma_n \tag{4-201}$$

The second sum becomes an integral in the limiting case, but we have to study the detailed dependence on $a$ of all quantities involved before we can determine this sum. When $a$ is large, the wave function $\tilde{\phi}(x;\varepsilon_k)$ takes the form

$$\tilde{\phi}(x;\varepsilon_k)=A_k\sin\left[\frac{(2m\varepsilon_k)^{1/2}x}{\hbar}+\alpha_k\right] \tag{4-202}$$

for large values of $x$. Here $A$ is the amplitude and $\alpha$ is a phase factor. For large values of $a$ the normalization function $N(a; \varepsilon_k)$ behaves asymptotically like

$$N(a;\varepsilon_k) \simeq \tfrac{1}{2}A_k^2 a \tag{4-203}$$

The values of the energy eigenvalues are determined by the condition that the wave function is zero for $x=a$. For large values of $a$ we may deduce from Eq. (4-202) that this condition is

$$(2m\varepsilon_k)^{1/2}a+\alpha_k\hbar = k\hbar\pi \tag{4-204}$$

with $k$ a positive integer. The difference $\Delta\varepsilon_k$ between $\varepsilon_{k+1}$ and $\varepsilon_k$ can be obtained as

$$\Delta\varepsilon_k = \frac{(2m\varepsilon_k)^{1/2}\hbar\pi}{ma} \tag{4-205}$$

when $k$ is sufficiently large. We now rewrite Eq. (4-198) as

$$\psi(x) = \sum_n c_n\tilde{\phi}_n(x) + \sum_k \frac{c(a;\varepsilon_k)}{N(a;\varepsilon_k)\Delta\varepsilon_k}\tilde{\phi}(x;\varepsilon_k)\Delta\varepsilon_k \tag{4-206}$$

It follows now from Eqs. (4-203) and (4-205) that the product $N(a; \varepsilon_k)\Delta\varepsilon_k$ tends to a finite limit when $a$ tends to infinity:

$$\lim_{a\to\infty} N(a;\varepsilon_k)\Delta\varepsilon_k = \frac{A_k^2(2m\varepsilon_k)^{1/2}\hbar\pi}{2m} \tag{4-207}$$

We are still free to choose the normalizing factor $A$, and we do so in such a way that the limit (4-207) becomes unity. It follows now that in the limit for $a$ tending to infinity the expansion (4-206) becomes

$$\psi(x) = \sum_n \gamma_n\phi_n(x) + \int_0^\infty \gamma(E)\phi(x;E)\,dE \tag{4-208}$$

with

$$\gamma_n = \langle \phi_n(x)|\psi(x)\rangle$$

$$\gamma(E) = \int_{-\infty}^{\infty} \phi^*(x;E)\psi(x)\,dx \tag{4-209}$$

From this expansion we can derive an orthogonality relation for the functions $\phi(x; E)$. We choose $\psi_o(x)$ orthogonal to the eigenfunctions $\phi_n(x)$ of the discrete states, and we obtain

$$\psi_o(x) = \int_0^\infty \gamma(E)\phi(x;E)\,dE \tag{4-210}$$

If we substitute this back into the second Eq. (4-209), we obtain

$$\gamma(E)=\int_{-\infty}^{\infty}\gamma(E')\,dE'\int_{-\infty}^{\infty}\phi^*(x;E)\phi(x;E')\,dx \tag{4-211}$$

Let us now compare this with Eq. (4-45), which we rewrite as

$$\gamma(E)=\int_{-\infty}^{\infty}\delta(E-E')\gamma(E')\,dE' \tag{4-212}$$

Obviously it may be concluded that

$$\int_{-\infty}^{\infty}\phi^*(x;E)\phi(x;E')\,dx=\delta(E-E') \tag{4-213}$$

This is known as the improper orthogonality of continuum eigenfunctions.

## Problems

**1** Write the operators for $x$, $y$, $z$, $p_x$, $p_y$, and $p_z$ and derive the commutators between all pairs of these operators.

**2** Show that

$$[x^n, p_x]=i\hbar n x^{n-1}$$

if $n$ is a positive integer.

**3** Show that the Laplace operator $\Delta$ and the operators $p_x$, $p_y$, and $p_z$ are all Hermitian operators for normalizable functions.

**4** Write the operators for $M_x$, $M_y$, and $M_z$ and derive the commutators between all pairs of these operators. (Here **M** stands for the angular momentum.)

**5** Show that the operators $M_z$ and $M^2$ commute with one another.

**6** The Schrödinger equation for a particle on a circle of radius $R$ is

$$-\frac{\hbar^2}{2mR^2}\frac{\partial^2\psi}{\partial\alpha^2}=E\psi$$

where $\alpha$ is the polar angle. Derive the eigenvalues and eigenfunctions. Hint: The eigenvalues are determined from the condition that the wave function is single-valued.

**7** Determine the eigenvalues and eigenfunctions of the system that is described by the potential

$$
\begin{aligned}
V(x) &= \infty \qquad && x \leq 0 \\
V(x) &= 0 && 0 < x < a \\
V(x) &= U && x \geq a
\end{aligned}
$$

for energies $E$ that are smaller than $U$, $E \leq U$.

**8** Calculate the transmission coefficient for tunneling of an electron through a rectangular potential barrier with a height of 2 eV and a width of 0.5 Å when the electron has successive energy values 0.1 eV, 1 eV, and 1.8 eV.

**9** Calculate the transmission coefficient for tunneling of a proton through a rectangular potential barrier with a height of 100 $cm^{-1}$ and a width of 0.1 Å when the proton has an energy of 50 $cm^{-1}$ and of 90 $cm^{-1}$.

**10** Consider an electron moving in a finite potential well of depth $U$ and length $a$. How many discrete energy levels are there if $U = 1$ eV and $a = 1$ Å? How many are there if $U = 1000$ $cm^{-1}$ and $a = 1$ Å. How many are there if $U = 100$ $cm^{-1}$ and $a = 0.5$ Å?

## Recommended Reading

The material in this chapter can be found in most text books of quantum mechanics. We list one general text on quantum mechanics, namely Schiff's book, as well as Kramer's book, which treats eigenfunction expansions very well. In addition we mention the German book by Flügge and Marschall, since this is a collection of quantum mechanics problems, many of them dealing with step function potentials. Finally we list the book by Titchmarsh, since this is the standard book on eigenfunction expansions, it is a highly sophisticated mathematical treatment.

L. I. Schiff, *Quantum Mechanics*, 3rd. ed., McGraw-Hill Book Co., Inc., New York 1968.

H. A. Kramers, *Quantum Mechanics*, North-Holland Publishing Company, Amsterdam, 1958.

S. Flügge and H. Marschall, *Rechenmethoden der Quantentheorie*, 2nd. ed., Springer-Verlag, Berlin, 1952.

E. C. Titchmarsh, *Eigenfunction Expansions Associated with Second-Order Differential Equations*, Parts I and II, Clarendon Press, Oxford, 1962 and 1958.

CHAPTER FIVE

# Differential Equations and the Schrödinger Equation

## 5-1 Introduction

In the previous chapter we discussed various solutions of the Schrödinger equation for one-dimensional systems where the potential function is a step function. In those cases, the solution of the differential equation was straightforward. In the present chapter, we derive the solution of the Schrödinger equation for various systems where the solution of the differential equation is more challenging. In order to discuss those situations we must present certain aspects of the theory of differential equations.

We shall see that certain types of differential equations occur in many physical problems and that they have been studied in great detail. Their solutions are known as special functions, and many of them have been named after the mathematicians who have studied their behavior and their properties. For example, we have Bessel functions, Legendre functions, Hankel functions, Gegenbauer functions, and so on. We also have the hypergeometric function, the confluent hypergeometric function, and so on.

We discuss the solution of the Schrödinger equation for three different systems together with the corresponding special functions. They are the harmonic oscillator and the Hermite polynomials, the Morse potential and the confluent hypergeometric function, and the rigid rotor together with the Legendre functions and the spherical harmonic functions. We also derive the eigenfunctions and eigenvalues of the angular momentum operators.

We dedicate a separate chapter, Chapter 6, to discussion of the hydrogen atom and of a particle in a central force field. The special functions that are relevant to this problem are the confluent hypergeometric function and the spherical harmonics function. They will be treated in this chapter.

First we discuss the classification and some general properties of differential equations and then we present the various special functions and the corresponding quantum mechanical problems.

## 5-2 Classification of Differential Equations

An equation that contains a dependent variable $u$, one or more independent variables $x_1, x_2, \ldots, x_n$, and the derivatives of $u$ with respect to the $x_i$ is called a differential equation. When the equation contains more than one independent variable and various partial derivatives, it is called a partial differential equation. When there is only one independent variable, we speak of an ordinary differential equation. We consider only ordinary differential equations.

Differential equations are classified according to their order and to their degree. The order of the equation is equal to the order of the highest derivative in the equation. The degree is equal to the power of the highest derivative in the equation after all fractional powers have been removed. For example,

$$\frac{d^2u}{dx^2}+5x\left(\frac{du}{dx}\right)^{1/2}+u=x^2 \tag{5-1}$$

is a quadratic, second-order equation.

Obviously there are so many different types of equations that it is impossible to discuss them all. We limit ourselves to those which occur most frequently in quantum mechanical problems and to a few that give us some general insight. Our first restriction is that we consider only linear equations of the type

$$f_n(x)\frac{d^nu}{dx^n}+f_{n-1}(x)\frac{d^{n-1}u}{dx^{n-1}}+\cdots+f_1(x)\frac{du}{dx}+f_o(x)u=g(x) \tag{5-2}$$

The most interesting equations of this type are second-order equations. This is because first-order equations can be solved in a straightforward manner according to standard procedures, and equations of higher than second-order cannot be solved analytically. Therefore, we limit ourselves to second-order equations of the type

$$\frac{d^2u}{dz^2}+p(z)\frac{du}{dz}+q(z)u=0 \tag{5-3}$$

The inhomogeneous equations, where the right side is equal to $g(z)$ instead of zero, as in Eq. (5-3), are not discussed here, since they seldom occur in quantum mechanical problems and since their solutions are customarily derived from the corresponding homogeneous equations.

Finally, we impose some restrictions on the functions $p(z)$ and $q(z)$. First we consider all values of the variable $z$ for which $p(z)$ and $q(z)$ are finite and continuous and have well-defined derivatives; these are called ordinary points of the differential equation (5-3). All other points are called singular points. Let us now take one such singular point, say $c$, and let us consider the functions $(z-c)p(z)$ and $(z-c)^2q(z)$. If both these functions are finite and continuous and if they have well-defined derivatives in $c$, then we call $c$ a regular point of the differential equation, otherwise $c$ is called an irregular point. The restriction that we impose is that the differential equation (5-3) has no irregular points in the region where we solve it. If these conditions are satisfied, we can solve Eq. (5-3) according to the method that we outline in Section 5-3.

We wish to discuss here a simple example of Eq. (5-3), since we used the results in previous chapters. If $p(z)$ and $q(z)$ are both constants, the equation becomes

$$\frac{d^2u}{dz^2}+p\frac{du}{dz}+qu=0 \tag{5-4}$$

It is easily verified that the function $e^{\lambda z}$ is a solution of this equation if $\lambda$ satisfies the condition

$$\lambda^2+p\lambda+q=0 \tag{5-5}$$

In general, this quadratic equation in $\lambda$ has two different roots $\lambda_1$ and $\lambda_2$. Equation (5-4) has two solutions $e^{\lambda_1 z}$ and $e^{\lambda_2 z}$. Its general solution $u(z)$ can be written in the form

$$u(z)=A_1e^{\lambda_1 z}+A_2e^{\lambda_2 z} \tag{5-6}$$

with $A_1$ and $A_2$ being arbitrary parameters.

We will see that, in general, Eq. (5-3) has two different, linearly independent solutions and that its general solution is a linear combination of these two specific solutions. It is to be expected that a second-order differential equation has two undetermined parameters because we get two integration parameters if we integrate twice.

Let us now return to Eq. (5-4) and ask what happens when the two roots $\lambda_1$ and $\lambda_2$ happen to coincide. In this case the equation is

$$\frac{d^2u}{dz^2}-2\lambda\frac{du}{dz}+\lambda^2u=0 \tag{5-7}$$

We substitute

$$u(z)=v(z)e^{\lambda z} \tag{5-8}$$

and we find that the equation for $v(z)$ is

$$\frac{d^2v}{dz^2}=0 \tag{5-9}$$

so that

$$u(z)=(Az+B)e^{\lambda z} \tag{5-10}$$

## 5-3 Series Expansion Method for Solving Differential Equations

It can be shown that the second-order differential equation of Eq. (5-3) can be solved by substituting a power series expansion of $u(z)$ into it if the functions $p(z)$ and $q(z)$ satisfy the conditions that were outlined in Section 5-2. If we are interested in the behavior of the solution in the vicinity of a regular point $c$, we ought to expand in terms of powers of $(z-c)$, but this is not a serious restriction, since it can be shown that the power series expansion represents the solution in other regions also, although the convergence may be slower. We think it may be helpful to discuss first the solution of a specific differential equation by means of the series expansion method to show how it works. As our example we choose the equation

$$x\frac{d^2u}{dx^2}+(c-x)\frac{du}{dx}-au=0 \tag{5-11}$$

which is the equation for the confluent hypergeometric function. Our choice has a dual purpose. The confluent hypergeometric series is one of the three special functions that we discuss in this chapter and we derive some of its properties by solving Eq. (5-11).

In order to solve Eq. (5-11), we substitute the power series expansion for $u(x)$:

$$u(x)=x^{\rho}\sum_{n=0}^{\infty} b_n x^n \tag{5-12}$$

The parameter $\rho$ is unknown for the time being, but it is chosen in such a way that the coefficient $b_o$ is different from zero. Substitution of Eq. (5-12) into Eq. (5-11) gives

$$\sum_{n=0}^{\infty}(n+\rho)(n+\rho+c-1)b_n x^{n+\rho-1}-\sum_{n=0}^{\infty}(n+\rho+a)b_n x^{n+\rho}=0 \tag{5-13}$$

which can also be written as

$$\rho(\rho+c-1)b_o x^{\rho-1}+\sum_{n=0}^{\infty}\left[(n+\rho+1)(n+\rho+c)b_{n+1}-(n+\rho+a)b_n\right]x^{n+\rho}=0 \tag{5-14}$$

The left side of Eq. (5-14) is zero for any value of $x$, and therefore the coefficient of each power of $x$ must be zero. The lowest power of $x$ is $(\rho-1)$, and if we equate the corresponding coefficient to zero, we find

$$\rho(\rho+c-1)b_o=0 \tag{5-15}$$

Since $b_o$ is always different from zero, the solutions of this equation are

$$\rho=0$$
$$\rho=1-c \tag{5-16}$$

Let us first consider the solution $\rho=0$. If we substitute this back into Eq. (5-14), we obtain the equation

$$(n+1)(n+c)b_{n+1}-(n+a)b_n=0 \tag{5-17}$$

or

$$cb_1=ab_o$$
$$2(c+1)b_2=(a+1)b_1 \tag{5-18}$$
$$3(c+2)b_3=(a+2)b_2 \qquad \text{and so on}$$

Since $b_o$ seems to be an arbitrary parameter, we take it equal to unity and find

$$b_o=1 \qquad b_1=\frac{a}{c}$$
$$b_2=\frac{a(a+1)}{1\cdot 2\cdot c(c+1)} \qquad b_3=\frac{a(a+1)(a+2)}{3!c(c+1)(c+2)} \quad \text{and so on} \tag{5-19}$$

The corresponding solution $u_1(x)$ is

$$u_1(x)=1+\frac{a}{c}\frac{x}{1!}+\frac{a(a+1)}{c(c+1)}\frac{x^2}{2!}+\frac{a(a+1)(a+2)}{c(c+1)(c+2)}\frac{x^3}{3!}+\cdots \tag{5-20}$$

Now we consider the second solution $\rho=1-c$. Substitution into Eq. (5-14) gives

$$(n+1)(n+2-c)b'_{n+1}=(n+1+a-c)b'_n \tag{5-21}$$

This leads to the solution

$$u_2(x)=x^{1-c}+\frac{1+a-c}{2-c}\frac{x^{2-c}}{1!}+\frac{(1+a-c)(2+a-c)}{(2-c)(3-c)}\frac{x^{3-c}}{2!}+\cdots \tag{5-22}$$

The general solution of Eq. (5-11) is thus

$$u(x)=A_1u_1(x)+A_2u_2(x) \tag{5-23}$$

Now that we have gained some insight into the series expansion method, we outline it in general. For the sake of convenience we take a differential equation that has $x=0$ as a regular point, and we write it as

$$x^2\frac{d^2u}{dx^2}+xf(x)\frac{du}{dx}+g(x)u=0 \tag{5-24}$$

The functions $f(x)$ and $g(x)$ are expanded as a power series in $x$:

$$f(x)=\sum_{n=0}^{\infty} a_n x^n \qquad g(x)=\sum_{n=0}^{\infty} b_n x^n \tag{5-25}$$

and we substitute the power series for $u(x)$:

$$u(x)=x^{\rho}\sum_{n=0}^{\infty} c_n x^n \tag{5-26}$$

Again, we substitute Eqs. (5-25) and (5-26) into Eq. (5-24), and we set the various coefficients of $x^{\rho+n}$ equal to zero. The lowest power of $x$, which is $x^{\rho}$, gives

$$(\rho^2+a_o\rho-\rho+b_o)c_o=0 \tag{5-27}$$

This is a quadratic equation in $\rho$, which generally has two different roots $\rho_1$ and $\rho_2$. The equations for the higher powers of $x$ lead to relationships between the coefficients $c_n$ that can be solved successively. If we substitute $\rho_1$ and solve the equations, we obtain the solution $u_1(x)$, and from $\rho_2$ we derive the solution $u_2(x)$. The final solution $u(x)$ is again a linear combination of $u_1(x)$ and $u_2(x)$.

When the two roots $\rho_1$ and $\rho_2$ are equal to each other, this method yields only one specific solution. In this case we have to follow a different procedure in order to derive the second solution. However, this is a rather complicated problem, and since we do not encounter it in our applications in this book, we do not discuss it. Moreover, if $\rho_1-\rho_2$ is equal to an integer, we may also obtain only one solution by the procedure above.

## 5-4 The Confluent Hypergeometric Function

The first special function that we discuss is the confluent hypergeometric function because it is related directly to the differential equation of the previous section. The function is denoted by the symbol ${}_1F_1(a; b; x)$, and it is defined by the series expansion

$${}_1F_1(a\colon b; x)=1+\frac{a}{b}\frac{x}{1!}+\frac{a(a+1)}{b(b+1)}\frac{x^2}{2!}+\frac{a(a+1)(a+2)}{b(b+1)(b+2)}\frac{x^3}{3!}+\cdots \tag{5-28}$$

Since this expression is identical with Eq. (5-20), we can conclude immediately that the function ${}_1F_1(a;\, b;\, x)$ is one of the solutions of the differential equation (5-11), that is,

$$x\frac{d^2y}{dx^2}+(b-x)\frac{dy}{dx}-ay=0 \tag{5-29}$$

The second solution of the differential equation is given by Eq. (5-22), and we write it as

$$y_2=x^{1-b}\left[1+\frac{1+a-b}{2-b}\frac{x}{1!}+\frac{(1+a-b)(2+a-b)}{(2-b)(3-b)}\frac{x^2}{2!}+\cdots\right] \tag{5-30}$$

Clearly this second solution can be represented as

$$y_2=x^{1-b}{}_1F_1(a-b+1;2-b;x) \tag{5-31}$$

First we consider the asymptotic behavior of the two solutions $y_1={}_1F_1(a;\, b;\, x)$ and $y_2$ for very small and for very large values of the variable $x$. Clearly,

$$y_1(0)={}_1F_1(a;\, b;0)=1 \tag{5-32}$$

The other solution $y_2$ behaves asymptotically as $x^{1-b}$ for $x\rightarrow 0$.

For larger values of $x$ the function ${}_1F_1(a;\, b;\, x)$ behaves asymptotically as $e^x$. For large values of $x$ the differential equation (5-29) can be approximated as

$$\frac{d^2y}{dx^2}-\frac{dy}{dx}=0 \tag{5-33}$$

and its solution is $e^x$. An alternative argument considers the ratio between the coefficient $c_{n+1}$ of $x^{n+1}$ and the coefficient $c_n$ of $x^n$ in Eq. (5-28). We find

$$\frac{c_{n+1}}{c_n}=\frac{a+n}{b+n}\cdot\frac{1}{n+1} \tag{5-34}$$

The limit of this ratio when $n$ tends to infinity is the same as the corresponding limit for the power series expansion of $e^x$. Consequently the function ${}_1F_1(a;\, b;\, x)$ behaves asymptotically as $e^x$ for large values of $x$.

An interesting property of the confluent hypergeometric function can be derived by substituting

$$y(x)=e^x w(x) \tag{5-35}$$

into the differential equation (5-29). The equation for $w(x)$ becomes

$$x\frac{d^2w}{dx^2}+(b+x)\frac{dw}{dx}+(b-a)w=0 \tag{5-36}$$

If we introduce the new variable $t=-x$, the equation

$$t\frac{d^2w}{dt^2}+(b-t)\frac{dw}{dt}-(b-a)w=0 \tag{5-37}$$

is again the differential equation for the confluent hypergeometric series. Its two solution are

$$\begin{aligned} w_1(t)&={}_1F_1(b-a;b;t)\\ w_2(t)&=t^{1-b}{}_1F_1(1-a;2-b;t) \end{aligned} \tag{5-38}$$

If we substitute this back into Eq. (5-35), we obtain four solutions to the original Eq. (5-29), namely

$$\begin{aligned} y_1&={}_1F_1(a;b;x)\\ y_2&=x^{1-b}{}_1F_1(a-b+1;2-b;x)\\ y_3&=e^x{}_1F_1(b-a;b;-x)\\ y_4&=x^{1-b}e^x{}_1F_1(1-a;2-b;-x) \end{aligned} \tag{5-39}$$

A second-order differential equation can have only two linearly independent solutions and therefore we should have two linear relations between the four solutions of Eq. (5-39). From the behavior of the solutions in the vicinity of the point $x=0$, it follows that

$$y_1=y_3 \qquad y_2=y_4 \tag{5-40}$$

or

$${}_1F_1(a;b;x)=e^x{}_1F_1(b-a;b;-x) \tag{5-41}$$

This is known as Kummer's transformation.

Various special functions are closely related to the confluent hypergeometric function. We discuss one of them, the Whittaker function, because it is relevant to the Schrödinger equation of the hydrogen atom.

We consider the differential equation

$$\frac{d^2z}{dx^2}+\left(-\frac{1}{4}+\frac{k}{x}+\frac{1-4m^2}{4x^2}\right)z=0 \tag{5-42}$$

If we attempt to solve this equation by means of the series expansion method, and if we substitute

$$z=x^{\rho}\sum_{n=0}^{\infty}c_nx^n \tag{5-43}$$

then the equation for $\rho$ becomes

$$\rho(\rho-1)+\tfrac{1}{4}-m^2=0$$

$$\rho-\tfrac{1}{2}=\pm m \tag{5-44}$$

Hence we substitute

$$z(x)=x^{m+1/2}w(x) \tag{5-45}$$

into Eq. (5-42). The new equation becomes

$$\frac{d^2w}{dx^2}+\frac{1+2m}{x}\frac{dw}{dx}+\left(\frac{k}{x}-\frac{1}{4}\right)w=0 \tag{5-46}$$

Next we substitute

$$w(x)=e^{-x/2}y \tag{5-47}$$

and we obtain

$$x\frac{d^2y}{dx^2}+(1+2m-x)\frac{dy}{dx}-\left(m+\frac{1}{2}-k\right)y=0 \tag{5-48}$$

Clearly one of the two solutions of Eq. (5-48) is the function

$$y_1={}_1F_1(m-k+\tfrac{1}{2},2m+1;x) \tag{5-49}$$

The corresponding solution $z_1$ of the differential equation (5-42) is defined as the Whittaker function $M_{k,m}(x)$. By making use of Eqs. (5-45) and (5-47) we find

$$z_1=M_{k,m}(x)=x^{m+1/2}e^{-x/2}{}_1F_1(m-k+\tfrac{1}{2},2m+1;x) \tag{5-50}$$

A second solution of Eq. (5-48) is found by making use of Eq. (5-39),

$$y_2=x^{-2m}{}_1F_1(-m-k+\tfrac{1}{2},-2m+1;x) \tag{5-51}$$

The corresponding solution $z_2$ of Eq. (5-42) is given by

$$z_2=x^{-m+1/2}e^{-x/2}{}_1F_1(-m-k+\tfrac{1}{2},-2m+1;x) \tag{5-52}$$

By comparing this with Eq. (5-50) we find that

$$z_2=M_{k,-m}(x) \tag{5-53}$$

As we mentioned before, the Whittaker functions play a role in the solution of the Schrödinger equation for the hydrogen atom and for the Morse potential.

## 5-5 Orthogonal Polynomials

In the following section we discuss the properties of various special functions that fall in the general category of orthogonal polynomials. In particular, we present the Hermite polynomials, the Laguerre polynomials, and the Legendre polynomials. Even though these functions are quite different from one another, they have certain common features and they exhibit the same type of properties.

First, every class of orthogonal polynomials consists of a set of polynomials $w_n(x)$ with $n=0,1,2,3,\ldots$, and so on, which are orthogonal with respect to a certain interval and a certain density function $\rho(x)$. We can write this as

$$\int_a^b w_n(x) w_m(x) \rho(x)\, dx = \lambda_n \delta_{n,m} \tag{5-54}$$

This means that the integral is zero when $n \neq m$ and it is nonzero and equal to a normalization constant $\lambda_n$ when the two polynomials are equal.

Second, all orthogonal polynomials may be defined by means of a generating function $K(x, h)$. This generating function is expanded as a power series in the variable $h$,

$$K(x,h) = \sum_{n=0}^{\infty} w_n(x) h^n \tag{5-55}$$

and the coefficients in this power series expansion are defined as the orthogonal polynomials $w_n(x)$.

Third, it is always possible to derive recurrence relations from the generating functions. These are relations between $w_n$, $w_{n+1}$, $w_{n-1}$, and their derivatives. These recurrence relations in combination with the orthogonality properties are useful for the evaluation of integrals involving $w_n$.

Finally, most orthogonal polynomials are solutions of second-order differential equations and they may be considered as special functions. It should be noted that the three types of polynomials that we discuss in this chapter may all be considered special cases of the confluent hypergeometric function. That is why we decided to discuss the latter function first.

## 5-6 Hermite Polynomials

Hermite polynomials can be defined by means of a generating function $K(x, h)$ that has the form

$$K(x,h) = \exp(2hx - h^2) \tag{5-56}$$

Here the variable $x$ can take all real values. We expand $K(x, h)$ as a power series in $h$:

$$K(x,h)=\sum_{n=0}^{\infty}\frac{H_n(x)}{n!}h^n \tag{5-57}$$

and we define the coefficient $H_n(x)$ as the Hermite polynomial in $x$ of order $n$.

By differentiating Eq. (5-57) $n$ times with respect to $h$, we can easily show that

$$H_n(x)=\left[\frac{\partial^n}{\partial h^n}K(x,h)\right]_{h=0} \tag{5-58}$$

We can rewrite this equation in a more useful form by observing that $K(x, h)$ can be represented also as

$$K(x,h)=e^{x^2}e^{-(x-h)^2} \tag{5-59}$$

Consequently we have

$$H_n(x)=e^{x^2}\left[\frac{\partial^n}{\partial h^n}e^{-(x-h)^2}\right]_{h=0} \tag{5-60}$$

We observe that

$$\frac{\partial^n}{\partial h^n}e^{-(x-h)^2}=(-1)^n\frac{\partial^n}{\partial x^n}e^{-(x-h)^2} \tag{5-61}$$

so that

$$\left[\frac{\partial^n}{\partial h^n}e^{-(x-h)^2}\right]_{h=0}=(-1)^n\left[\frac{\partial^n}{\partial x^n}e^{-(x-h)^2}\right]_{h=0} \tag{5-62}$$

and

$$H_n(x)=(-1)^n e^{x^2}\frac{d^n}{dx^n}e^{-x^2} \tag{5-63}$$

Each function $H_n(x)$ is a polynomial in $x$, as the name indicates, and the first few polynomials are, according to Eq. (5-63),

$$\begin{aligned} H_0(x)&=1\\ H_1(x)&=2x\\ H_2(x)&=4x^2-2\\ H_3(x)&=8x^3-12x \end{aligned} \tag{5-64}$$

Hermite polynomials of different order and their derivatives are related to one another by equations that are known as recurrence relations. They are derived from the generating function $K(x, h)$ of Eq. (5-56) in combination with Eq. (5-57). If we differentiate $K(x, h)$ with respect to $h$ we find that

$$\frac{\partial K}{\partial h} = -2(h-x)K \tag{5-65}$$

Substitution of Eq. (5-57) gives

$$\sum_{n=0}^{\infty} \left[ H_{n+1}(x) - 2xH_n(x) + 2nH_{n-1}(x) \right] \frac{h^n}{n!} = 0 \tag{5-66}$$

This expansion is identically zero for all values of $h$, and therefore each coefficient of the power series expansion must be zero. This leads to our first recurrence relation:

$$H_{n+1}(x) - 2xH_n(x) + 2nH_{n-1}(x) = 0 \tag{5-67}$$

A second recurrence relation is derived in a similar way by differentiating Eq. (5-56) with respect to $x$:

$$\frac{\partial K}{\partial x} = 2hK \tag{5-68}$$

Substitution of Eq. (5-57) gives

$$\frac{dH_n(x)}{dx} = 2nH_{n-1}(x) \tag{5-69}$$

It can be shown from the properties of the generating function that the functions $H_n(x)e^{-x^2/2}$ form an orthonormal set in the interval from $-\infty$ to $+\infty$. We evaluate the integrals

$$I_{n,m} = \int_{-\infty}^{\infty} H_n(x)H_m(x)e^{-x^2}\,dx \tag{5-70}$$

by considering the expression

$$e^{-x^2}K(x, h)K(x, k) = e^{-x^2} \sum_{n=0}^{\infty} \sum_{m=0}^{\infty} H_n(x)H_m(x) \frac{h^n k^m}{n!m!} \tag{5-71}$$

Integration of both sides of Eq. (5-71) gives

$$\int_{-\infty}^{\infty} K(x, h)K(x, k)e^{-x^2}\,dx = \sum_{n=0}^{\infty} \sum_{m=0}^{\infty} I_{n,m} \frac{h^n k^m}{n!m!} \tag{5-72}$$

The integral on the left is evaluated by substituting Eq. (5-56):

$$\int_{-\infty}^{\infty} K(x,h)K(x,k)e^{-x^2}\,dx=\int_{-\infty}^{\infty} e^{-(x^2-2hx-2kx+h^2+k^2)}\,dx$$

$$=e^{2hk}\int_{-\infty}^{\infty} e^{-(x-h-k)^2}\,d(x-h-k)=e^{2hk}\sqrt{\pi} \tag{5-73}$$

where we make use of the result of Eq. (3-109). Substitution of this result into Eq. (5-72) gives

$$\sum_{n=0}^{\infty}\sum_{m=0}^{\infty} I_{n,m}\frac{h^n k^m}{n!m!}=\sqrt{\pi}\,e^{2hk}=\sqrt{\pi}\sum_{n=0}^{\infty}\frac{(2hk)^n}{n!} \tag{5-74}$$

Since this expression is an identity in $h$ and $k$, it follows that

$$I_{n,m}=0 \qquad n\neq m$$

$$I_{n,n}=2^n n!\sqrt{\pi} \tag{5-75}$$

Let us now derive the differential equation that is satisfied by $H_n(x)$. To this end we consider the function

$$w(x)=Ce^{-x^2} \tag{5-76}$$

and we notice that it satisfies the differential equation

$$\frac{dw}{dx}+2xw=0 \tag{5-77}$$

for all values of the arbitrary constant $C$. We differentiate this expression $(n+1)$ times and obtain

$$\frac{d^{n+2}w}{dx^{n+2}}+2x\frac{d^{n+1}w}{dx^{n+1}}+2(n+1)\frac{d^n w}{dx^n}=0 \tag{5-78}$$

It follows that the function

$$q(x)=\frac{d^n w}{dx^n}=C\frac{d^n}{dx^n}e^{-x^2} \tag{5-79}$$

satisfies the equation

$$\frac{d^2q}{dx^2}+2x\frac{dq}{dx}+2(n+1)q=0 \tag{5-80}$$

A comparison of Eqs. (5-63) and (5-79) shows that $H_n(x)$ can be written as

$$H_n(x)=e^{x^2}q(x) \tag{5-81}$$

if we take the constant $C$ equal to $(-1)^n$. Therefore, if we substitute

$$q(x)=y(x)e^{-x^2} \tag{5-82}$$

into Eq. (5-80), we obtain the differential equation for $H_n(x)$. This substitution leads to

$$\frac{d^2y}{dx^2}-2x\frac{dy}{dx}+2ny=0 \tag{5-83}$$

We have shown that for a non-negative integer $n$ the Hermite polynomial $H_n(x)$ is a solution of Eq. (5-83), but this covers only a small fraction of the solutions of the differential equations. In the first place, Eq. (5-83) also has solutions for noninteger values of $n$, and second, it has two possible solutions for each value of $n$. We wish to investigate some properties of all these solutions, and therefore we have to find the general solution of Eq. (5-83). This can be achieved by reducing Eq. (5-83) to the differential equation for the confluent hypergeometric function.

We introduce the new variable

$$t=x^2 \tag{5-84}$$

and we have

$$\begin{aligned}\frac{dy}{dx}&=2x\frac{dy}{dt}\\ \frac{d^2y}{dx^2}&=2\frac{dy}{dt}+4x^2\frac{d^2y}{dt^2}\end{aligned} \tag{5-85}$$

Substitution into Eq. (5-83) gives

$$t\frac{d^2y}{dt^2}+\left(\frac{1}{2}-t\right)\frac{dy}{dt}+\frac{n}{2}y=0 \tag{5-86}$$

This equation is identical with Eq. (5-29) for the confluent hypergeometric function, and its two solutions are given by Eqs. (5-28) and (5-31):

$$\begin{aligned}y_1&={}_1F_1\left(-\frac{n}{2};\frac{1}{2};x^2\right)\\ y_2&=x\,{}_1F_1\left(-\frac{n}{2}+\frac{1}{2};\frac{3}{2};x^2\right)\end{aligned} \tag{5-87}$$

We have shown in Eq. (5-33) that these functions behave asymptotically as $\exp(x^2)$ for large values of $x$. However, it can be seen from the definition (5-28) of the confluent hypergeometric function that it reduces to a finite polynomial if the parameter $a$ is a negative integer $a=0,-1,-2,-3,\ldots$, and so on.

It can be seen that, in general, the two solutions $y_1$ and $y_2$ behave asymptotically as $\exp(x^2)$. If $n$ is an even positive integer, $n=2k$, the first solution $y_1$ reduces to a finite polynomial:

$$y_1 = {}_1F_1(-k; \tfrac{1}{2}; x^2) \tag{5-88}$$

and the second solution $y_2$ behaves asymptotically as $\exp(x^2)$. If $n$ is an odd positive integer, $n=2k+1$, then the second solution $y_2$ reduces to a finite polynomial:

$$y_2 = x\,{}_1F_1(-k; \tfrac{3}{2}; x^2) \tag{5-89}$$

and the first solution $y_1$ behaves asymptotically as $\exp(x^2)$. In both cases, the solution $y_1$ of Eq. (5-88) and the solution $y_2$ of Eq. (5-89) must be proportional to the Hermite polynomials $H_{2k}(x)$ or $H_{2k+1}(x)$, respectively. The proportionality constant can be derived by substituting $x=0$ in the expansion (5-57) and in Eq. (5-88). It is found that

$$H_{2k}(x) = (-1)^k (2k)!(k!)^{-1}\,{}_1F_1(-k; \tfrac{1}{2}; x^2) \tag{5-90}$$

A similar expression can be derived for $H_{2k+1}(x)$.

The various results that we derived in this section will be used for the solution of the Schrödinger equation of the harmonic oscillator. This problem is discussed in the following section.

## 5-7 The Harmonic Oscillator

A particle moving in a one-dimensional quadratic potential field with a minimum is called a harmonic oscillator. If we take the minimum of the potential as the zero point for the energy and the position of the minimum as the origin of the coordinate system, we can write the potential function as

$$V(x) = \tfrac{1}{2}kx^2 \tag{5-91}$$

where $k$ is the force constant. The Schrödinger equation is

$$-\frac{\hbar^2}{2m}\frac{d^2\phi}{dx^2} + \frac{1}{2}kx^2\phi = E\phi \tag{5-92}$$

We introduce a new parameter

$$\varepsilon = \frac{2mE}{\hbar^2} \tag{5-93}$$

so that the equation becomes

$$\frac{d^2\phi}{dx^2} + \left(\varepsilon - \frac{km}{\hbar^2}x^2\right)\phi = 0 \tag{5-94}$$

Now we change from the variable $x$ to a new variable $y$ by way of the transformation

$$x = \sqrt{\alpha}\, y \qquad \alpha^2 = \frac{\hbar^2}{km} \tag{5-95}$$

The equation then becomes

$$\frac{d^2\phi}{dy^2} + (\alpha\varepsilon - y^2)\phi = 0 \tag{5-96}$$

In order to solve an equation of this kind, it is often useful to determine the asymptotic behavior of its solution first. We consider the function

$$g(y) = e^{\pm y^2/2} \tag{5-97}$$

which has a second derivative

$$g''(y) = (y^2 \pm 1)e^{\pm y^2/2} \tag{5-98}$$

The differential equation for $g$ is therefore

$$g'' - (\pm 1 + y^2)g = 0 \tag{5-99}$$

For very large values of $y$ the term $\pm 1$ in Eq. (5-99) becomes negligible, as does the term $\alpha\varepsilon$ in Eq. (5-96), so that Eqs. (5-96) and (5-99) become identical for large values of $y$. We can therefore conclude that for large values of $y$ the solutions of Eq. (5-96) behave asymptotically like $\exp(\pm y^2/2)$. The solution that behaves like $\exp(y^2/2)$ is unsuitable, since it tends to infinity when $y$ tends to plus or minus infinity. Consequently the desired solution of Eq. (5-96) behaves asymptotically like $\exp(-y^2/2)$, and in order to solve the equation, we substitute

$$\phi(y) = w(y)e^{-y^2/2} \tag{5-100}$$

into Eq. (5-96). Then the differential equation for $w(y)$ is

$$\frac{d^2w}{dy^2}-2y\frac{dw}{dy}+(\alpha\varepsilon-1)w=0 \tag{5-101}$$

Let us compare this with Eq. (5-83) for the Hermite polynomial $H_n(y)$, which is

$$\frac{d^2w}{dy^2}-2y\frac{dw}{dy}+2nw=0 \tag{5-102}$$

Equations (5-101) and (5-102) become identical if we take

$$\alpha\varepsilon-1=2n \tag{5-103}$$

We saw in Section 5-6 that for integer non-negative values of $n$ one of the two solutions of Eq. (5-102) is a Hermite polynomial $H_n(y)$. The other solution and all solutions for noninteger values of $n$ behave asymptotically like $\exp(y^2)$ and they are not allowed since they would cause the wave function $\phi(y)$ to behave asymptotically like $\exp(y^2/2)$, according to Eq. (5-100). We find, therefore, that we obtain acceptable solutions of the Schrödinger equation if we impose the condition

$$\alpha\varepsilon=2n+1 \qquad n=0,1,2,3,4,\ldots, \text{ and so on} \tag{5-104}$$

The energy eigenvalues are therefore

$$E_n=(n+\tfrac{1}{2})\hbar\omega \qquad n=0,1,2,\ldots, \text{ and so on} \tag{5-105}$$

according to Eqs. (5-93), (5-95), and (5-104). Here we have introduced the angular frequency

$$\omega=\left(\frac{k}{m}\right)^{1/2} \tag{5-106}$$

It follows from Eq. (5-100) and from the considerations above that the corresponding eigenfunctions are

$$\phi_n=C_nH_n(y)e^{-y^2/2} \qquad y=\alpha^{-1/2}x \tag{5-107}$$

The normalization constants $C_n$ are determined from the condition

$$C_nC_n^*\int_{-\infty}^{\infty}H_n(y)H_n(y)e^{-y^2}\,dx=C_nC_n^*\sqrt{\alpha}\int_{-\infty}^{\infty}H_n(y)H_n(y)e^{-y^2}\,dy=1 \tag{5-108}$$

The value of the integral is reported in Eq. (5-75), and the substitution leads to

$$C_n C_n^* \sqrt{\alpha}\, 2^n n! \sqrt{\pi} = 1 \tag{5-109}$$

or

$$C_n = (\pi\alpha)^{-1/4} (2^n \cdot n!)^{-1/2} \tag{5-110}$$

In Eq. (5-64) we reported the explicit forms of some of the Hermite polynomials. We use these expressions to plot the normalized eigenfunctions versus the variable $y$ in Fig. 5-1 for some of the lower eigenvalues.

It may be interesting to draw a comparison between the quantum mechanical description of the harmonic oscillator and the classical description we presented in Section 1-3. In Fig. 5-2 we show the potential curve of the harmonic oscillator together with a horizontal line at the level of the lowest energy eigenvalue $E_o$. In the region inside the potential curve the energy $E_o$ is larger than the potential energy, the difference being the kinetic energy of the particle. In the region outside the potential curve $E_o$ is smaller than the potential energy; in this region the kinetic energy should have a negative value. This is clearly not allowed in classical mechanics. The points of intersection between the vertical line, $+x_o$ and $-x_o$, are the classical turning points of the harmonic oscillator. Their values are given by

$$\tfrac{1}{2}\hbar\omega = \tfrac{1}{2}kx_o^2$$

$$x_o^2 = \frac{\hbar}{\sqrt{km}} = \alpha \tag{5-111}$$

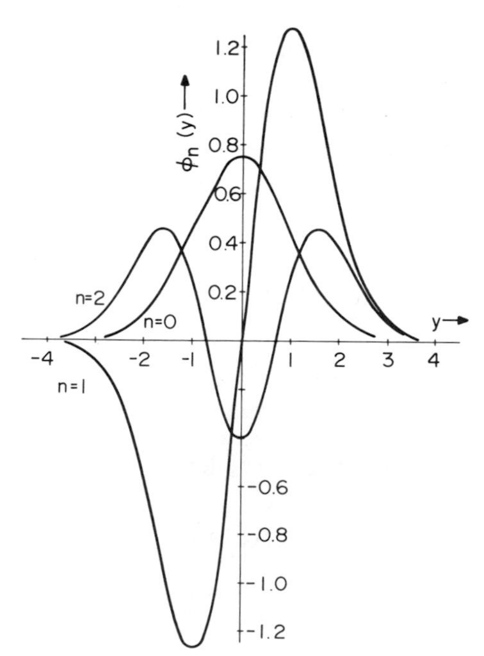


**Fig. 5-1** The normalized eigenfunctions corresponding to the eigenstates $n=0$, $n=1$, and $n=2$ of the harmonic oscillator.

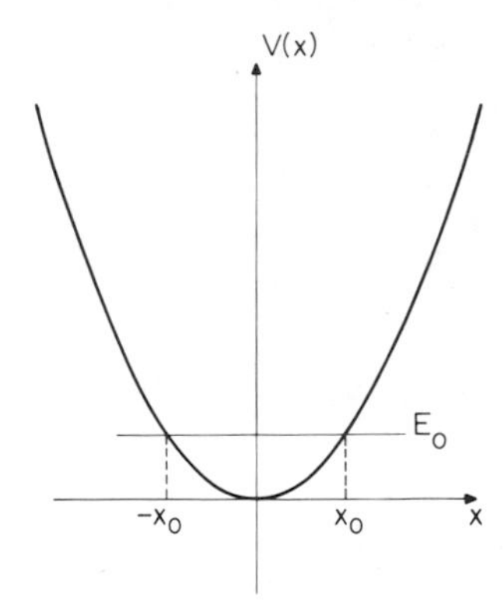


**Fig. 5-2** The potential function $V(x)$ and the lowest energy eigenvalue $E_0$ of the harmonic oscillator. The figure is used to determine the unit of length $x_0$ of the harmonic oscillator according to Eq. (5-11).

According to Eq. (5-95), this is exactly the unit of length that we use in our quantum mechanical description.

It can be seen in Fig. 5-1*a* that the wave function $\phi_o(y)$ is nonzero outside the points $\pm 1$, and it follows that there is a finite, nonzero probability of finding the particle in the areas $y<-1$ and $y>1$. This result is in marked contrast with the classical description where the motion is confined to the region $-1 \leq y \leq 1$ if the total energy is equal to $E_o$.

We are interested also in comparing our present results with the matrix representation of the harmonic oscillator. In Section 2-9 we described Heisenberg's matrix mechanics and we mentioned that the harmonic oscillator was described by this model. We use the results of this section together with the properties of Hermite polynomials in order to evaluate the quantities

$$(x)_{n,m} = \langle \phi_n(x) | x | \phi_m(x) \rangle \tag{5-112}$$

and

$$(p)_{n,m} = -i\hbar(\delta)_{n,m}$$

$$(\delta)_{n,m} = \left\langle \phi_n(x) \middle| \frac{d}{dx} \middle| \phi_m(x) \right\rangle \tag{5-113}$$

We write Eq. (5-112) as

$$(x)_{n,m} = C_n C_m \int_{-\infty}^{\infty} H_n(y) x H_m(y) e^{-y^2}\, dx$$

$$= \left( \frac{\pi}{\alpha} 2^{n+m} \cdot n! \cdot m! \right)^{-1/2} \int_{-\infty}^{\infty} H_n(y) y H_m(y) e^{-y^2}\, dy \tag{5-114}$$

According to Eq. (5-67), we have

$$yH_m(y) = mH_{m-1}(y) + \tfrac{1}{2} H_{m+1}(y) \tag{5-115}$$

Substitution into Eq. (5-114) gives

$$(x)_{n,m}=\left(\frac{\pi}{\alpha}\cdot 2^{n+m}n!m!\right)^{-1/2}\left(mI_{n,m-1}+\tfrac{1}{2}I_{n,m+1}\right)$$

with

$$I_{n,m}=\int_{-\infty}^{\infty}H_n(y)H_m(y)e^{-y^2}\,dy=\left(2^n\cdot n!\sqrt{\pi}\right)\delta_{n,m} \qquad (5\text{-}116)$$

We obtain

$$\begin{aligned}(x)_{n,n+1}&=\left[\tfrac{1}{2}\alpha(n+1)\right]^{1/2}\\(x)_{n,n-1}&=\left[\tfrac{1}{2}\alpha n\right]^{1/2}\\(x)_{n,m}&=0 \qquad \text{if} \qquad m\neq n\pm 1\end{aligned} \qquad (5\text{-}117)$$

Equation (5-113) can be written as

$$\begin{aligned}(\delta)_{n,m}&=C_nC_m\int_{-\infty}^{\infty}H_n(y)e^{-y^2/2}\frac{d}{dy}\left[H_m(y)e^{-y^2/2}\right]dy\\&=C_nC_m\int_{-\infty}^{\infty}\left[H_n(y)H_m'(y)-H_n(y)yH_m(y)\right]e^{-y^2}\,dy\end{aligned} \qquad (5\text{-}118)$$

Substitution of the recurrence Eqs. (5-67) and (5-69) gives

$$(\delta)_{n,m}=(\pi\alpha\cdot 2^{n+m}\cdot n!m!)^{-1/2}\left(mI_{n,m-1}-\tfrac{1}{2}I_{n,m+1}\right) \qquad (5\text{-}119)$$

It follows from Eq. (5-116) that

$$\begin{aligned}(\delta)_{n,n+1}&=\left(\frac{n+1}{2\alpha}\right)^{1/2}\\(\delta)_{n,n-1}&=-\left(\frac{n}{2\alpha}\right)^{1/2}\\(\delta)_{n,m}&=0 \qquad \text{if} \qquad m\neq n\pm 1\end{aligned} \qquad (5\text{-}120)$$

Finally, we wish to derive a general property of the eigenfunctions. It can be seen in Fig. 5-1 that the various eigenfunctions are either symmetric or antisymmetric in the coordinate. We show that this is true in general for all systems where the potential function $V(x)$ is symmetric in $x$, that is,

$$V(x)=V(-x) \qquad (5\text{-}121)$$

We write the Schrödinger equation as

$$\left[-\frac{\hbar^2}{2m}\frac{d^2}{dx^2}+V(x)\right]\phi(x)=E\phi(x) \tag{5-122}$$

Let us now consider an eigenvalue $E_n$ that has only one corresponding eigenfunction $\phi_n(x)$. We then have

$$\left[-\frac{\hbar^2}{2m}\frac{d^2}{dx^2}+V(x)\right]\phi_n(x)=E_n\phi_n(x) \tag{5-123}$$

and also

$$\left[-\frac{\hbar^2}{2m}\frac{d^2}{dx^2}+V(-x)\right]\phi_n(-x)=E_n\phi_n(-x) \tag{5-124}$$

Because of Eq. (5-121) we can also write Eq. (5-124) as

$$\left[-\frac{\hbar^2}{2m}\frac{d^2}{dx^2}+V(x)\right]\phi_n(-x)=E_n\phi_n(-x) \tag{5-125}$$

Let us now compare Eqs. (5-123) and (5-125). It follows that not only $\phi_n(x)$, but also $\phi_n(-x)$, is an eigenfunction belonging to $E_n$. However, since $E_n$ can have only one eigenfunction, $\phi_n(x)$ and $\phi_n(-x)$ must be proportional to each other, or

$$\phi_n(-x)=\rho\phi_n(x) \tag{5-126}$$

If we apply Eq. (5-126) twice, we find that

$$\phi_n(-x)=\rho\phi_n(x)=\rho^2\phi_n(-x) \tag{5-127}$$

or

$$\rho^2=1 \tag{5-128}$$

Obviously $\rho=\pm1$, so that there are two possibilities, namely

$$\phi_n(x)=\phi_n(-x) \tag{5-129}$$

or

$$\phi_n(x)=-\phi_n(-x) \tag{5-130}$$

In the first case we call $\phi_n(x)$ symmetric in $x$, and in the second case $\phi_n(x)$ is antisymmetric in $x$.

## 5-8 Laguerre Polynomials

The Laguerre polynomials $L_n(x)$ can be derived from the generating function

$$K(x,h)=\frac{\exp[-xh(1-h)^{-1}]}{1-h} \tag{5-131}$$

If $|h|<1$ then the generating function (5-131) can be expanded as a power series in $h$. The Laguerre polynomials $L_n(x)$ are then defined as

$$K(x,h)=\sum_{n=0}^{\infty} L_n(x)h^n \tag{5-132}$$

It is possible to derive a differential expression for $L_n(x)$, analogous to Eq. (5-63) for the Hermite polynomials. We have

$$L_n(x)=\frac{1}{n!}\left[\frac{\partial^n K(x,h)}{\partial h^n}\right]_{h=0} \tag{5-133}$$

We write the generating function as

$$K(x,h)=e^x\frac{\exp[-x/(1-h)]}{1-h}=e^x\sum_{k=0}^{\infty}\frac{(-x)^k}{(1-h)^{k+1}k!} \tag{5-134}$$

It is easily seen that

$$\frac{\partial^n K(x,h)}{\partial h^n}=e^x\sum_{k=0}^{\infty}\frac{(k+n)!}{k!}\frac{(-1)^k x^k}{(1-h)^{k+n+1}k!}$$

or, according to Eq. (5-133),

$$L_n(x)=\frac{e^x}{n!}\sum_{k=0}^{\infty}\frac{(-1)^k(k+n)!x^k}{(k!)^2} \tag{5-135}$$

It can be shown also that

$$\frac{d^n}{dx^n}(x^n e^{-x})=\frac{d^n}{dx^n}\sum_{k=0}^{\infty}\frac{(-1)^k x^{k+n}}{k!}=\sum_{k=0}^{\infty}\frac{(-1)^k(k+n)!x^k}{(k!)^2} \tag{5-136}$$

By comparing Eqs. (5-135) and (5-136) we find that

$$L_n(x)=\frac{e^x}{n!}\frac{d^n}{dx^n}(x^n e^{-x}) \tag{5-137}$$

The first few Laguerre polynomials can be derived from Eq. (5-137). They are

$$
\begin{aligned}
L_0(x)&=1\\
L_1(x)&=1-x\\
L_2(x)&=1-2x+\tfrac{1}{2}x^2\\
L_3(x)&=1-3x+\tfrac{3}{2}x^2-\tfrac{1}{6}x^3
\end{aligned}
\tag{5-138}
$$

Various recurrence relations can be derived from the generating function. By differentiating with respect to $x$ or $h$, respectively, we find

$$
\begin{aligned}
(1-h)\frac{\partial K}{\partial x}+hK&=0\\
(1-h)^2\frac{\partial K}{\partial h}+xK-(1-h)K&=0
\end{aligned}
\tag{5-139}
$$

From the first Eq. (5-139) we derive that

$$\frac{dL_n}{dx}-\frac{dL_{n-1}}{dx}+L_{n-1}=0 \tag{5-140}$$

and from the second Eq. (5-139) we derive that

$$(n+1)L_{n+1}-(2n+1)L_n+nL_{n-1}+xL_n=0 \tag{5-141}$$

By differentiating Eq. (5-141) and by making use of Eq. (5-140) we can derive a third recurrence relation:

$$x\frac{dL_n}{dx}=nL_n-nL_{n-1} \tag{5-142}$$

The orthogonality properties of the Laguerre polynomials are easily derived from the generating function. We have

$$
\begin{aligned}
&\int_0^\infty e^{-x}K(x,h)K(x,k)\,dx\\
&=(1-h)^{-1}(1-k)^{-1}\\
&\times\int_0^\infty \exp\left[\{-x(1-h)(1-k)-xh(1-k)-xk(1-h)\}(1-h)^{-1}(1-k)^{-1}\right]dx\\
&=(1-h)^{-1}(1-k)^{-1}\int_0^\infty \exp\left[-x(1-hk)(1-h)^{-1}(1-k)^{-1}\right]dx\\
&=(1-hk)^{-1}=\sum_{n=0}^{\infty}(hk)^n
\end{aligned}
\tag{5-143}
$$

It follows from Eq. (5-132) that

$$\int_0^\infty e^{-x}K(x,h)K(x,k)\,dx=\sum_{n=0}^{\infty}\sum_{m=0}^{\infty}h^nk^m\int_0^\infty e^{-x}L_n(x)L_m(x)\,dx \quad (5\text{-}144)$$

By comparing the results of Eqs. (5-143) and (5-144) we find that

$$\int_0^\infty L_n(x)L_n(x)e^{-x}\,dx=1$$

$$\int_0^\infty L_n(x)L_m(x)e^{-x}\,dx=0 \qquad n\neq m \quad (5\text{-}145)$$

The differential equation for the Laguerre polynomials is easily derived from the expression (5-137). First we note that the function

$$g(x)=x^ne^{-x} \quad (5\text{-}146)$$

satisfies the equation

$$x\frac{dg}{dx}+(x-n)g=0 \quad (5\text{-}147)$$

By differentiating this equation $(n+1)$ times we obtain

$$x\frac{d^2w}{dx^2}+(1+x)\frac{dw}{dx}+(n+1)w=0 \quad (5\text{-}148)$$

as the equation for the function

$$w=\frac{d^ng}{dx^n} \quad (5\text{-}149)$$

Finally, we substitute

$$w=e^{-x}y \quad (5\text{-}150)$$

and we obtain

$$x\frac{d^2y}{dx^2}+(1-x)\frac{dy}{dx}+ny=0 \quad (5\text{-}151)$$

as the equation for the Laguerre polynomials.

Equation (5-151) is identical with the differential equation for the confluent hypergeometric function, Eq. (5-29). The two solutions of Eq. (5-29) reduce to the same form, namely

$$y = {}_1F_1(-n; 1; x) \tag{5-152}$$

Clearly this becomes a finite polynomial for integer values of the parameter $n$.

In addition to the Laguerre polynomials we may encounter the associated Laguerre polynomials $L_n^k(x)$. We define the latter polynomials as

$$L_n^k(x) = (-1)^k \frac{d^k}{dx^k} L_{n+k}(x) \tag{5-153}$$

The latter polynomials can also be expressed in terms of the confluent hypergeometric function.

It follows from the definition (5-28) that

$$\frac{d}{dx}\,{}_1F_1(a; b; x) = \frac{a}{b}\,{}_1F_1(a+1; b+1; x) \tag{5-154}$$

Hence

$$\frac{d}{dx} L_{n+k}(x) = \frac{d}{dx}\,{}_1F_1(-n-k; 1; x) = -\frac{(n+k)}{1}\,{}_1F_1(-n-k+1; 2; x) \tag{5-155}$$

and

$$L_n^k(x) = \frac{(n+k)!}{n!k!}\,{}_1F_1(-n; 1+k; x) \tag{5-156}$$

Obviously the differential equation for the associated Laguerre polynomial $L_n^k(x)$ is

$$x\frac{d^2y}{dx^2} + (1+k-x)\frac{dy}{dx} + ny = 0 \tag{5-157}$$

We should mention that the associated Laguerre polynomials $L_n^m(x)$ can be defined also by means of a generating function:

$$(1-h)^{-m-1}\exp\left[-xh(1-h)^{-1}\right] = \sum_{n=0}^{\infty} L_n^m(x)h^n \tag{5-158}$$

The associated Laguerre polynomials have certain properties that are analogous to the properties of the Laguerre polynomials and that are derived in analogous ways. For example, we have

$$L_n^m(x) = \frac{x^{-m}e^x}{n!}\frac{d^n}{dx^n}(x^{n+m}e^{-x}) \tag{5-159}$$

analogous to Eq. (5-137). By means of an argument that is similar to Eq. (5-143) we can also show that

$$\int_0^\infty x^m e^{-x} L_n^m(x) L_n^m(x)\, dx = \frac{(n+m)!}{n!} \tag{5-160}$$

and

$$\int_0^\infty x^m e^{-x} L_n^m(x) L_{n'}^m(x)\, dx = 0 \qquad \text{if} \qquad n \neq n' \tag{5-161}$$

These various relations will prove to be helpful in solving the Schrödinger equation for the hydrogen atom in the next chapter. We also use some of them for solving the Morse potential problem in Section 5-9.

## 5-9 The Morse Potential

The vibrational motion of a diatomic molecule is sometimes approximated by the harmonic oscillator potential that we discussed in Section 5-7. A more realistic model is the Morse potential sketched in Fig. 5-3. The analytical form of the potential proposed by Morse is

$$V(r) = D\{1 - \exp[-\alpha(r - r_o)]\}^2 \tag{5-162}$$

This potential has a minimum value 0 for $r = r_o$; for large values of $r$ it approaches $D$ asymptotically and for negative values of $r$ it becomes very large. The corresponding Schrödinger equation is given by

$$-\frac{\hbar^2}{2\mu}\frac{d^2\psi}{dr^2} + D\{1 - \exp[-\alpha(r - r_o)]\}^2\psi = E\psi \tag{5-163}$$

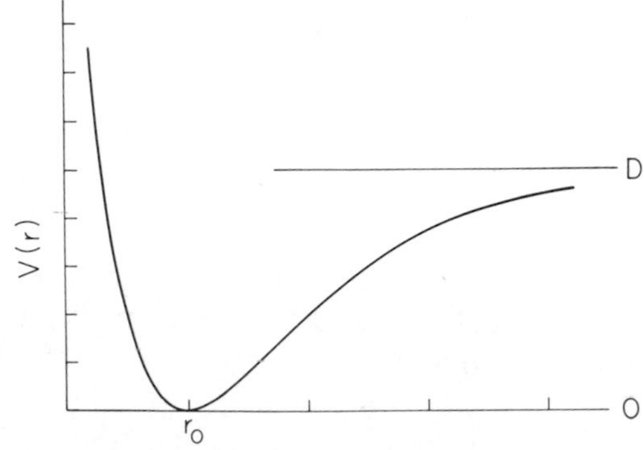


**Fig. 5-3** The potential function corresponding to the Morse potential of Eq. (5-162).

It may be anticipated that the system has a finite number of discrete energy eigenvalues for $E<D$ and a set of continuum eigenvalues for $E>D$.

In order to solve Eq. (5-163) we first introduce the new variable

$$x=\exp[-\alpha(r-r_o)] \tag{5-164}$$

The new equation becomes

$$x^2\frac{d^2\psi}{dx^2}+x\frac{d\psi}{dx}-\frac{2\mu(D-E)}{\hbar^2\alpha^2}\psi-\frac{2\mu D}{\hbar^2\alpha^2}(x^2-2x)\psi=0 \tag{5-165}$$

For convenience, we define

$$\lambda^2=\frac{2\mu(D-E)}{\hbar^2\alpha^2} \qquad \gamma^2=\frac{2\mu D}{\hbar^2\alpha^2} \tag{5-166}$$

so that Eq. (5-165) can be written as

$$x^2\frac{d^2\psi}{dx^2}+x\frac{d\psi}{dx}-(\gamma^2x^2-2\gamma^2x+\lambda^2)\psi=0 \tag{5-167}$$

This equation can be reduced to the confluent hypergeometric equation. We substitute

$$\psi=x^\lambda w \tag{5-168}$$

and we obtain

$$x\frac{d^2w}{dx^2}+(2\lambda+1)\frac{dw}{dx}-\gamma^2(x-2)w=0 \tag{5-169}$$

Next we substitute

$$w(x)=e^{-\gamma x}u(x) \tag{5-170}$$

The new equation becomes

$$x\frac{d^2u}{dx^2}+(2\lambda+1-2\gamma x)\frac{du}{dx}+(2\gamma^2-2\lambda\gamma-\gamma)u=0 \tag{5-171}$$

Finally, we introduce the new variable

$$t=2\gamma x \tag{5-172}$$

and we obtain

$$t\frac{d^2u}{dt^2}+(2\lambda+1-t)\frac{du}{dt}+\left(\gamma-\lambda-\frac{1}{2}\right)u=0 \tag{5-173}$$

This is identical with Eq. (5-29) for the confluent hypergeometric function. The two solutions are

$$u_1 = {}_1F_1(-\gamma+\lambda+\tfrac{1}{2}; 2\lambda+1; t)$$

$$u_2 = t^{-2\lambda}{}_1F_1(-\gamma-\lambda+\tfrac{1}{2}; -2\lambda+1; t) \tag{5-174}$$

We should remember that the variable $x$ (or the variable $t$) ranges from zero to infinity. The second solution $u_2$ is inadmissible because it tends to infinity when $t$ tends to zero. The first solution $u_1$ leads to a solution

$$\psi_1 = x^{\lambda}e^{-\gamma x}{}_1F_1(-\gamma+\lambda+\tfrac{1}{2}; 2\lambda+1; 2\gamma x) \tag{5-175}$$

of the original Eq. (5-167). This solution behaves asymptotically as $\exp(\gamma x)$ for large values of $x$, according to Eqs. (5-33) and (5-34). It follows that the solution is acceptable only if the confluent hypergeometric function reduces to a finite polynomial. In other words, if

$$\lambda+\tfrac{1}{2}-\gamma=-n \qquad n=0,1,2,3,\ldots \tag{5-176}$$

The energy eigenvalues $E_n$ are given by

$$\frac{2\mu(D-E)}{\hbar^2\alpha^2} = \lambda^2 = (\gamma-n-\tfrac{1}{2})^2 = \gamma^2 - 2\gamma(n+\tfrac{1}{2}) + (n+\tfrac{1}{2})^2 \tag{5-177}$$

or

$$E_n = \frac{\hbar^2\alpha^2\gamma(n+\frac{1}{2})}{\mu} - \frac{\hbar^2\alpha^2(n+\frac{1}{2})^2}{2\mu} \tag{5-178}$$

It may be helpful to write this result in a different form so that we can compare it with the results of the harmonic oscillator. We define the force constant $k$ of the Morse potential as

$$k = \left(\frac{\partial^2 V}{\partial r^2}\right)_{r_o} = 2\alpha^2 D \tag{5-179}$$

and we define the angular frequency $\omega_o$ as

$$\omega_o = \left(\frac{k}{\mu}\right)^{1/2} = \left(\frac{2\alpha^2 D}{\mu}\right)^{1/2} = \frac{\hbar\alpha^2\gamma}{\mu} \tag{5-180}$$

Consequently we can write Eq. (5-178) as

$$E_n = \hbar\omega_o\left[(n+\tfrac{1}{2}) - (2\gamma)^{-1}(n+\tfrac{1}{2})^2\right] \tag{5-181}$$

The corresponding eigenfunctions are derived from Eq. (5-175) by substituting Eq. (5-176):

$$\psi_n(x) = x^{\gamma-n-1/2} e^{-\gamma x} {}_1F_1(-n; 2\gamma - 2n; 2\gamma x) \tag{5-182}$$

It is possible to express this either in terms of the Whittaker functions of Eq. (5-50) or in terms of the associated Laguerre polynomials of Eq. (5-156), but there does not seem to be any useful purpose in these transformations.

Finally, we wish to determine the number of discrete energy levels. This number $n_o$ is determined from Eq. (5-181) by taking

$$\frac{\partial E}{\partial n} = 0 \qquad n_o + \frac{1}{2} = \gamma \tag{5-183}$$

We can derive from Eqs. (5-166) and (5-180) that

$$n_o \approx \gamma = \frac{2D}{\hbar \omega_o} \tag{5-184}$$

For the vibrational motion of a typical diatomic molecule this number is of the order of 10 to 50.

## 5-10 Legendre Polynomials

The Legendre polynomials $P_n(x)$ are derived from the generating function

$$K(x,h) = (1 - 2xh + h^2)^{-1/2} \tag{5-185}$$

If the variable $h$ is sufficiently small, that is, $|h| < 1$, then the generating function can be expanded as a power series in $h$, and the Legendre polynomials $P_n(x)$ are defined as the coefficients in this expansion.

$$K(x,h) = \sum_{n=0}^{\infty} P_n(x) h^n \tag{5-186}$$

In order to derive explicit expressions for the Legendre polynomials we first observe that $(1-y)^{-1/2}$ can be expanded as

$$(1-y)^{-1/2} = 1 + \frac{1}{2}y + \frac{1}{2}\cdot\frac{3}{2}\cdot\frac{y^2}{2!} + \frac{1}{2}\cdot\frac{3}{2}\cdot\frac{5}{2}\cdot\frac{y^3}{3!} + \frac{1}{2}\cdot\frac{3}{2}\cdot\frac{5}{2}\cdot\frac{7}{2}\cdot\frac{y^4}{4!} + \cdots \tag{5-187}$$

if $|y| < 1$. The product $1\cdot3\cdot5\cdot7\ldots(2n-1)$ can be written as

$$1\cdot3\cdot5\cdot7\ldots(2n-1) = \frac{1\cdot2\cdot3\cdot4\cdot5\ldots2n}{2^n\cdot1\cdot2\cdot3\cdot4\ldots n} = \frac{(2n)!}{2^n n!} \tag{5-188}$$

so that Eq. (5-187) can be reduced to

$$(1-y)^{-1/2}=\sum_{n=0}^{\infty}\frac{(2n)!y^n}{2^{2n}\cdot(n!)^2} \tag{5-189}$$

The generating function $K(x, h)$ can therefore be expanded as

$$K(x,h)=\sum_{n=0}^{\infty}\frac{(2n)!h^n(2x-h)^n}{2^{2n}\cdot(n!)^2} \tag{5-190}$$

Let us now expand each product $(2x-h)^n$ as

$$(2x-h)^n=\sum_{m=0}^{n}\frac{(-1)^m n!(2x)^{n-m}h^m}{m!(n-m)!} \tag{5-191}$$

and substitute the result into Eq. (5-190):

$$K(x,h)=\sum_{n=0}^{\infty}\frac{(2n)!h^n}{n!}\sum_{m=0}^{n}\frac{(-1)^m h^m x^{n-m}}{2^{n+m}m!(n-m)!} \tag{5-192}$$

The Legendre polynomial $P_k(x)$ of order $k$ is now obtained from Eq. (5-192) by collecting all terms that contain the factor $h^k$. These terms are obtained by taking $n=k$ and $m=0$, $n=k-1$ and $m=1$, $n=k-2$ and $m=2$, and so on. Hence we find $P_k(x)$ by substituting the index $k-r$ for $n$ and the number $r$ for $m$ into Eq. (5-192) and by summing over $r$ instead of over $n$ and $m$:

$$P_k(x)=\sum_{r=0}^{\alpha}\frac{(-1)^r(2k-2r)!x^{k-2r}}{w^k r!(k-r)!(k-2r)!} \tag{5-193}$$

The maximum value $\alpha$ that the summation index $r$ can assume is determined by the condition that in Eq. (5-192) $m \leq n$. If we substitute $r$ for $m$ and $k-r$ for $n$, we find

$$2r \leq k \tag{5-194}$$

Since $r$ also has to be an integer, it follows that $\alpha=\frac{1}{2}k$ when $k$ is even and $\alpha=\frac{1}{2}(k-1)$ when $k$ is odd.

We can derive from Eq. (5-193) that the first few Legendre polynomials have the form

$$
\begin{aligned}
&P_0(x)=1 \\
&P_1(x)=x \\
&P_2(x)=(\tfrac{1}{2})(3x^2-1) \\
&P_3(x)=(\tfrac{1}{2})(5x^3-3x) \\
&P_4(x)=(\tfrac{1}{8})(35x^4-30x^2+3) \\
&P_5(x)=(\tfrac{1}{8})(63x^5-70x^3+15x)
\end{aligned}
\tag{5-195}
$$

We show that Eq. (5-193) is equivalent to a different expression for $P_k(x)$ which is more concise and which is known as Rodrigues' formula. For this purpose we observe that

$$(x^2-1)^k=\sum_{r=0}^{k}\frac{(-1)^r k!x^{2k-2r}}{r!(k-r)!} \tag{5-196}$$

If we differentiate this expression $k$ times, we find that

$$\frac{d^k}{dx^k}(x^2-1)^k=\sum_{r=0}^{\alpha}\frac{(-1)^r k!(2k-2r)!x^{k-2r}}{r!(k-r)!(k-2r)!} \tag{5-197}$$

A comparison with Eq. (5-193) leads to Rodrigues' formula:

$$P_k(x)=\frac{1}{2^k\cdot k!}\frac{d^k}{dx^k}(x^2-1)^k \tag{5-198}$$

Let us now derive some of the recurrence formulas for Legendre polynomials of different orders and their derivatives. By differentiating Eq. (5-185) with respect to $h$, we find that

$$(1-2hx+h^2)\frac{\partial K}{\partial h}=(x-h)K \tag{5-199}$$

If we now substitute the expansion of Eq. (5-186) into this result, we find that

$$(n+1)P_{n+1}(x)-(2n+1)xP_n(x)+nP_{n-1}(x)=0 \tag{5-200}$$

A second recurrence relation is obtained if we differentiate Eq. (5-185) with respect to $x$ and to $h$. From a comparison of the two results it follows that

$$h\frac{\partial K}{\partial h}=(x-h)\frac{\partial K}{\partial x} \tag{5-201}$$

Substitution of Eq. (5-186) now yields

$$x\frac{dP_n(x)}{dx}-\frac{dP_{n-1}(x)}{dx}=nP_n(x) \tag{5-202}$$

By combining Eqs. (5-200) and (5-202), we derive three additional recurrence formulas:

$$\begin{aligned} P'_{n+1}(x)-xP'_n(x)&=(n+1)P_n(x)\\ P'_{n+1}(x)-P'_{n-1}(x)&=(2n+1)P_n(x)\\ (x^2-1)P'_n(x)&=nxP_n(x)-nP_{n-1}(x) \end{aligned} \tag{5-203}$$

The differential equation for $P_n(x)$ is easily derived with the aid of Rodrigues' formula. We first consider the function

$$q(x)=C(1-x^2)^n \tag{5-204}$$

and we observe that it satisfies the equation

$$(1-x^2)\frac{dq}{dx}+2nxq=0 \tag{5-205}$$

for all values of the arbitrary parameter $C$. Differentiating Eq. (5-205) $(n+1)$ times gives

$$(1-x^2)\frac{d^{n+2}q}{dx^{n+2}}-2x\frac{d^{n+1}q}{dx^{n+1}}+n(n+1)\frac{d^nq}{dx^n}=0 \tag{5-206}$$

If we now introduce the function $g(x)$, which is

$$g(x)=C\frac{d^nq}{dx^n}=C'P_n(x) \tag{5-207}$$

according to Eq. (5-198), then we find the required equation

$$(1-x^2)\frac{d^2g}{dx^2}-2x\frac{dg}{dx}+n(n+1)g=0 \tag{5-208}$$

for $P_n(x)$.

We know that for a non-negative integer $n$ the Legendre polynomial $P_n(x)$ is one of the solutions of the equations. Let us now attempt to derive the general solution of Eq. (5-208) for arbitrary values of $n$. Since the parameter $\rho$ is equal to zero or unity in the present case, we substitute

$$g(x)=\sum_{k=0}^{\infty} a_k x^k \tag{5-209}$$

into Eq. (5-208) and obtain

$$a_{k+2}=\frac{k(k+1)-n(n+1)}{(k+1)(k+2)}a_k=\frac{(k-n)(k+n+1)}{(k+1)(k+2)}a_k \tag{5-210}$$

This leads to the following general solution of Eq. (5-208):

$$\begin{aligned} g(x)=a_0\bigg[1+&\frac{(-n)(1+n)}{2!}x^2+\frac{(-n)(2-n)(1+n)(3+n)}{4!}x^4+\cdots \\ &+\frac{(-n)(2-n)(4-n)\cdots(2s-2-n)(n+1)(n+3)\cdots(2s-1+n)}{(2s)!}x^{2s}+\cdots\bigg] \\ +a_1x\bigg[1+&\frac{(1-n)(2+n)}{3!}x^2+\frac{(1-n)(3-n)(2+n)(4+n)}{5!}x^4+\cdots \\ &+\frac{(1-n)(3-n)\cdots(2s-1-n)(2+n)(4+n)\cdots(2s+n)}{(2s+1)!}x^{2s}+\cdots\bigg] \end{aligned} \tag{5-211}$$

When $n$ is not an integer both solutions of Eq. (5-211) have the form of infinite power series in $x$. In quantum mechanical problems we are interested in the behavior of the power series for real values of the variable in the interval $-1\le x\le 1$, and we want to know whether the solutions have any singularities in this interval. This reduces to the question of whether the two power series are convergent for all values of $x$ in the interval $-1\le x\le 1$.

It would lead us too far to discuss the various convergence tests for the series of Eq. (5-211). We just mention that the series converges when $|x|\le 1$, but that it diverges for $x=\pm 1$. The relevant convergence test is known as Raabe's test. In order to obtain a solution of Eq. (5-208) that is finite in the whole interval $-1\le x\le 1$ we must impose the condition that $n$ is an integer. If $n$ is even, the first solution of Eq. (5-211) becomes a finite polynomial $P_n(x)$ that is finite over the whole interval and the second solution is infinite for $x=+1$. If $n$ is odd, the second solution becomes $P_n(x)$ and the first solution has singularities for $x=\pm 1$. When $n$ is negative we obtain nothing new; it is easily seen that Eq. (5-208) remains the same when we replace $n$ by $-n-1$. Therefore, we need only consider positive values of $n$. We can conclude, therefore, that the only solutions of Eq. (5-208) that are finite in the interval $-1 \leq x \leq 1$ are the Legendre polynomials $P_n(x)$. This conclusion is very important in quantum mechanical applications.

In order to derive the orthogonality properties of the Legendre polynomials we again evaluate the integral

$$\int_{-1}^{1} K(x,h)K(x,k)\,dx = \int_{-1}^{1} (1-2xh+h^2)^{-1/2}(1-2xk+k^2)^{-1/2}\,dx \tag{5-212}$$

This is a standard integral. If we define

$$R(x) = a + bx + cx^2 \tag{5-213}$$

it can be shown that

$$\int R^{-1/2}\,dx = c^{-1/2}\log\left[2\sqrt{cR} + 2cx + b\right] \tag{5-214}$$

By using Eq. (5-214) it can be shown that

$$\int_{-1}^{1} (1-2xh+h^2)^{-1/2}(1-2xk+k^2)^{-1/2}$$

$$= \frac{1}{\sqrt{hk}}\log\frac{1+\sqrt{hk}}{1-\sqrt{hk}} = \sum_{n=0}^{\infty} \frac{2}{2n+1}(hk)^n \tag{5-215}$$

From the definition of the Legendre polynomials it follows then that

$$\sum_{n=0}^{\infty}\sum_{m=0}^{\infty} h^n k^m \int_{-1}^{1} P_n(x)P_m(x)\,dx = \sum_{n=0}^{\infty} 2(2n+1)^{-1}(hk)^n \tag{5-216}$$

or

$$\int_{-1}^{1} P_n(x)P_m(x)\,dx = 0 \qquad n \neq m$$

$$\int_{-1}^{1} P_n(x)P_n(x)\,dx = \frac{2}{2n+1} \tag{5-217}$$

In addition to Legendre polynomials we will encounter the associated Legendre functions $P_n^m(x)$, which are defined as

$$P_n^m(x) = (1-x^2)^{m/2}\frac{d^m}{dx^m}P_n(x) \tag{5-218}$$

Here $n$ and $m$ are non-negative integers, and $n \geq m$. We limit our discussion to the behavior of the functions in the interval $-1 \leq x \leq 1$, since this is the interval that we consider in quantum mechanical applications. The associated Legendre functions satisfy a differential equation that is derived from Eq. (5-208) for the Legendre functions. If we differentiate this equation $m$ times, we find

$$(1-x^2)\frac{d^{m+2}g}{dx^{m+2}} - 2x(m+1)\frac{d^{m+1}g}{dx^{m+1}} + (n-m)(n+m+1)\frac{d^m g}{dx^m} = 0 \quad (5\text{-}219)$$

where

$$g = P_n(x) \quad (5\text{-}220)$$

If we introduce the function

$$q(x) = \frac{d^m}{dx^m} g(x) \quad (5\text{-}221)$$

it satisfies the equation

$$(1-x^2)\frac{d^2 q}{dx^2} - 2x(m+1)\frac{dq}{dx} + (n-m)(n+m+1)q = 0 \quad (5\text{-}222)$$

It follows from a comparison of Eqs. (5-218) and (5-221) that the associated Legendre function $P_n^m(x)$ is identical to the function $w(x)$ defined as

$$q(x) = (1-x^2)^{-m/2} w(x) \quad (5\text{-}223)$$

Substitution of Eq. (5-223) into Eq. (5-222) leads, therefore, to the desired differential equation for $P_n^m(x)$:

$$(1-x^2)\frac{d^2 w}{dx^2} - 2x\frac{dw}{dx} + \left[ n(n+1) - \frac{m^2}{1-x^2} \right] w = 0 \quad (5\text{-}224)$$

In future applications we must know the integral properties of the associated Legendre polynomials, in particular the integrals

$$I_{n,n'}^m = \int_{-1}^{1} P_n^m(x) P_{n'}^m(x)\, dx \quad (5\text{-}225)$$

First, we show that these integrals are zero when $n \neq n'$. From the differential equation (5-224) we can derive the following two equations:

$$P_{n'}^m(1-x^2)\frac{d^2P_n^m}{dx^2}-2xP_{n'}^m\frac{dP_n^m}{dx}+\left[n(n+1)-\frac{m^2}{1-x^2}\right]P_n^mP_{n'}^m=0$$

$$P_n^m(1-x^2)\frac{d^2P_{n'}^m}{dx^2}-2xP_n^m\frac{dP_{n'}^m}{dx}+\left[n'(n'+1)-\frac{m^2}{1-x^2}\right]P_n^mP_{n'}^m=0 \quad (5\text{-}226)$$

The first of these equations is the differential equation for $P_n^m$, multiplied by $P_{n'}^m$, and the second is the differential equation for $P_{n'}^m$, multiplied by $P_n^m$. If we subtract the second equation from the first and integrate, we obtain

$$\int_{-1}^{1}\frac{d}{dx}\left\{(1-x^2)\left[P_{n'}^m\frac{dP_n^m}{dx}-P_n^m\frac{dP_{n'}^m}{dx}\right]\right\}dx$$

$$+\left[n(n+1)-n'(n'+1)\right]\int_{-1}^{1}P_n^mP_{n'}^m\,dx=0 \quad (5\text{-}227)$$

Since the first integral is zero, we find that

$$\int_{-1}^{1}P_n^m(x)P_{n'}^m(x)\,dx=0 \qquad n\neq n' \quad (5\text{-}228)$$

In order to evaluate the integral of Eq. (5-225) for $n=n'$, we start with the definition of Eq. (5-218), which we write as

$$P_n^{m+1}(x)=(1-x^2)^{(m+1)/2}\frac{d^{m+1}}{dx^{m+1}}P_n(x) \quad (5\text{-}229)$$

By differentiating Eq. (5-218) once, we find

$$\frac{dP_n^m(x)}{dx}=(1-x^2)^{m/2}\frac{d^{m+1}}{dx^{m+1}}P_n(x)-mx(1-x^2)^{(m-2)/2}\frac{d^m}{dx^m}P_n(x) \quad (5\text{-}230)$$

The combination of Eqs. (5-229) and (5-230) leads to

$$P_n^{m+1}(x)=(1-x^2)^{1/2}\frac{dP_n^m}{dx}+mx(1-x^2)^{-1/2}P_n^m(x) \quad (5\text{-}231)$$

The integral $I_{n,n}^m$ is evaluated from the above recurrence formula (5-231). If we square and integrate this expression, we obtain

$$\int_{-1}^{1}\left[P_n^{m+1}(x)\right]^2dx=\int_{-1}^{1}(1-x^2)\left[\frac{dP_n^m}{dx}\right]^2dx$$

$$+2m\int_{-1}^{1}xP_n^m\frac{dP_n^m}{dx}dx+m^2\int_{-1}^{1}\frac{x^2(P_n^m)^2}{1-x^2}dx \quad (5\text{-}232)$$

By partial integration of the first and second integral on the right side, we obtain

$$\int_{-1}^{1}\left[P_n^{m+1}(x)\right]^2 dx=-\int_{-1}^{1}P_n^m(x)\frac{d}{dx}\left[(1-x^2)\frac{dP_n^m}{dx}\right]dx$$
$$-m\int_{-1}^{1}\left[P_n^m(x)\right]^2 dx+m^2\int_{-1}^{1}\frac{x^2}{1-x^2}\left[P_n^m(x)\right]^2 dx \tag{5-233}$$

It follows from the differential equation (5-224) that

$$\frac{d}{dx}\left[(1-x^2)\frac{dP_n^m}{dx}\right]=(1-x^2)\frac{d^2P_n^m}{dx^2}-2x\frac{dP_n^m}{dx}$$
$$=\frac{m^2}{1-x^2}P_n^m(x)-n(n+1)P_n^m(x) \tag{5-234}$$

Substitution of this result into Eq. (5-233) gives

$$\int_{-1}^{1}\left[P_n^{m+1}(x)\right]^2 dx=\left[n(n+1)-m-m^2\right]\int_{-1}^{1}\left[P_n^m(x)\right]^2 dx$$
$$=(n-m)(n+m+1)\int_{-1}^{1}\left[P_n^m(x)\right]^2 dx \tag{5-235}$$

If we repeat this procedure a number of times, we find that

$$\int_{-1}^{1}\left[P_n^m(x)\right]^2 dx=(n-m+1)(n-m+2)$$
$$\cdots(n)\cdot(n+m)(n+m-1)\cdots(n+1)\int_{-1}^{1}\left[P_n(x)\right]^2 dx \tag{5-236}$$

Substitution of Eq. (5-217) finally gives

$$\int_{-1}^{1}\left[P_n^m(x)\right]^2 dx=\frac{(n+m)!}{(n-m)!}\,\frac{2}{2n+1} \tag{5-237}$$

In quantum mechanical problems we often encounter the functions

$$S_{n,m}(\theta,\phi)=P_n^{|m|}(\cos\theta)e^{im\phi} \tag{5-238}$$

which are defined for $0\leq\phi\leq2\pi$ and for $0\leq\theta\leq\pi$. We assume that they are defined in this way on the surface of a sphere with constant radius. If we integrate the product of two such functions over the surface of the sphere, we can write this integral as

$$\int_0^{2\pi}\int_0^{\pi} S^*_{n',m'}(\theta,\phi)S_{n,m}(\theta,\phi)\sin\theta\, d\theta\, d\phi$$

$$=\int_{-1}^{1} P_n^{|m|}(x)P_{n'}^{|m'|}(x)\,dx\int_0^{2\pi} e^{i(m-m')\phi}\,d\phi \quad (5\text{-}239)$$

The integral over $\phi$ is zero if $m\neq m'$, and if $m=m'$, the integral over $\theta$ is zero unless $n=n'$. It follows, therefore, from Eq. (5-237) that

$$\int_0^{2\pi} d\phi\int_0^{\pi} S^*_{n',m'}(\theta,\phi)S_{n,m}(\theta,\phi)\sin\theta\, d\theta=\frac{4\pi}{2n+1}\frac{(n+|m|)!}{(n-|m|)!}\delta_{n,n'}\delta_{m,m'}$$

(5-240)

Obviously, the functions

$$Y_{n,m}(\theta,\phi)=\left[\frac{2n+1}{4\pi}\frac{(n-|m|)!}{(n+|m|)!}\right]^{1/2} P_n^{|m|}(\cos\theta)e^{im\phi} \quad (5\text{-}241)$$

form an orthonormal set of functions on the spherical surface. They are known as spherical harmonics, and they play an important role in many quantum mechanical problems.

## 5-11 The Rigid Rotor

We define a rigid rotor as two particles of masses $m_1$ and $m_2$ which are connected by a rigid weightless rod of length $R$ and which move in a potential field that is everywhere zero. We impose the restriction that the center of gravity of the system has to remain at rest. It is convenient to make use of the results that we obtained in Section 1–5 in order to derive the Hamiltonian. According to Eq. (1–94) the Hamiltonian of two particles moving in a potential field $V(\mathbf{r})=0$ can be written as

$$H=\frac{P^2}{2(m_1+m_2)}+\frac{p^2}{2\mu} \quad (5\text{-}242)$$

where

$$\mu=\frac{m_1 m_2}{m_1+m_2} \quad (5\text{-}243)$$

Here **P** is the momentum of the center of gravity, and **p** is associated with the relative motion of the particles; in Section 1–5 it was defined as

$$\mathbf{p}=\mu\frac{d\mathbf{r}}{dt} \qquad \mathbf{r}=\mathbf{r}_1-\mathbf{r}_2 \tag{5-244}$$

if $\mathbf{r}_1$ and $\mathbf{r}_2$ are the positions of the two particles, respectively.

It follows that the motion of two particles, moving in a zero potential field in such a way that their center of gravity remains at rest, is described by a Schrödinger equation

$$-\frac{\hbar^2}{2\mu}\left(\frac{\partial^2}{\partial x^2}+\frac{\partial^2}{\partial y^2}+\frac{\partial^2}{\partial z^2}\right)\psi(x,y,z)=E\psi(x,y,z) \tag{5-245}$$

Before we consider the restriction that the distance $r$ has to remain equal to $R$, we transform Eq. (5-245) to the polar coordinates $(r,\theta,\phi)$, which are defined as

$$\begin{aligned} x&=r\sin\theta\cos\phi \\ y&=r\sin\theta\sin\phi \\ z&=r\cos\theta \end{aligned} \tag{5-246}$$

This transformation is discussed in Appendix A, and the result is given by Eq. (A-15). Substitution into Eq. (5-245) gives

$$\left(\frac{\partial^2}{\partial r^2}+\frac{2}{r}\frac{\partial}{\partial r}+\frac{1}{r^2}\frac{\partial^2}{\partial\theta^2}+\frac{\cos\theta}{r^2\sin\theta}\frac{\partial}{\partial\theta}+\frac{1}{r^2\sin^2\theta}\frac{\partial^2}{\partial\phi^2}\right)\psi(r,\theta,\phi)$$
$$+\frac{2\mu E}{\hbar^2}\psi(r,\theta,\phi)=0 \tag{5-247}$$

Now we impose the restriction that $r$ has to be equal to a constant $R$. We can then omit the differentiations with respect to $r$ and replace $r$ by $R$ in the other terms. This gives

$$\left(\frac{\partial^2}{\partial\theta^2}+\frac{\cos\theta}{\sin\theta}\frac{\partial}{\partial\theta}+\frac{1}{\sin^2\theta}\frac{\partial^2}{\partial\phi^2}\right)\psi(\theta,\phi)+\lambda\psi(\theta,\phi)=0 \tag{5-248}$$

with

$$\lambda=\frac{2\mu R^2E}{\hbar^2} \tag{5-249}$$

This is the Schrödinger equation for the rigid rotor, which we take as the starting point for our considerations.

In order to solve Eq. (5-248) we write the function $\psi(\theta,\phi)$ in the form

$$\psi(\theta,\phi)=f(\theta)g(\phi) \tag{5-250}$$

Substitution into Eq. (5-248) gives

$$\frac{1}{f(\theta)}\left(\frac{\partial^2}{\partial\theta^2}+\frac{\cos\theta}{\sin\theta}\frac{\partial}{\partial\theta}\right)f(\theta)+\frac{1}{g(\phi)\sin^2\theta}\frac{\partial^2 g}{\partial\phi^2}+\lambda=0 \tag{5-251}$$

which can be separated to

$$\frac{d^2g}{d\phi^2}=-m^2g(\phi) \tag{5-252a}$$

$$\frac{d^2f}{d\theta^2}+\frac{\cos\theta}{\sin\theta}\frac{df}{d\theta}+\left(\lambda-\frac{m^2}{\sin^2\theta}\right)f(\theta)=0 \tag{5-252b}$$

Here $m$ is an arbitrary real parameter.

The solutions of Eq. (5-252a) can all be written in the form

$$g(\phi)=e^{im\phi} \tag{5-253}$$

if we let $m$ take both positive and negative values. Let us now investigate for which values of $m$ we obtain acceptable wave functions. We mentioned in Section 3–10 that the general conditions for the wave functions are that they and their derivatives are everywhere finite, continuous, and single-valued. The essential condition that we have to worry about in the present situation is that $g(\phi)$ is single-valued. It can be seen that the sets of polar coordinates $(\theta,\phi)$ and $(\theta,\phi+2\pi)$ present the same orientation of the rigid rotor, and if we wish the wave function to be single-valued, we have to impose the condition

$$g(\phi)=g(\phi+2\pi) \tag{5-254}$$

or

$$e^{2\pi im\phi}=1 \tag{5-255}$$

It follows that the allowed values of $m$ are

$$m=0,\pm 1,\pm 2,\pm 3,\ldots, \text{ and so on} \tag{5-256}$$

Let us now consider the differential equation (5-252b), with $m$ given by Eq. (5-256). We introduce the new variable

$$s=\cos\theta \tag{5-257}$$

and observe that

$$\frac{df}{d\theta} = -\sin\theta \frac{df}{ds} \tag{5-258}$$

$$\frac{d^2f}{d\theta^2} = \sin^2\theta \frac{d^2f}{ds^2} - \cos\theta \frac{df}{ds}$$

Equation (5-252b) then becomes

$$(1-s^2)\frac{d^2f}{ds^2} - 2s\frac{df}{ds} + \left[\lambda - \frac{m^2}{1-s^2}\right]f = 0 \tag{5-259}$$

We notice that this is identical to the differential equation (5-224) for the associated Legendre polynomials $P_n^m(x)$:

$$(1-x^2)\frac{d^2w}{dx^2} - 2x\frac{dw}{dx} + \left[n(n+1) - \frac{m^2}{1-x^2}\right]w = 0 \tag{5-260}$$

if we take

$$\lambda = n(n+1) \tag{5-261}$$

If we take $m$ equal to zero, Eq. (5-260) reduces to the differential equation (5-208) for the Legendre polynomials. We showed in Section 5–10 that the solutions of Eq. (5-208) are usually infinite for $x = \pm 1$. It can be shown in a similar way that for $m \neq 0$ the solutions of Eq. (5-260) are usually infinite for $x = \pm 1$ and that, only if $n$ is a non-negative integer and $n \geq |m|$, one of the two solutions, namely $P_n^{|m|}(x)$, is finite for $x = \pm 1$. Hence the condition that the wave function is acceptable is given by Eq. (5-261), with $n$ a non-negative integer and $n \geq |m|$. The energy eigenvalues of the system are obtained by combining Eqs. (5-249) and (5-261):

$$E_n = n(n+1)\frac{\hbar^2}{2\mu R^2} \qquad n = 0, 1, 2, 3, \ldots, \text{ and so on} \tag{5-262}$$

The corresponding eigenfunctions are

$$\psi_{n,m}(\theta,\phi) = P_n^{|m|}(\cos\theta)e^{im\phi} = Y_{n,m}(\theta,\phi)$$
$$m = -n, -n+1, \ldots, -1, 0, 1, 2,, \ldots, n \tag{5-263}$$

where the functions $Y_{n,m}(\theta,\phi)$ are the spherical harmonics that were defined in Eq. (5-241). They are orthonormal functions on the surface of a sphere.

We see that there is more than one eigenfunction belonging to an eigenvalue; to be exact, the eigenvalue $E_n$ has $(2n+1)$ eigenfunctions. We speak here of a degeneracy or of a $(2n+1)$-fold degenerate eigenvalue. The consequences of degeneracies have been discussed in detail in Chapter 4.

It should be mentioned here that the eigenvalues and eigenfunctions of the rigid rotor are the same as those of the angular momentum operator $M^2$. As such, they play a very important role in the quantum theory of atomic systems and the above results will prove to be useful when we discuss the hydrogen atom in the following chapter.

## Problems

**1** Find the general solution of the equation

$$\frac{d^2y}{dx^2}+6\frac{dy}{dx}+9y=0$$

**2** Prove that

$$\frac{d}{dx}\,{}_1F_1(a;c;x)=\frac{a}{c}\,{}_1F_1(a+1;c+1;x)$$

**3** Evaluate the integral

$$\int_0^x e^t\,{}_1F_1(a;c;-t)\,dt$$

by expressing the result in terms of the confluent hypergeometric function.

**4** Solve by means of the series expansion method the differential equation

$$\frac{d^2y}{dx^2}-x^2\frac{dy}{dx}-xy=0$$

Also, rewrite the equation in terms of the new variable $t=x^3$ and express the solutions in terms of the confluent hypergeometric function.

**5** The function ${}_1F_1(b+2;b;x)$ is zero for two values of $x$, namely $x_1$ and $x_2$. Determine these two values, expressed in terms of the parameter $b$.

**6** Solve the differential equation

$$(1-x^2)\frac{d^2y}{dx^2}-2x\frac{dy}{dx}+\frac{3}{4}y=0$$

by means of the series expansion method.

**7** Determine the values of the constants $C_n$ for which the functions

$$g_n(x)=C_n H_n(x)\exp\left(\frac{-x^2}{2}\right)$$

form an orthonormal set in the interval $-\infty<x<\infty$.

**8** Calculate the integrals

$$\int_{-\infty}^{\infty} g_n(x)x^2 g_n(x)\,dx$$

$$\int_{-\infty}^{\infty} g_n(x)\left(\frac{d^2}{dx^2}\right)g_n(x)\,dx$$

for the functions $g_n(x)$ defined in Problem 5–7.

**9** Show that the functions $g_n(x)$ defined in Problems 5–7 satisfy the following relations:

$$\left[\frac{d}{dx}-x\right]g_n(x)=A_n g_{n+1}(x)$$

$$\left[\frac{d}{dx}+x\right]g_n(x)=B_n g_{n-1}(x)$$

and determine the constants $A_n$ and $B_n$.

**10** Show that

$$\int_0^1 x^2 P_{n+1}(x)P_{n-1}(x)\,dx=\frac{n(n+1)}{(2n-1)(2n+1)(2n+3)}$$

**11** Show that

$$\int_{-1}^{1}(1-x^2)\left[\frac{dP_n}{dx}\right]^2 dx=\frac{2n(n+1)}{2n+1}$$

**12** Show that

$$x\frac{dP_n}{dx}=nP_n(x)+(2n-3)P_{n-2}(x)+(2n-7)P_{n-4}(x)+\cdots$$

**13** Show explicitly that

$$\int_{-1}^{1}(1-2xh+h^2)^{1/2}(1-2xk+k^2)^{1/2}\,dx=\frac{1}{\sqrt{hk}}\log\frac{1+\sqrt{hk}}{1-\sqrt{hk}}$$

**14** Show that for sufficiently small values of $|x|$ and $|h|$ we have

$$\frac{1-h^2}{(1-2hx+h^2)^{3/2}} = \sum_{n=0}^{\infty} (2n+1)h^n P_n(x)$$

**15** Prove that

$$\int_{-1}^{1} \frac{1-h^2}{(1-2xh+h^2)^{3/2}(1-2xk+k^2)^{1/2}} dx$$

$$= \int_{-1}^{1} \frac{1-k^2}{(1-2xh+h^2)^{1/2}(1-2xk+k^2)^{3/2}} dx = \frac{2}{1-hk}$$

**16** Show that for positive integers $n$

$$P_n(x) = \sum_{k=0}^{n} \frac{(-1)^k (n+k)!}{(n-k)!k!k!2^{k+1}} \left[(1-x)^k + (-1)^n (1+x)^k\right]$$

**17** Derive the differential equation of the associated Laguerre polynomial $L_n^k(x)$ from its definition (5-153) by differentiating the equation for the polynomial $L_n(x)$.

**18** Derive from the generating function of the associated Laguerre polynomials that

$$n! \int_0^{\infty} x^m e^{-x} L_n^m(x) L_n^m(x)\, dx = (n+m)!$$

**19** Determine the eigenvalues and eigenfunctions of the two-dimensional rigid rotor. This is the system composed of two particles of masses $m_1$ and $m_2$ connected by a rigid weightless rod of length $R$ and moving in two dimensions such that the center of gravity of the system remains at rest.

**20** If $\psi_n(x)$ is the eigenfunction of the $n$th state of a harmonic oscillator and if it is zero for the values $x = x_1, x = x_2, \ldots, x = x_n$, evaluate $\Sigma_i x_i^2$.

**21** Evaluate the five lowest rotational energies ($J = 1, 2, 3, 4, 5$) for the molecules $H_2$, $HF$, and $F_2$ by treating them as rigid rotors and by neglecting the masses of the electrons.

**22** To a first approximation we can represent the vibrational motion of a diatomic molecule $AB$ as a harmonic oscillator. The oscillations are described by a displacement coordinate $q = R - R_o$, where $R$ is the internuclear distance and $R_o$ is its equilibrium value, and by a reduced mass

$\mu_{AB}=m_A m_B(m_A+m_B)^{-1}$, where $m_A$ and $m_B$ are the nuclear masses. It has been found experimentally that the energy differences between the vibrational ground states and the first excited states for $N_2^{14}$, for $H^1I^{127}$, and for $I^{127}Cl^{35}$ are 2359.61 $cm^{-1}$, 2309.5 $cm^{-1}$, and 384.18 $cm^{-1}$, respectively. Calculate for each of these molecules the force constant $k$ of the harmonic motion and the root-mean-square deviation from equilibrium, that is, the square root of the expectation value of $q^2$, for the ground state of the harmonic oscillator.

**23** The electronic ground state of the $H_2$ molecule has a dissociation energy of $D=4.747$ eV. The energy difference between the lowest vibrational level and the first excited vibrational level is 4395.2 $cm^{-1}$. Construct the corresponding Morse potential and predict the number of stationary state energy levels.

**24** In the case of the $HCl^{35}$ molecule the observed vibrational transitions $0\rightarrow 1$ and $1\rightarrow 2$ are 2885.9 $cm^{-1}$ and 5668.0 $cm^{-1}$, respectively. Assuming that the motion is harmonic, calculate the zero-point energy of the corresponding harmonic oscillator. Next, represent the motion by a Morse potential and derive the corresponding zero-point energy. Compare the two different theoretical results.

**25** Assume that an electron is allowed to move on a sphere of radius $a_0$, where $a_0$ is the Bohr radius. Derive its eigenvalues and eigenfunctions and express the excitation energy between the lowest and the first excited states in terms of the Rydberg constant $R=e^2/2a_0(a_0=\hbar^2/me^2)$.

## Recommended Reading

We list a few books on the theory of differential equations and on special functions. Whittaker and Watson's text is a classic, and the Bateman series is a good reference source. The other books are classical treatments of the specific functions that they discuss.

E. T. Whittaker and G. N. Watson, *A Course of Modern Analysis*, Cambridge University Press, Cambridge, 1902.

Sir Harold Jeffreys and B. S. Jeffreys, *Methods of Mathematical Physics*, Cambridge University Press, Cambridge, 1966.

Bateman Manuscript Project, *Higher Transcendental Functions*, Vols. 1,2,3, McGraw-Hill Book Company, Inc., New York, 1953.

E. W. Hobson, *The Theory of Spherical and Ellipsoidal Harmonics*, Cambridge University Press, Cambridge, 1931.

L. J. Slater, *Confluent Hypergeometric functions*, Cambridge University Press, Cambridge, 1960.

P. M. Morse and H. Feshbach, *Methods of Theoretical Physics*, McGraw-Hill Book Company, Inc., New York 1953.

CHAPTER SIX

# The Hydrogen Atom

## 6-1 Introduction

The hydrogen atom is a special case of a particle of mass $\mu$ in a three-dimensional central force field. A central force field is characterized by a potential function that depends only on the distance between the particle and the origin. If we employ polar coordinates $(r, \theta, \phi)$, then $V$ does not depend on the angles $\theta$ and $\phi$. It is a function of $r$ only and we write it as $V(r)$. The Hamiltonian is

$$H=-\frac{\hbar^2}{2\mu}\Delta+V(r)=-\frac{\hbar^2}{2\mu}\left(\frac{\partial^2}{\partial x^2}+\frac{\partial^2}{\partial y^2}+\frac{\partial^2}{\partial z^2}\right)+V(r) \tag{6-1}$$

and the corresponding Schrödinger equation is

$$\left[\frac{\partial^2}{\partial x^2}+\frac{\partial^2}{\partial y^2}+\frac{\partial^2}{\partial z^2}+\frac{2\mu}{\hbar^2}(E-V)\right]\Psi(r,\theta,\phi)=0 \tag{6-2}$$

We discussed the classical description of the hydrogen atom in Section 1-5. The hydrogen atom consists of two particles: a heavy proton with mass $M$, coordinate $\mathbf{r}_n$, and momentum $\mathbf{p}_n$ and a much lighter electron with mass $m$, coordinate $\mathbf{r}_e$, and momentum $\mathbf{p}_e$. The potential energy $V$ depends on the distance $|\mathbf{r}_e-\mathbf{r}_n|$ only, and the Hamiltonian is given by

$$H=\frac{p_n^2}{2M}+\frac{p_e^2}{2m}-\frac{e^2}{|\mathbf{r}_e-\mathbf{r}_n|} \tag{6-3}$$

The last term in Eq. (6-3) represents the Coulomb interaction between the electron and the proton.

We showed in Section 1-5 how the hydrogen atom can be solved classically by transforming to a new set of coordinates and momenta; we use the same transformations in our quantum mechanical description. Our new coordinates are the position $\mathbf{R}$ of the center of gravity, defined as

$$(M+m)\mathbf{R}=M\mathbf{r}_n+m\mathbf{r}_e \tag{6-4}$$

and the relative position of the electron with respect to the nucleus, defined as

$$\mathbf{r} = \mathbf{r}_e - \mathbf{r}_n \tag{6-5}$$

The momenta $\mathbf{p}_n$ and $\mathbf{p}_e$ are defined as

$$\mathbf{p}_n = M\frac{d\mathbf{r}_n}{dt} \qquad \mathbf{p}_e = m\frac{d\mathbf{r}_e}{dt} \tag{6-6}$$

and we introduce the new momenta

$$\mathbf{P} = (m+M)\frac{d\mathbf{R}}{dt} \qquad \mathbf{p} = \mu\frac{d\mathbf{r}}{dt} \tag{6-7}$$

with

$$\mu = \frac{mM}{m+M} \tag{6-8}$$

It has been shown in Section 1-5 that the Hamiltonian (6-3) of the hydrogen atom can be expressed as

$$H = \frac{P^2}{2(M+m)} + \frac{p^2}{2\mu} - \frac{e^2}{r} \tag{6-9}$$

in terms of the new coordinates **R** and **r**. The first term of Eq. (6-9) represents the motion of the total hydrogen atom, considered as a free particle. The other terms

$$\mathcal{H} = \frac{p^2}{2\mu} - \frac{e^2}{r} \tag{6-10}$$

represent the motion of the electron relative to the nucleus. It is easily seen that this is a special case of a particle in a central force field since the potential function depends on the coordinate $r$ only.

The Schrödinger equation of a particle in a central force field can be separated into two equations. The first equation deals with the angular part of the eigenfunction and the second equation deals with the radial motion. This separation may be understood better by considering the properties of the angular momentum and by using the properties of commuting operators that we discussed in Section 4-3.

The angular momentum **M** of a particle in a central force field is defined as

$$\mathbf{M} = \mathbf{r} \times \mathbf{p} \tag{6-11}$$

and its three components are given by

$$M_x = yp_z - zp_y$$
$$M_y = zp_x - xp_z$$
$$M_z = xp_y - yp_x \qquad (6\text{-}12)$$

In quantum mechanics the angular momentum components can be represented as operators. In Section 3-9 we showed that

$$p_x = h\sigma_x = \frac{\hbar}{i}\frac{\partial}{\partial x} \qquad \text{and so on} \qquad (6\text{-}13)$$

Substitution into Eq. (6-12) gives

$$M_x = \frac{\hbar}{i}\left( y\frac{\partial}{\partial z} - z\frac{\partial}{\partial y} \right)$$
$$M_y = \frac{\hbar}{i}\left( z\frac{\partial}{\partial x} - x\frac{\partial}{\partial z} \right)$$
$$M_z = \frac{\hbar}{i}\left( x\frac{\partial}{\partial y} - y\frac{\partial}{\partial x} \right) \qquad (6\text{-}14)$$

It is easily verified that

$$M_x V(r) = \frac{\hbar}{i}\left( y\frac{\partial V}{\partial z} - z\frac{\partial V}{\partial y} \right) = \frac{\hbar}{i}\left( \frac{yz}{r} - \frac{zy}{r} \right)\frac{\partial V}{\partial r} = 0 \qquad (6\text{-}15)$$

or

$$[M_x, V] = M_x V - V M_x = 0 \qquad (6\text{-}16)$$

Since $M_x$ commutes also with the kinetic energy part of the Hamiltonian of Eq. (6-1), we have

$$[M_x, H] = 0 \qquad (6\text{-}17)$$

if $H$ represents a particle moving in a central force field.

It can be shown also that

$$[M_y, H] = [M_z, H] = [M_x, H] = 0 \qquad (6\text{-}18)$$

Consequently, we also have

$$[M^2, H] = 0 \qquad (6\text{-}19)$$

where

$$M^2 = M_x^2 + M_y^2 + M_z^2 \tag{6-20}$$

We have shown in Section 4-3 that commuting operators have the same eigenfunctions. Since the operators $H$ and $M^2$ commute, they must have the same set of eigenfunctions. Clearly, a study of the eigenvalues and eigenfunctions of the various angular momentum operators should be helpful in deriving the eigenvalues and eigenfunctions of a particle in a central force field. Therefore, we first discuss the quantum mechanical description of the angular momentum before we tackle the hydrogen atom. We shall see that the eigenfunctions of the operator $M^2$ constitute the angular parts of the eigenfunctions of the hydrogen atom. We derive this connection from purely mathematical considerations and from the commutator relations of the various operators.

## 6-2 Eigenvalues and Eigenfunctions of the Angular Momentum

The eigenvalues and eigenfunctions of the angular momentum operators can be derived by introducing polar coordinates:

$$\begin{aligned} x &= r\sin\theta\cos\phi \\ y &= r\sin\theta\sin\phi \\ z &= r\cos\theta \end{aligned} \tag{6-21}$$

However, before presenting this derivation we show the connection between the angular momentum problem and the rigid rotor that we discussed in Section 5-11.

A rigid rotor is defined as two particles of masses $m_1$ and $m_2$ that are connected by a rigid weightless rod of length $r$. In order to describe its motion, we choose the origin as the center of gravity of the two particles, and we denote the position and momentum of the first particle by $\mathbf{r}_1$ and $\mathbf{p}_1$ and the position and momentum of the second particle by $\mathbf{r}_2$ and $\mathbf{p}_2$.

We have

$$\mathbf{r}_1 = \frac{m_2}{m_1+m_2}\mathbf{r} = \frac{\mu}{m_1}\mathbf{r} \qquad \mathbf{r}_2 = -\frac{\mu}{m_2}\mathbf{r} \tag{6-22}$$

where $\mu$ is again the reduced mass. The velocities $\mathbf{v}_1$ and $\mathbf{v}_2$ of the two particles can be expressed in terms of the angular velocity $\boldsymbol{\omega}$, which is the same for both particles. It follows that

$$\begin{aligned} v_1 &= \omega r_1 = \left(\frac{\mu}{m_1}\right)\omega r \\ p_1 &= m_1 v_1 = \mu\omega r \\ p_2 &= m_2 v_2 = \mu\omega r \end{aligned} \tag{6-23}$$

The angular momentum $M$ of the rigid rotor is given by

$$M = p_1 r_1 + p_2 r_2 = \frac{\mu^2 \omega r^2}{m_1} + \frac{\mu^2 \omega r^2}{m_2} = \mu \omega r^2$$

$$M^2 = \mu^2 \omega^2 r^4 \tag{6-24}$$

The energy is

$$E = \frac{m_1 v_1^2}{2} + \frac{m_2 v_2^2}{2} = \frac{\mu^2 \omega^2 r^2}{2}\left(\frac{1}{m_1} + \frac{1}{m_2}\right) = \frac{\mu \omega^2 r^2}{2} \tag{6-25}$$

Obviously,

$$E = \frac{M^2}{2\mu r^2} \tag{6-26}$$

In the rigid rotor the distance $r$ between the two mass points is a constant, and it follows that the operator $H$ representing the energy and the operator $M^2$ representing the angular momentum should have the same eigenfunctions. According to Eqs. (5-262) and (5-263) the energy eigenfunctions and eigenvalues are given by

$$H_{\text{op}} Y_{l,m}(\theta, \phi) = l(l+1) \frac{\hbar^2}{2\mu r^2} Y_{l,m}(\theta, \phi) \tag{6-27}$$

It follows from Eq. (6-26) that the eigenvalues and eigenfunctions of the angular momentum operator $(M^2)_{\text{op}}$ are given by

$$(M^2)_{\text{op}} Y_{l,m}(\theta, \phi) = l(l+1)\hbar^2 Y_{l,m}(\theta, \phi) \tag{6-28}$$

Of course, we have proved the validity of this result [Eq. (6-28)] only for the case of the rigid rotor, but we suspect that it may be valid also in other cases. In order to derive the eigenvalues and eigenfunctions of the operator $(M^2)_{\text{op}}$ we transform the various expressions for the angular momentum components in terms of polar coordinates. By substituting Eqs. (6-21) into Eq. (6-14) we obtain

$$M_x = i\hbar\left(\sin\phi \frac{\partial}{\partial\theta} + \cot\theta \cos\phi \frac{\partial}{\partial\phi}\right)$$

$$M_y = i\hbar\left(-\cos\phi \frac{\partial}{\partial\theta} + \cot\theta \sin\phi \frac{\partial}{\partial\phi}\right)$$

$$M_z = -i\hbar \frac{\partial}{\partial\phi} \tag{6-29}$$

By combining these results we obtain

$$(M^2)_{op}=M_x^2+M_y^2+M_z^2=-\hbar^2\left(\frac{\partial^2}{\partial\theta^2}+\frac{\cos\phi}{\sin\phi}\frac{\partial}{\partial\phi}+\frac{1}{\sin^2\theta}\frac{\partial^2}{\partial\phi^2}\right) \quad (6\text{-}30)$$

It follows from Eq. (5-247) that the operator $H_{op}$ for the rigid rotor is given by

$$H_{op}=-\frac{\hbar^2}{2\mu r^2}\left(\frac{\partial^2}{\partial\theta^2}+\frac{\cos\theta}{\sin\theta}\frac{\partial}{\partial\theta}+\frac{1}{\sin^2\theta}\frac{\partial^2}{\partial\phi^2}\right) \quad (6\text{-}31)$$

Obviously,

$$(M^2)_{op}=2\mu r^2 H_{op} \quad (6\text{-}32)$$

In other words, the operator $(M^2)_{op}$ is proportional to the Hamiltonian of the rigid rotor and the two operators have the same set of eigenfunctions. The expressions (6-28) for the eigenvalues and eigenfunctions of the operator $(M^2)_{op}$ are generally valid.

We are interested also in deriving the eigenvalues and eigenfunctions of the operator $M_z$ of Eq. (6-29). This operator depends on the polar angle $\phi$ only, and the eigenvalue equation has the form

$$\frac{\hbar}{i}\frac{\partial\chi}{\partial\phi}=\lambda\chi \quad (6\text{-}33)$$

The solutions have the form

$$\chi(\phi)=\exp\left(\frac{i\lambda\phi}{\hbar}\right) \quad (6\text{-}34)$$

The function $\chi(\phi)$ must be single-valued. This means that the exponential must have the form $e^{im\phi}$ with $m$ an integer. The result is

$$(M_z)_{op}e^{im\phi}=m\hbar e^{im\phi} \qquad m=0,\pm1,\pm2,\ldots,\text{ and so on} \quad (6\text{-}35)$$

It follows that the functions $Y_{l,m}(\theta,\phi)$, the eigenfunctions of the operator $(M^2)_{op}$, are also eigenfunctions of the operator $(M_z)_{op}$. We defined the spherical harmonics in Eq. (5-241) of Section 5-10, and it may be recalled that the definition is

$$Y_{l,m}(\theta,\phi)=\left[\frac{2l+1}{4\pi}\frac{(l-|m|)!}{(l+|m|)!}\right]^{1/2}P_l^{|m|}(\cos\theta)e^{im\phi} \quad (6\text{-}36)$$

The $\phi$-dependent part of this function is an eigenfunction of $(M_z)_{op}$. Clearly we can multiply this eigenfunction by an arbitrary function of the angle $\theta$. We have, therefore,

$$(M_z)_{op} Y_{l,m}(\theta,\phi) = m\hbar Y_{l,m}(\theta,\phi) \tag{6-37}$$

in addition to

$$(M^2)_{op} Y_{l,m}(\theta,\phi) = l(l+1)\hbar^2 Y_{l,m}(\theta,\phi) \tag{6-38}$$

The possible values of the quantum numbers $l$ and $m$ are

$$l=0,1,2,3,\ldots \qquad m=0,\pm 1,\pm 2,\ldots,\pm l \tag{6-39}$$

We look further into the relation between Eq. (6-37) and (6-38) when we discuss the commutator relations between the various angular momentum operators in the next section. Here, we just point out that the operator $(M^2)_{op}$ represents the square of the length of the angular momentum vector and the operator $(M_z)_{op}$ represents the projection on the $Z$ axis of an angular momentum vector of given length. Clearly, the projection should be smaller than the length of the vector, hence the condition

$$|m| \leq l \tag{6-40}$$

In the following Section 6-3 we derive various commutator relations between the angular momentum operators, and we use those relations as a basis for deriving additional properties of the angular momentum.

## 6-3 Commutation Relations

The operators representing the angular momentum components $M_x$, $M_y$, and $M_z$ are described in Eq. (6-14). The operator $M^2$ is defined by Eq. (6-20). It may be recalled that the commutator between two operators $\Omega_1$ and $\Omega_2$ is defined as

$$[\Omega_1,\Omega_2] = \Omega_1\Omega_2 - \Omega_2\Omega_1 \tag{6-41}$$

as we discussed in Section 4-2.

We first derive the commutation relations between the various components $M_i$. We have

$$\begin{aligned}[M_x, M_y] &= (i\hbar)^2\left[\left(y\frac{\partial}{\partial z} - z\frac{\partial}{\partial y}\right)\left(z\frac{\partial}{\partial x} - x\frac{\partial}{\partial z}\right)\right.\\ &\qquad \left. - \left(z\frac{\partial}{\partial x} - x\frac{\partial}{\partial z}\right)\left(y\frac{\partial}{\partial z} - z\frac{\partial}{\partial y}\right)\right]\\ &= (ih)^2\left(y\frac{\partial}{\partial x} - x\frac{\partial}{\partial y}\right)\\ &= i\hbar M_z \end{aligned} \tag{6-42}$$

Therefore, we find

$$[M_x, M_y] = i\hbar M_z$$

$$[M_y, M_z] = i\hbar M_x$$

$$[M_z, M_x] = i\hbar M_y \tag{6-43}$$

Obviously, we also have

$$[M_x, M_x] = [M_y, M_y] = [M_z, M_z] = 0 \tag{6-44}$$

In order to derive the commutation relations involving $M^2$ we note that two arbitrary operators $\Omega$ and $\Lambda$ obey the relation

$$[\Omega^2, \Lambda] = \Omega^2\Lambda - \Lambda\Omega^2 = \Omega\Omega\Lambda - \Omega\Lambda\Omega + \Omega\Lambda\Omega - \Lambda\Omega\Omega = \Omega[\Omega, \Lambda] + [\Omega, \Lambda]\Omega \tag{6-45}$$

We can use this equation to derive that

$$\begin{aligned}[M^2, M_x] &= M_x[M_x, M_x] + [M_x, M_x]M_x + M_y[M_y, M_x] \\ &\quad + [M_y, M_x]M_y + M_z[M_z, M_x] + [M_x, M_x]M_z \\ &= -i\hbar M_y M_z - i\hbar M_z M_y + i\hbar M_z M_y + i\hbar M_y M_z = 0\end{aligned} \tag{6-46}$$

In the same way it can be shown that

$$[M^2, M_y] = [M^2, M_z] = 0 \tag{6-47}$$

We introduce a new pair of operators, namely

$$M_1 = M_x + iM_y$$

$$M_{-1} = M_x - iM_y \tag{6-48}$$

It can be derived that

$$\begin{aligned}M_1 M_{-1} &= (M_x + iM_y)(M_x - iM_y) \\ &= M_x^2 + M_y^2 + i(M_y M_x - M_x M_y) \\ &= M^2 - M_z^2 - i[M_x, M_y] \\ &= M^2 - M_z^2 + \hbar M_z \\ M_{-1} M_1 &= (M_x - iM_y)(M_x + iM_y) \\ &= M^2 - M_z^2 - \hbar M_z\end{aligned} \tag{6-49}$$

Furthermore, we have

$$[M_z, M_1] = [M_z, M_x] + i[M_z, M_y]$$
$$= i\hbar M_y + \hbar M_x = \hbar M_1$$
$$[M_z, M_{-1}] = [M_z, M_x] - i[M_z, M_y]$$
$$= i\hbar M_y - \hbar M_x = -\hbar M_{-1} \tag{6-50}$$

We use the above commutation relations for a discussion of the angular momentum eigenvalues and eigenfunctions. First we note that the two operators $M^2$ and $M_z$ commute and that they must have the same eigenfunctions according to Section 4-3. We can interpret this property by considering the results of the previous section. Here it follows from Eq. (6-35) that the eigenvalues and eigenfunctions of $M_z$ are given by

$$M_z e^{im\phi} = m\hbar e^{im\phi} \tag{6-51}$$

The operator $M_z$ depends only on the angle $\phi$, and we can multiply its eigenfunctions by an arbitrary function of another variable such as the polar angle $\theta$,

$$M_z f(\theta) e^{im\phi} = m\hbar f(\theta) e^{im\phi} \tag{6-52}$$

The eigenfunctions of $M^2$ are described by Eq. (6-36) and they are derived by taking the functions $f(\theta)$ as the associated Legendre function $P_l^{|m|}(\cos\theta)$.

We can also reverse the argument. If we denote the normalized spherical harmonics of Eq. (6-36) by $\psi_{l,m}$, then the eigenfunctions of $M^2$ are given by

$$M^2 \psi_{l,m} = l(l+1)\hbar^2 \psi_{l,m} \qquad m = 0, \pm 1, \pm 2, \ldots, \pm l \tag{6-53}$$

Each eigenvalue is $(2l+1)$-fold degenerate. We now have the situation described by Eq. (4-83). Since $M_z$ and $M^2$ commute we have, according to Eq. (4-84),

$$M_z \psi_{l,m} = \sum_{m'} c_{m,m'}^{(l)} \psi_{l,m'} \tag{6-54}$$

The eigenfunctions of $M_z$ are obtained as specific linear combinations of the functions $\psi_{l,m}$.

It turns out that we have been lucky because the set of functions $\psi_{l,m}$ that we selected, the spherical harmonics functions of Eq. (6-36), are already eigenfunctions of $M_z$. In other words, the matrix $c_{m,m'}^{(l)}$ is diagonal and we have

$$M_z \psi_{l,m} = m\hbar \psi_{l,m} \qquad m = 0, \pm 1, \pm 2, \ldots, \pm l \tag{6-55}$$

We have seen in Eqs. (6-46) and (6-47) that the operators $M_x$ or $M_y$ also commute with $M^2$, and we may ask what would happen if we consider $M_x$ or $M_y$ instead of the operator $M_z$. Clearly, we would arrive at equations that are analogous to Eq. (6-54), namely

$$M_x\psi_{l,m}=\sum_{m'}a_{m,m'}^{(l)}\psi_{l,m'}$$

$$M_y\psi_{l,m}=\sum_{m'}b_{m,m'}^{(l)}\psi_{l,m'} \tag{6-56}$$

However, the set of spherical harmonics functions that are eigenfunctions of $M_z$ are not eigenfunctions of $M_x$ or $M_y$, and this means that the matrices $a_{m,m'}$ or $b_{m,m'}$ are no longer diagonal. We have described in Section 4-3 that the eigenfunctions of $M_x$ or $M_y$ are obtained by means of unitary transformations that diagonalize the matrices $a_{m,m'}$ or $b_{m,m'}$. Naturally, the eigenvalues of the three operators $M_x$, $M_y$, or $M_z$ should be the same, but their eigenfunctions should be different linear combinations of the set of functions $\psi_{l,m}$.

The specific form of the matrices $a_{m,m'}$ or $b_{m,m'}$ can be derived from Eq. (6-29) by using the recurrence relations of the associated Legendre polynomials, but this is a very cumbersome procedure. An easier and more elegant derivation is based on the commutator relations.

We first set out to derive the matrices corresponding to the operators $M_1$ and $M_{-1}$:

$$M_1\psi_{l,m}=\sum_{m'}A_{m,m'}\psi_{l,m'}$$

$$M_{-1}\psi_{l,m}=\sum_{m'}B_{m,m'}\psi_{l,m'} \tag{6-57}$$

It follows from the first commutator relation (6-50) that

$$[M_z,M_1]-\hbar M_1=0 \tag{6-58}$$

or

$$\langle\psi_{l,m'}|[M_z,M_1]-\hbar M_1|\psi_{l,m}\rangle=0 \tag{6-59}$$

for all values of $m$ and $m'$. Since the functions $\psi_{l,m}$ and $\psi_{l,m'}$ are eigenfunctions of $M_z$, according to Eq. (6-55), we have

$$(m'-m-1)\hbar\langle\psi_{l,m'}|M_1|\psi_{l,m}\rangle=0 \tag{6-60}$$

Clearly,

$$\langle\psi_{l,m'}|M_1|\psi_{l,m}\rangle=0 \tag{6-61}$$

unless

$$m' = m + 1 \tag{6-62}$$

Because of the orthonormality of the set of functions $\psi_{l,m}$ we have

$$M_1 \psi_{l,m} = A_m \psi_{l,m+1} \tag{6-63}$$

In a similar fashion, it can be derived from the second recurrence relation (6-50) that

$$M_{-1} \psi_{l,m} = B_m \psi_{l,m-1} \tag{6-64}$$

We can derive two additional relations by taking the complex conjugates of these two equations. The complex conjugate of Eq. (6-63) gives

$$M_{-1} \psi_{l,-m} = A_m^* \psi_{l,-m-1} \tag{6-65}$$

The complex conjugate of Eq. (6-64) is

$$M_1 \psi_{l,-m} = B_m^* \psi_{l,-m+1} \tag{6-66}$$

By combining the equations we obtain

$$B_{-m} = A_m^* \qquad A_{-m} = B_m^* \tag{6-67}$$

Additional relations for $A_m$ and $B_m$ can be derived from Eq. (6-49). We have

$$\begin{aligned} M_1 M_{-1} \psi_{l,m} &= A_{m-1} B_m \psi_{l,m} = \left(M^2 - M_z^2 + \hbar M_z\right) \psi_{l,m} \\ &= \hbar^2 \left[ l(l+1) - m^2 + m \right] \psi_{l,m} \\ M_{-1} M_1 \psi_{l,m} &= B_{m+1} A_m \psi_{l,m} = \left(M^2 - M_z^2 - \hbar M_z\right) \psi_{l,m} \\ &= \hbar^2 \left[ l(l+1) - m^2 - m \right] \psi_{l,m} \end{aligned} \tag{6-68}$$

Consequently

$$\begin{aligned} A_{m-1} B_m &= \hbar^2 \left[ l(l+1) - m^2 + m \right] \\ A_m B_{m+1} &= \hbar^2 \left[ l(l+1) - m^2 - m \right] \end{aligned} \tag{6-69}$$

It can be seen that both Eqs. (6-67) and (6-69) are satisfied if we take

$$\begin{aligned} A_m &= \hbar \left[ l(l+1) - m(m+1) \right]^{1/2} \\ B_m &= \hbar \left[ l(l+1) - m(m-1) \right]^{1/2} \end{aligned} \tag{6-70}$$

The matrices for $M_x$ and $M_y$ are easily derived from Eqs. (6-48) and (6-70). We have

$$M_x\psi_{l,m}=\tfrac{1}{2}(M_1+M_{-1})\psi_{l,m}=\tfrac{1}{2}(A_m\psi_{l,m+1}+B_m\psi_{l,m-1})$$
$$=\left(\frac{\hbar}{2}\right)[l(l+1)-m(m+1)]^{1/2}\psi_{l,m+1}$$
$$+\left(\frac{\hbar}{2}\right)[l(l+1)-m(m-1)]^{1/2}\psi_{l,m-1}$$
$$M_y\psi_{l,m}=\left(\frac{1}{2i}\right)(M_1-M_{-1})\psi_{l,m}=\left(\frac{1}{2i}\right)(A_m\psi_{l,m+1}-B_m\psi_{l,m-1})$$
$$=\left(\frac{-i\hbar}{2}\right)[l(l+1)-m(m+1)]^{1/2}\psi_{l,m+1}$$
$$+\left(\frac{i\hbar}{2}\right)[l(l+1)-m(m-1)]^{1/2}\psi_{l,m-1} \tag{6-71}$$

The matrix representation of the angular momentum operators $M_x$ and $M_y$ are easily derived from the above two expressions.

Finally, we discuss the relation between the angular momentum eigenfunctions and the eigenvalue problem of a particle in a central force field. According to Eq. (6-18), the Hamiltonian $H$ of a particle in a central force field commutes with the operator $M^2$, and we can conclude again that the two operators have the same eigenfunctions. We have seen that the operator $M^2$ depends on the two polar angles $\theta$ and $\phi$ and not on the variable $r$. Its eigenfunctions are the spherical harmonics functions $\psi_{l,m}(\theta,\phi)$, defined in Eq. (6-36). Obviously, we can multiply these eigenfunctions by an arbitrary function of $r$ since the operator is independent of $r$; hence we have

$$M^2 g(r)Y_{l,m}(\theta,\phi)=l(l+1)\hbar^2 g(r)Y_{l,m}(\theta,\phi) \tag{6-72}$$

Since $M^2$ and $H$ commute we suspect that the eigenfunctions $\Psi(r,\theta,\phi)$ of the operator $H$ can also be written in the form

$$\Psi(r,\theta,\phi)=g_l(r)Y_{l,m}(\theta,\phi) \tag{6-73}$$

It is easily proved in the following section that this assumption is indeed correct.

## 6-4 A Particle in a Central Force Field

The Schrödinger equation of a particle in a central force field is given by Eq. (6-2). In order to solve the equation we introduce the polar coordinates that we defined in Eq. (6-21). We must express the Laplace operator $\Delta$ in terms of these polar coordinates. The derivation is given in Appendix A and the result is

$$\Delta = \frac{\partial^2}{\partial x^2} + \frac{\partial^2}{\partial y^2} + \frac{\partial^2}{\partial z^2}$$

$$= \frac{\partial^2}{\partial r^2} + \frac{2}{r}\frac{\partial}{\partial r} + \frac{1}{r^2}\frac{\partial^2}{\partial \theta^2} + \frac{\cos\theta}{r^2 \sin\theta}\frac{\partial}{\partial \theta} + \frac{1}{r^2 \sin^2\theta}\frac{\partial^2}{\partial \phi^2} \tag{6-74}$$

If we compare this with Eq. (6-30) for the operator $M^2$ in terms of polar coordinates, we see that the angle-dependent part of the operator $\Delta$ is proportional to the operator $M^2$. We can write the Hamiltonian $H$ of a particle in a central force field as

$$H = -\frac{\hbar^2}{2\mu}\Delta + V(r)$$

$$= -\frac{\hbar^2}{2\mu}\left(\frac{\partial^2}{\partial r^2} + \frac{2}{r}\frac{\partial}{\partial r}\right) + \frac{M^2}{2\mu r^2} + V(r) \tag{6-75}$$

where we have substituted Eq. (6-30) for the operator $M^2$.

The Schrödinger equation of a particle in a central force field is given by

$$H\Psi(r,\theta,\phi) = E\Psi(r,\theta,\phi) \tag{6-76}$$

It is easily seen that the eigenfunctions of this equation can be taken as the functions (6-73) because substitution of these functions give

$$Hg_l(r)Y_{l,m}(\theta,\phi) = Y_{l,m}(\theta,\phi)\left[-\frac{\hbar^2}{2\mu}\frac{\partial^2 g_l}{\partial r^2} - \frac{\hbar^2}{2\mu}\frac{2}{r}\frac{\partial g_l}{\partial r}\right]$$

$$+ \frac{g_l(r)}{2\mu r^2}\left[M^2 Y_{l,m}(\theta,\phi)\right] + V(r)g_l(r)Y_{l,m}(\theta,\phi)$$

$$= -\frac{\hbar^2}{2\mu}Y_{l,m}(\theta,\phi)\left[\frac{\partial^2 g_l}{\partial r^2} + \frac{2}{r}\frac{\partial g_l}{\partial r} - \frac{l(l+1)g_l}{r^2}\right]$$

$$+ Vg_l Y_{l,m}(\theta,\phi) \tag{6-77}$$

By substituting this into the Schrödinger equation (6-76) we obtain

$$-\frac{\hbar^2}{2\mu}\left(\frac{\partial^2 g_l}{\partial r^2} + \frac{2}{r}\frac{\partial g_l}{\partial r} - \frac{l(l+1)g_l}{r^2}\right) + V(r)g_l = Eg_l \tag{6-78}$$

as the radial Schrödinger equation.

It follows thus that the angular part of the eigenfunction is independent of the form of the potential function $V(r)$; in other words, this angular part is the same for every central force field. We proved this already in the previous section from the commutation properties of the operators $H$ and $M^2$ and we have proved it again from the detailed form of the operators.

The hydrogen atom is a special case of this general situation and we discuss it in the following section.

## 6-5 The Hydrogen Atom

We have shown in Section 6-1, Eq. (6-10), that the hydrogen atom is a special case of a particle in a central force field, namely the case where the potential function $V(r)$ is the Coulomb potential,

$$V(r) = -\frac{e^2}{r} \tag{6-79}$$

Substitution into the radial Schrödinger equation (6-78) gives

$$-\frac{\hbar^2}{2\mu}\left[\frac{\partial^2 g_l}{\partial r^2} + \frac{2}{r}\frac{\partial g_l}{\partial r} - \frac{l(l+1)g_l}{r^2}\right] - \frac{e^2 g_l}{r} = E g_l \tag{6-80}$$

as the Schrödinger equation for the hydrogen atom.

Before we attempt to solve the equation we introduce new units of length and energy. We denote our new unit of length by $a$ and we take our unit of energy as $(e^2/a)$. In terms of these units the equation becomes

$$-\frac{\hbar^2}{2\mu a^2}\cdot\frac{a}{e^2}\left[\frac{\partial^2 g_l}{\partial r^2} + \frac{2}{r}\frac{\partial g_l}{\partial r} - \frac{l(l+1)g_l}{r^2}\right] - \frac{g_l}{r} = \varepsilon g_l \tag{6-81}$$

Obviously, if we take

$$a = a_o = \frac{\hbar^2}{\mu e^2} \qquad \varepsilon_o = \frac{e^2}{a_o} \tag{6-82}$$

as our units of length and energy, the equation takes a particularly simple form

$$-\frac{1}{2}\frac{d^2 g_l}{dr^2} - \frac{1}{r}\frac{dg_l}{dr} + \frac{l(l+1)g_l}{2r^2} - \frac{g_l}{r} = \varepsilon g_l \tag{6-83}$$

In order to solve Eq. (6-83) we make the following two substitutions

$$\varepsilon = -\frac{1}{2\rho^2} \qquad r = \frac{\rho t}{2} \tag{6-84}$$

The result is

$$\frac{d^2g_l}{dt^2}+\frac{2}{t}\frac{dg_l}{dt}+\left[-\frac{1}{4}+\frac{\rho}{t}-\frac{l(l+1)}{t^2}\right]g_l=0 \tag{6-85}$$

In our classical description of the hydrogen atom in Section 1-5 we showed that the energy of the system is negative for bound states. We assume therefore that the stationary states of the hydrogen atom, corresponding to the bound states, have negative energies. For these states the parameter $\rho$ of Eq. (6-84) is real, and it must be positive if we take both variables $r$ and $t$ to be positive. We shall see at the end of this chapter that the continuum states of the hydrogen atom are derived by taking $\rho$ to be imaginary.

The differential equation (6-85) bears a close resemblance to the differential equation (5-42) for the Whittaker function. We transform Eq. (6-85) to the latter form by substituting

$$q_l(t)=tg_l(t) \qquad \frac{d^2q_l}{dt^2}=t\frac{d^2g_l}{dt^2}+2\frac{dg_l}{dt} \tag{6-86}$$

We obtain

$$\frac{d^2q_l}{dt^2}+\left[-\frac{1}{4}+\frac{\rho}{t}+\frac{1-4(l+\frac{1}{2})^2}{4t^2}\right]q_l=0 \tag{6-87}$$

The two solutions of this equation are described by Eqs. (5-50) and (5-52); they are

$$q_l=M_{\rho,l+1/2}(t)=t^{l+1}\exp\left(\frac{-t}{2}\right){}_1F_1(l+1-\rho;2l+2;t) \tag{6-88}$$

and

$$q_l=M_{\rho,-l-1/2}(t)=t^{-l}\exp\left(\frac{-t}{2}\right){}_1F_1(-l-\rho;-2l;t) \tag{6-89}$$

The second solution, Eq. (6-89), is not allowed because it leads to a radial function $g(t)$ which is infinite at the origin. The first solution, Eq. (6-88), gives the radial function

$$g_l(t)=t^l\exp\left(\frac{-t}{2}\right){}_1F_1(l+1-\rho;2l+2;t) \tag{6-90}$$

We discussed in Section 5-4 that the confluent hypergeometric function behaves asymptotically as $e^t$ for large values of the parameter $t$. In general, we find that the solution $g_l(t)$ does not give a physically acceptable wave function.

Only in those cases where the confluent hypergeometric function reduces to a finite polynomial do we obtain a solution that is normalizable. This happens when the parameter $a$ of Eq. (5-28) is a nonpositive integer; in other words, when

$$l+1-\rho=-\nu \qquad \nu=0,1,2,3,\ldots, \text{ and so on} \tag{6-91}$$

The eigenvalues of the hydrogen atom are therefore determined from the condition

$$\rho=l+1+\nu \qquad \nu=0,1,2,3,\ldots, \text{ and so on} \tag{6-92}$$

The corresponding eigenfunctions $g_{l,\nu}(r)$ are

$$g_{l,\nu}(t)=t^l e^{-t/2}{}_1F_1(-\nu;2l+2;t)$$

$$t=\frac{2r}{l+1+\nu} \tag{6-93}$$

The complete set of eigenfunctions $\Psi_{\nu,l,m}(r,\theta,\phi)$ of the hydrogen atom is now obtained by combining Eqs. (6-73) and (6-93):

$$\Psi_{\nu,l,m}(r,\theta,\phi)=g_{l,\nu}(t)Y_{l,m}(\theta,\phi) \tag{6-94}$$

where the allowed values for $\nu$, $l$, and $m$ are

$$\begin{aligned}\nu&=0,1,2,3,\ldots\\ l&=0,1,2,3,\ldots\\ m&=0,\pm1,\pm2,\pm3,\ldots,\pm l\end{aligned} \tag{6-95}$$

This set of indices is known as the quantum numbers, describing the states of the hydrogen atom.

Although these quantum numbers describe all eigenfunctions of the hydrogen atom, it is customary to use a different set. We can write

$$\rho=n \qquad n=l+1,\, l+2,\, l+3,\ldots \tag{6-96}$$

$$l=0,1,2,3,4,\ldots$$

The condition here is that $n \geq l+1$, but this can also be satisfied by taking

$$\begin{aligned}n&=1,2,3,4,\ldots\\ l&=0,1,2,3,\ldots,n-1\\ m&=0,\pm1,\pm2,\ldots,\pm l\end{aligned} \tag{6-97}$$

Now the energy eigenvalues are, according to Eqs. (6-82) and (6-84),

$$E_n = -\frac{1}{2n^2}\varepsilon_o \qquad \varepsilon_o = \frac{e^2}{a_o} \tag{6-98}$$

and the corresponding eigenfunctions are

$$\Psi_{n,l,m}(r,\theta,\phi) = t^l e^{-t/2}{}_1F_1(l+1-n;2l+2;t)Y_{l,m}(\theta,\phi)$$

$$t = \frac{2r}{n} \tag{6-99}$$

The allowed values of the quantum numbers $n$, $l$, and $m$ are prescribed by Eq. (6-97).

In order to understand the physical significance of the quantum numbers we collect the results of Eqs. (6-52), (6-72), and (6-98):

$$H\Psi_{n,l,m}(r,\theta,\phi) = \left(\frac{-\varepsilon_o}{2n^2}\right)\Psi_{n,l,m}(r,\theta,\phi)$$

$$M^2\Psi_{n,l,m}(r,\theta,\phi) = l(l+1)\hbar^2\Psi_{n,l,m}(r,\theta,\phi)$$

$$M_z\Psi_{n,l,m}(r,\theta,\phi) = m\hbar\Psi_{n,l,m}(r,\theta,\phi) \tag{6-100}$$

First, we consider the quantum number $n$. This describes the energy of the system. A particular eigenvalue $E_n$ is usually degenerate because the other quantum numbers $l$ and $m$ can take the values

$$l = 0,1,2,3,\ldots,n-1$$

$$m = 0,\pm 1,\pm 2,\ldots,\pm l \tag{6-101}$$

The order $\rho_n$ of the degeneracy of $E_n$ is given by

$$\rho_n = 1+3+5+7+\cdots(2n-1) = n^2 \tag{6-102}$$

The second quantum number $l$ describes the magnitude of the angular momentum in a given stationary state. If we know both the energy and the magnitude of the angular momentum of the system, we still have a degeneracy of order $(2l+1)$ because of the orientation of the angular momentum. The latter orientation is related to the quantum number $m$ which represents the projection of the angular momentum along the $Z$ axis. The three quantum numbers $n$, $l$, and $m$ together give a unique description of each stationary state by labeling the values of the energy, the angular momentum magnitude, and the angular momentum direction. The above description is, of course, related to the commutator relations between the three operators $H$, $M^2$, and $M_z$, which we discussed in Sections 6-1 and 6-3.

## 6-6 The Behavior of the Hydrogen Atom Eigenfunctions

It is useful to gain some insight into the general properties of the eigenfunctions and eigenvalues of the hydrogen-like atoms. These are the series of atoms and ions H, $He^+$, $Li^{2+}$, and so on. In general, we consider the system of a nucleus $A$ with charge $Z_A e$ and mass $m_A$ and an electron with charge $(-e)$ and mass $m_e$. We show that the eigenfunctions and eigenvalues of all these systems can be described by Eqs. (6-98) and (6-99) by introducing the proper units of length and energy.

To avoid carrying too many numerical factors in atomic calculations, it is convenient to make use of the atomic units introduced by Hartree. The units of length and energy are defined as

$$a_o = \frac{\hbar^2}{m_e e^2} \qquad \varepsilon_o = 2R_\infty \qquad R_\infty = \frac{e^2}{2a_o} \tag{6-103}$$

Here $a_o$ is called the Bohr radius, and its magnitude is $a_o = 5.2917 \times 10^{-9}$ cm. The quantity $R$ is known as the Rydberg constant for infinite mass, and its magnitude is 109,727.3 $cm^{-1}$. The energy unit $\varepsilon_o$ expressed in terms of electron volts is $\varepsilon_o = 27.21$ ev.

In order to describe a specific atom $A$ consisting of a nucleus, with mass $m_A$ and charge $Z_A e$, and one electron, we introduce the quantities

$$a_A = \frac{\hbar^2}{\mu_A Z_A e^2} = \frac{m_A + m_e}{Z_A m_A} a_o \tag{6-104}$$

and

$$\varepsilon_A = \frac{Z_A e^2}{a_A} = \frac{Z_A^2 m_A}{m_A + m_e} \varepsilon_o \tag{6-105}$$

as units of length and energy, respectively. We also introduce the Rydberg constant $R_A$ for atom $A$:

$$\varepsilon_A = 2Z_A^2 R_A \qquad R_A = \frac{m_A R_\infty}{m_A + m_e} \tag{6-106}$$

We see that $R_A$ approaches $R_\infty$ when $m_A$ tends to infinity; this is why the notation $R_\infty$ is used.

The reduced mass $\mu_A$ of ion $A$ is defined as

$$\mu_A = \frac{m_A m_e}{m_A + m_e} \tag{6-107}$$

and its Hamiltonian $H_A$ is given by

$$H_A = -\frac{\hbar^2}{2\mu_A}\left(\frac{\partial^2}{\partial x^2}+\frac{\partial^2}{\partial y^2}+\frac{\partial^2}{\partial z^2}\right)-\frac{Ze^2}{r} \tag{6-108}$$

If we introduce $a_A$ as unit of length and $\varepsilon_A = Ze^2/a_A$ as unit of energy, the Hamiltonian becomes

$$H_A = -\frac{\hbar^2}{2\mu_A a_A Ze^2}\left(\frac{\partial^2}{\partial x^2}+\frac{\partial^2}{\partial y^2}+\frac{\partial^2}{\partial z^2}\right)-\frac{1}{r} \tag{6-109}$$

Clearly, if we take

$$\frac{\hbar^2}{\mu_A a_A Ze^2}=1 \qquad a_A = \frac{\hbar^2}{\mu_A Ze^2} \tag{6-110}$$

the Hamiltonian assumes the same form as the hydrogen atom Hamiltonian of the previous section. The only difference is that the units of length and energy change from one system to the next. In the case of the hydrogen atom they were defined in Eq. (6-82) and in the more general case they are defined in Eqs. (6-104) and (6-105). The analytical form of the eigenvalues and eigenfunctions is the same for all different ions and atoms as long as we use the generalized units $a_A$ and $\varepsilon_A$. In the remainder of this chapter, we will discuss the hydrogen atom, but it should be realized that this discussion applies to any ion or atom $A$ as we have defined above.

Each stationary state of the hydrogen atom is determined by the quantum numbers $n$, $l$, and $m$, but we usually describe them by means of a different notation, which is due to the old spectroscopic theories. First we look at the quantum number $l$, and if $l=0$, we speak of an $s$ state; for $l=1$ we have a $p$ state; $l=2$ is a $d$ state; $l=3$ is an $f$ state; $l=4$ is a $g$ state; and so on. The value of $n$ is given by writing it in front of these letters. The lowest states of the hydrogen atom are given in Table 6-1. The values of $m$ are indicated by adding subscripts. For example, the $2p$ states are written as $2p_1$, $2p_0$, and $2p_{-1}$ for $m=1$, $m=0$, and $m=-1$, respectively.

**Table 6-1**

| | | |
|---|---|---|
| $n=1$ | $l=0$ | $1s$ |
| $n=2$ | $l=0$ | $2s$ |
| $n=2$ | $l=1$ | $2p$ |
| $n=3$ | $l=0$ | $3s$ |
| $n=3$ | $l=1$ | $3p$ |
| $n=3$ | $l=2$ | $3d$ |
| and so on | | |

It is useful to know all wave functions for the states given in Table 6-1, since these functions play a role in many chemical problems. If we express $r$ in terms of the unit $a_A$, the radial wave functions are, according to Eq. (6-94),

$$g_{l,n}(r)=r^l \exp\left(-\frac{r}{n}\right) {}_1F_1\left(l+1-n;2l+2;2\frac{r}{n}\right) \tag{6-111}$$

It follows from the definition of the confluent hypergeometric series that

$$\begin{aligned}
g_{0,1}(r)&=A_{0,1}e^{-r}\\
g_{0,2}(r)&=A_{0,2}\left(1-\frac{1}{2}r\right)e^{-(r/2)}\\
g_{1,2}(r)&=A_{1,2}re^{-(r/2)}\\
g_{0,3}(r)&=A_{0,3}\left(1-\frac{2r}{3}+\frac{2r^2}{27}\right)e^{-(r/3)}\\
g_{1,3}(r)&=A_{1,3}r\left(1-\frac{r}{6}\right)e^{-(r/3)}\\
g_{2,3}(r)&=A_{2,3}r^2e^{-(r/3)}
\end{aligned} \tag{6-112}$$

The $A_{l,n}$ are normalization constants, which are determined by the condition

$$\int_0^\infty [g_{l,n}(r)]^2 r^2\,dr=1 \tag{6-113}$$

The normalized functions are therefore

$$\begin{aligned}
g_{0,1}(r)&=2e^{-r}\\
g_{0,2}(r)&=\frac{1}{2\sqrt{2}}(r-2)e^{-(r/2)}\\
g_{1,2}(r)&=\frac{1}{2\sqrt{6}}re^{-(r/2)}\\
g_{0,3}(r)&=\frac{2}{81\sqrt{3}}(2r^2-18r+27)e^{-(r/3)}\\
g_{1,3}(r)&=\frac{2\sqrt{2}}{81\sqrt{3}}(r^2-6r)e^{-(r/3)}\\
g_{2,3}(r)&=\frac{2\sqrt{2}}{81\sqrt{15}}r^2e^{-(r/3)}
\end{aligned} \tag{6-114}$$

The orthonormal spherical harmonics $Y_{1,m}(\theta,\phi)$ are, according to Eq. (5-241),

$$
\begin{aligned}
Y_{0,0}(\theta,\phi) &= \frac{1}{\sqrt{4\pi}} \\
Y_{1,0}(\theta,\phi) &= \left(\frac{3}{4\pi}\right)^{1/2} \cos\theta \\
Y_{1,1}(\theta,\phi) &= \left(\frac{3}{8\pi}\right)^{1/2} \sin\theta e^{i\phi} \\
Y_{1,-1}(\theta,\phi) &= Y^*_{1,1}(\theta,\phi) \\
Y_{2,0}(\theta,\phi) &= \frac{1}{2}\left(\frac{5}{4\pi}\right)^{1/2}(3\cos^2\theta - 1) \\
Y_{2,1}(\theta,\phi) &= \left(\frac{15}{8\pi}\right)^{1/2} \sin\theta\cos\theta e^{i\phi} \\
Y_{2,2}(\theta,\phi) &= \left(\frac{15}{32\pi}\right)^{1/2} \sin^2\theta e^{2i\phi} \\
Y_{2,-1}(\theta,\phi) &= Y^*_{2,1}(\theta,\phi) \\
Y_{2,-2}(\theta,\phi) &= Y^*_{2,2}(\theta,\phi)
\end{aligned}
\tag{6-115}
$$

We can now construct the orthonormal eigenfunctions for the states described in Table 6-1. The *s* states are all nondegenerate, and we have

$$
\begin{aligned}
\psi(1s) &= \frac{1}{\sqrt{\pi}} e^{-r} \\
\psi(2s) &= \frac{1}{4\sqrt{2\pi}}(r-2)e^{-(r/2)} \\
\psi(3s) &= \frac{1}{81\sqrt{3\pi}}(2r^2 - 18r + 27)e^{-(r/3)}
\end{aligned}
\tag{6-116}
$$

Each $p$ state is threefold degenerate, and the wave functions for the $2p$ and $3p$ states are

$$\psi(2p_1)=\frac{1}{8\sqrt{\pi}}r\sin\theta\, e^{i\phi}e^{-(r/2)}$$

$$\psi(2p_0)=\frac{1}{4\sqrt{2\pi}}r\cos\theta\, e^{-(r/2)}$$

$$\psi(2p_{-1})=\frac{1}{8\sqrt{\pi}}r\sin\theta\, e^{-i\phi}e^{-(r/2)}$$

$$\psi(3p_1)=\frac{1}{81\sqrt{\pi}}(r^2-6r)\sin\theta\, e^{i\phi}e^{-(r/3)} \tag{6-117}$$

$$\psi(3p_0)=\frac{\sqrt{2}}{81\sqrt{\pi}}(r^2-6r)\cos\theta\, e^{-(r/3)}$$

$$\psi(3p_{-1})=\frac{1}{81\sqrt{\pi}}(r^2-6r)\sin\theta\, e^{-i\phi}e^{-(r/3)}$$

The fivefold degenerate $3d$ state has the eigenfunctions

$$\psi(3d_2)=\frac{1}{81\sqrt{4\pi}}r^2\sin^2\theta\, e^{2i\phi}e^{-(r/3)}$$

$$\psi(3d_1)=\frac{1}{81\sqrt{\pi}}r^2\sin\theta\cos\theta\, e^{i\phi}e^{-(r/3)}$$

$$\psi(3d_0)=\frac{1}{81\sqrt{6\pi}}r^2(3\cos^2\theta-1)e^{-(r/3)} \tag{6-118}$$

$$\psi(3d_{-1})=\frac{1}{81\sqrt{\pi}}r^2\sin\theta\cos\theta e^{-i\phi}e^{-(r/3)}$$

$$\psi(3d_{-2})=\frac{1}{81\sqrt{4\pi}}r^2\sin^2\theta\, e^{-2i\phi}e^{-(r/3)}$$

We have already mentioned that a set of degenerate eigenfunctions can be replaced by a different set, which is obtained by taking linear combinations of the first set. Consequently, we can represent the hydrogen wave functions in a different way. It follows form Eq. (6-117) that $\psi(2p_0)$ can also be written as

$$(2p_0)=\frac{1}{4\sqrt{2\pi}}z\,e^{-(r/2)} \tag{6-119}$$

and we often find it denoted by $\psi(2p_z)$. By taking the sum and the difference of $\psi(2p_1)$ and $\psi(2p_{-1})$, we can obtain the functions $\psi(2p_x)$ and $\psi(2p_y)$, and we can replace the set of eigenfunctions $\psi(2p_m)$ of Eq. (6-117) by

$$\begin{aligned}\psi(2p_x)&=\frac{1}{4\sqrt{2\pi}}x\,e^{-(r/2)}\\ \psi(2p_y)&=\frac{1}{4\sqrt{2\pi}}y\,e^{-(r/2)}\\ \psi(2p_z)&=\frac{1}{4\sqrt{2\pi}}z\,e^{-(r/2)}\end{aligned} \tag{6-120}$$

By the same method we can replace the $3p$ eigenfunctions of Eq. (6-117) by the set

$$\begin{aligned}\psi(3p_x)&=\frac{\sqrt{2}}{81\sqrt{\pi}}(r-6)x\,e^{-(r/3)}\\ \psi(3p_y)&=\frac{\sqrt{2}}{81\sqrt{\pi}}(r-6)y\,e^{-(r/3)}\\ \psi(3p_z)&=\frac{\sqrt{2}}{81\sqrt{\pi}}(r-6)z\,e^{-(r/3)}\end{aligned} \tag{6-121}$$

Often we also find a different set of $3d$ functions instead of the $\psi(3d_m)$ of Eq. (6-118), namely

$$\psi(3d_{zz})=\frac{1}{81\sqrt{6\pi}}(r^2-3z^2)e^{-(r/3)}$$

$$\psi(3d_{xz})=\frac{\sqrt{2}}{81\sqrt{\pi}}xz\,e^{-(r/3)}$$

$$\psi(3d_{yz})=\frac{\sqrt{2}}{81\sqrt{\pi}}yz\,e^{-(r/3)}$$

$$\psi(3d_{xy}) = \frac{\sqrt{2}}{81\sqrt{\pi}} xy\, e^{-(r/3)}$$

$$\psi(3d_{x^2-y^2}) = \frac{1}{81\sqrt{2\pi}} (x^2 - y^2) e^{-(r/3)} \tag{6-122}$$

It is convenient to express the probability density functions for these states in terms of polar coordinates. If we define $P(r, \theta, \phi)\, dr\, d\theta\, d\phi$ as the probability that the polar coordinates of the electron are between $r$ and $r + dr$, $\theta$ and $\theta + d\theta$, and $\phi$ and $\phi + d\phi$, then we have

$$P(r, \theta, \phi) = \psi\psi^* r^2 \sin\theta \tag{6-123}$$

For a state $(n, l, m)$ we can write this as

$$P_{n,l,m} = [g_{l,n}(r)]^2 r^2 Y_{l,m}(\theta, \phi) Y^*_{l,m}(\theta, \phi) \sin\theta \tag{6-124}$$

that is, as the product of a radial function

$$R_{n,l} = [g_{l,n}(r)]^2 r^2 \tag{6-125}$$

and an angular function

$$F(\Omega) = Y_{l,m}(\theta, \phi) Y^*_{l,m}(\theta, \phi) \sin\theta \tag{6-126}$$

For the $s$ states the angular distribution function is a constant, and we only have to consider the radial distribution functions, which we plot in Fig. 6-1. It

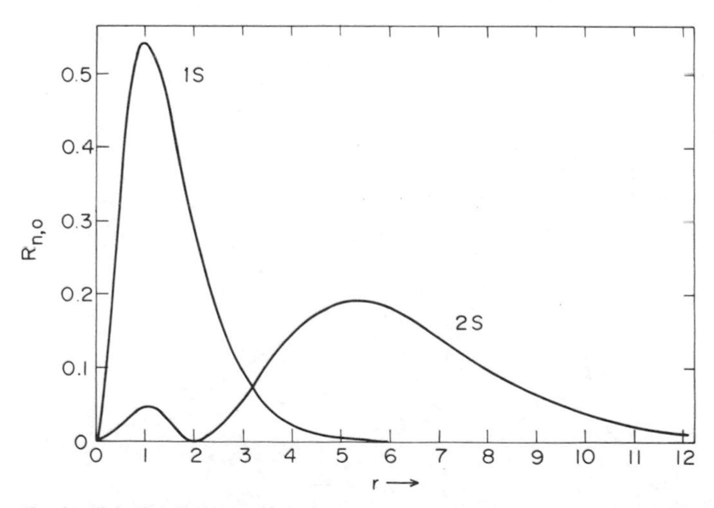


**Fig. 6-1** The radial distribution functions according to Eq. (6-125) for the $1s$ and the $2s$ eigenfunctions of the hydrogen atom.

may be useful to know that the expectation value of $r$, that is, the integral

$$(r)_n = \int [g_{0,n}(r)]^2 r^3\, dr \tag{6-127}$$

has the value

$$(r)_{n,s} = \frac{3n^2}{2} \tag{6-128}$$

It follows thus that the average distance between the electron and the nucleus increases quadratically with increasing values of the quantum number $n$.

The radial distribution functions for the $2p$ and $3p$ states are plotted in Fig. 6-2. In the case of $np$ states, $l=1$, the expectation values of $r$ are given by

$$(r)_{n,p} = \left(\frac{3n^2}{2}\right) - 1 \tag{6-129}$$

However, the angular distribution function now plays a very important role. For example, $(2p_z)$ is zero in the $xy$ plane, and for any given $r$ it has a maximum value on the $z$ axis. We may therefore say that $\psi(2p_z)$ is directed along the $z$ axis. In the same way we find that $\psi(2p_x)$ and $\psi(2p_y)$ are directed along the $x$ axis and $y$ axis, respectively.

In the case of the $3d$ functions the angular distribution is much more interesting than the radial dependence. It is easily verified that $\psi(3d_{xy})$ is zero in the planes $x=0$ and $y=0$ and that the planes $x=\pm y$ contain the relative maxima. The function $\psi(3d_{x^2-y^2})$ has the same behavior if we rotate 45° around the $z$ axis, and here the maxima are on the planes $x=0$ and $y=0$, and

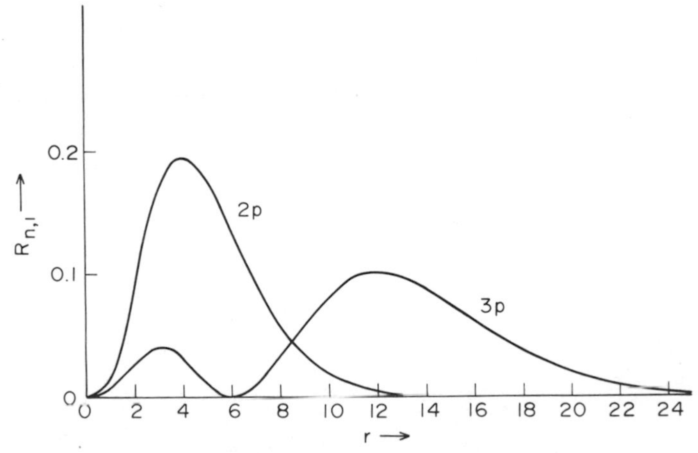


**Fig. 6-2** The radial distribution functions according to Eq. (6-125) for the $2p$ and $3p$ eigenfunctions of the hydrogen atom.

the wave function is zero when $x = \pm y$. Obviously, the functions $\psi(3d_{xz})$ and $\psi(3d_{yz})$ behave similarly. The function $\psi(3d_{zz})$ is zero when $3\cos^2\theta = 1$, that is, on the surface of a cone making an angle of $27\frac{1}{2}°$ with the $z$ axis and passing through the origin. The relative maxima of the function are in the plane $z = 0$ and along the $z$ axis.

## 6-7 The Continuum

When the energy $E$ of the hydrogen atom is positive, all energy values lead to allowed, unnormalizable wave functions that behave in an oscillatory fashion. We mentioned this when we introduced Eq. (6-84). In the subsequent discussion we pointed out that for real values of the parameter $\rho$ we obtain negative energies that lead to bound states. It should be noted that our subsequent mathematical analysis is valid for all possible values of the parameter $\rho$, both real and imaginary. In order to derive the quantum mechanical description of the continuum states, we substitute, therefore,

$$\rho = i\sigma \qquad \varepsilon = \frac{1}{2\sigma^2} \qquad r = \frac{i\sigma t}{2} \tag{6-130}$$

The solution of the Schrödinger equation is again given by Eq. (6-90), since we did not make use of the condition that $\rho$ has to be real in solving the equation. By combining Eq. (6-90) with Eq. (6-94) we find that the continuum eigenfunctions are given by

$$\Psi_{l,m}(r,\theta,\phi;\sigma) = r^l \exp\left(\frac{ir}{\sigma}\right) {}_1F_1\left(l+1-i\sigma; 2l+2; -\frac{2ir}{\sigma}\right) Y_{l,m}(\theta,\phi) \tag{6-131}$$

The continuum functions play an important role in the theoretical description of scattering, but we do not discuss this topic in this book.

## Problems

**1** The component of the angular momentum in a certain direction can be represented by the operator

$$M_p = \alpha M_x + \beta M_y + \gamma M_z$$

where $\alpha$, $\beta$, and $\gamma$ are direction cosines. Derive the eigenvalues and eigenfunctions of $M_p$ for the case in which $l = 1$. The eigenfunctions should be expressed in terms of the functions $\psi_{l,m}$ of Section 6-3.

**2** Derive the eigenvalues and eigenfunctions of the operator $M_x$ for $l = 2$.

3 Derive the eigenvalues and eigenfunctions of the operator

$$H=\left(\frac{D}{3}\right)(3M_z^2-M^2)-E\left(M_x^2-M_y^2\right)$$

for a system with $l=1$.

4 Express the radial wave functions $g_{n,l}(r)$ of the hydrogen atom in terms of the associated Laguerre polynomials of Eq. (5-153) and derive the validity of Eqs. (6-128) and (6-129) for the expectation values of $r$ from the properties of the associated Laguerre polynomials.

5 Derive the expectation values of $(1/r)$ and of $(r^2)$ with respect to the hydrogen atom eigenstates by means of the same procedure, that is, by using the integration properties of the associated Laguerre polynomials.

6 Consider a particle of mass $m$ in a spherical box where $V(r)=0$ if $r\leq R$ and $V(r)=\infty$ if $r>R$. Calculate the energy eigenvalues and the eigenfunctions for the case that the angular quantum number $l$ is equal to zero.

7 The eigenvalues and eigenfunctions of the three-dimensional harmonic oscillator

$$V(r)=\tfrac{1}{2}k\left(x^2+y^2+z^2\right)=\tfrac{1}{2}kr^2$$

are easily obtained by separating the Hamiltonian into an $x$-dependent, a $y$-dependent, and a $z$-dependent part. Show that the same results can be obtained by treating the problem as a particle in a central force field and by using polar coordinates.

8 Derive the detailed analytical expression for the normalized wave functions of the $4s$ and $5p$ states of the hydrogen atom.

9 The Rydberg constant $R_H$ for the hydrogen atom is

$$R_H=109{,}677.58 \text{ cm}^{-1}$$

Use this value to derive the lowest excitation energy $E(2p)-E(2s)$ of the ions $He^+$, $Li^{2+}$, and $Be^{3+}$.

## Recommended Reading

The hydrogen atom is discussed in every textbook on quantum mechanics. We quote Davydov's book because of its discussion of the angular momentum eigenvalue problem. The most extensive discussion is found in the text by Bethe and Salpeter, which is devoted exclusively to the hydrogen and helium atoms.

A. S. Davydov, *Quantum Mechanics*, Pergamon Press, Oxford, 1965.

H. A. Bethe and E. E. Salpeter, *Quantum Mechanics of One- and Two-Electron Atoms*, Springer-Verlag, Berlin, 1957.

CHAPTER SEVEN

# Approximate Methods in Quantum Mechanics

## 7-1 Introduction

We have mentioned that the various properties of an atomic or molecular system can all be predicted by solving the corresponding Schrödinger equation. Unfortunately, the exact solutions of the Schrödinger equation have been derived only for a few simple systems. Most of these systems have been discussed in the previous chapters. The most complex atomic or molecular system for which an exact solution can be obtained is the hydrogen molecular ion, $H_2^+$. We did not present this solution because the mathematics is fairly involved.

Obviously, we must resort to approximations if we wish to apply quantum mechanics to the description of atomic or molecular systems. The best known approximate procedures in quantum mechanics are perturbation theory and the various variational methods. The purpose of perturbation theory is the derivation of the eigenvalues and eigenfunctions of an operator $H$ if those of a similar operator $H_o$ are known. We start our discussions in this chapter by presenting the mathematical description of formal perturbation theory. However, if we want to apply perturbation theory, it becomes apparent that we need more than just the formal perturbation theory, because in most cases we deal with approximate solutions of both operators $H$ and $H_o$ and we must consider the relations between two different sets of approximate eigenvalues and eigenfunctions. We deal with these questions at the end of this chapter because the answers can be obtained by making use of variational procedures.

The variational principle makes it possible to derive approximate eigenvalues and eigenfunctions of a Hamiltonian. The accuracy of the results is, of course, related to the amount of effort that is invested in the problem. We show that the exact solutions can, in principle, be derived by means of variational procedures, but in practice, when a finite effort is expended, approximate results are obtained.

The various results of perturbation theory can also be derived from the variational principle and we discuss the relations between the two different approximate methods at the end of this chapter.

In summary, we present first the formal perturbation theory, then the variational procedures, and finally the relations between the two different methods.

## 7-2 Perturbation Theory for a Nondegenerate State

The goal of perturbation theory is to derive the eigenvalues and eigenfunctions of an operator $H$ if those of another operator $H_o$ are known. It is assumed here that $H_o$ is close to $H$; usually we write this as

$$H = H_o + \lambda H' \tag{7-1}$$

where $\lambda$ is a scaling parameter that is much smaller than unity. We do not attempt to specify how small $\lambda$ should be in order that perturbation theory is applicable. Usually this question is resolved by practical considerations. If the expansions in terms of $\lambda$ that we are going to use converge to our satisfaction, we conclude that $\lambda$ is sufficiently small, otherwise it is not.

We denote the known eigenvalues and eigenfunctions of $H_o$ by $\varepsilon_n$ and $\phi_n$, respectively:

$$H_o\phi_n = \varepsilon_n\phi_n \tag{7-2}$$

and the unknown eigenvalues and eigenfunctions of $H$ by $E_n$ and $\psi_n$:

$$H\psi_n = E_n\psi_n \tag{7-3}$$

The basic assumption of Rayleigh-Schrödinger perturbation theory is that we can expand $E_n$ and $\psi_n$ as

$$\begin{aligned} E_n &= \varepsilon_n + \sum_{k=1}^{\infty} \lambda^k E_{n,k} \\ \psi_n &= \phi_n + \sum_{k=1}^{\infty} \lambda^k \psi_{n,k} \end{aligned} \tag{7-4}$$

Here we bypass all complications that arise if the eigenvalue spectra of $H$ and $H_o$ are different. For the time being we also assume that the particular state for which we evaluate the perturbation has a nondegenerate eigenvalue.

Substitution of the series expansions (7-4) into the Schrödinger equation (7-3) gives

$$\left(H_o + \lambda H' - \varepsilon_n - \sum_{k=1}^{\infty} \lambda^k E_{n,k}\right)\left(\phi_n + \sum_{k=1}^{\infty} \lambda^k \psi_{n,k}\right) = 0 \tag{7-5}$$

We now set the coefficients of successive powers of $\lambda$ equal to zero in order to obtain successive approximations to $\psi_n$ and $E_n$. The first equation of this kind is

$$(H_o - \varepsilon_n)\phi_n = 0 \tag{7-6}$$

This equation is always satisfied because $\varepsilon_n$ is an eigenvalue of $H_o$ and $\phi_n$ is its corresponding eigenfunction. The second equation is

$$(H_o - \varepsilon_n)\psi_{n,1} + (H' - E_{n,1})\phi_n = 0 \tag{7-7}$$

The third equation is

$$(H_o - \varepsilon_n)\psi_{n,2} + (H' - E_{n,1})\psi_{n,1} - E_{n,2}\phi_n = 0 \tag{7-8}$$

The general equation is

$$(H_o - \varepsilon_n)\psi_{n,k} + (H' - E_{n,1})\psi_{n,k-1} - E_{n,2}\psi_{n,k-2} \cdots - E_{n,k-1}\psi_{n,1} - E_{n,k}\phi_n = 0 \tag{7-9}$$

Let us now attempt to solve these equations successively, starting with Eq. (7-7). If we multiply on the left by $\phi_n^*$ and integrate, we obtain

$$\langle \phi_n | H_o - \varepsilon_n | \psi_{n,1} \rangle + \langle \phi_n | H' - E_{n,1} | \phi_n \rangle = 0 \tag{7-10}$$

Since we suppose $H_o$ to be Hermitian, we have

$$\langle \phi_n | H_o - \varepsilon_n | \psi_{n,1} \rangle = \langle \psi_{n,1} | H_o - \varepsilon_n | \phi_n \rangle^* = 0 \tag{7-11}$$

because of Eq. (7-2). We find, therefore,

$$E_{n,1} = \langle \phi_n | H' | \phi_n \rangle \tag{7-12}$$

if we assume that the $\phi_n$ form an orthonormal set.

Now that we have derived an expression for $E_{n,1}$, we write Eq. (7-7) as

$$(H_o - \varepsilon_n)\psi_{n,1} = (E_{n,1} - H')\phi_n \tag{7-13}$$

Since we know all the quantities on the right side of the equation, this is an inhomogeneous differential equation in $\psi_{n,1}$ only. The corresponding homogeneous equation is

$$(H_o - \varepsilon_n)\psi = 0 \tag{7-14}$$

and we already know that its solution is $\phi_n$. According to the theory of inhomogeneous differential equations, we write the general solution of Eq. (7-13) as

$$\psi_{n,1} = \psi_{n,1}^o + \alpha\phi_n \tag{7-15}$$

where $\psi_{n,1}^{o}$ is a specific solution of Eq. (7-13) and $\alpha$ is an arbitrary parameter. It is now convenient to choose $\alpha$ in such a way that $\psi_{n,1}$ is orthogonal to $\phi_n$

$$\langle \psi_{n,1} | \phi_n \rangle = 0 \tag{7-16}$$

The specific form of $\psi_{n,1}$ obviously depends on $H_o$ and $H'$, so that we cannot derive the general solution. However, let us assume that we have obtained a solution $\psi_{n,1}$ of Eq. (7-13) that is orthogonal to $\phi_n$, and let us proceed to Eq. (7-8). Again we multiply on the left by $\phi_n^*$ and integrate, and we now find that

$$E_{n,2} = \langle \phi_n | H' | \psi_{n,1} \rangle \tag{7-17}$$

We now write Eq. (7-8) as

$$(H_o - \varepsilon_n)\psi_{n,2} = (E_{n,1} - H')\psi_{n,1} + E_{n,2}\phi_n \tag{7-18}$$

and we observe that all quantities on the right are known. Again we solve the differential equation in $\psi_{n,2}$, and we impose the condition that

$$\langle \psi_{n,2} | \phi_n \rangle = 0 \tag{7-19}$$

In this way $\psi_{n,2}$ is uniquely determined.

In general, we find that

$$E_{n,k} = \langle \phi_n | H' | \psi_{n,k-1} \rangle \tag{7-20}$$

if we see to it that all functions $\psi_{n,1}, \psi_{n,2}, \psi_{n,3}, \ldots, \psi_{n,k-1}$ are orthogonal to $\phi_n$.

In Section 7-3 we discuss a simple example of this method. We trust that this example will familiarize the reader with the procedures to be followed.

## 7-3 The Harmonic Oscillator in an Electric Field

We apply the approach above to the evaluation of the ground-state energy and eigenfunction of the harmonic oscillator in an electric field $F$. The Schrödinger equation for this system is

$$-\frac{\hbar^2}{2m}\frac{d^2\psi}{dx^2} + \frac{1}{2}kx^2\psi - eFx\psi = E\psi \tag{7-21}$$

From the considerations in Section 5-7 it is clear that the problem is considerably simplified if we introduce the quantity $\sqrt{\alpha}$ as the unit of length, where

$$\alpha = \left(\frac{\hbar^2}{km}\right)^{1/2} \tag{7-22}$$

and the quantity $(\hbar^2/2m\alpha)$ as the unit of energy. The equation then becomes

$$\left(-\frac{d^2}{dx^2}+x^2\right)\psi-\frac{2eF}{k\sqrt{\alpha}}x\psi=E\psi \tag{7-23}$$

If we now introduce the perturbation parameter $\mu$, we can write this as

$$(H_o+\mu H')\psi=E\psi$$

$$H_o=-\frac{d^2}{dx^2}+x^2$$

$$H'=-2x \tag{7-24}$$

$$\mu=\frac{eF}{k\sqrt{\alpha}}$$

The eigenvalues and eigenfunctions of the operator $H_o$ were obtained in Section 5-7, where we discussed the harmonic oscillator. They are

$$\varepsilon_n=2n+1 \qquad \phi_n=\frac{H_n(x)e^{-x^2/2}}{\left(\sqrt{\pi}\,2^n n!\right)^{1/2}} \tag{7-25}$$

Let us set out to calculate the perturbation to the ground state $n=0$. Here we have

$$\varepsilon_o=1 \qquad \phi_o=(\pi)^{-1/4}e^{-x^2/2} \tag{7-26}$$

First we calculate the energy correction $E_{0,1}$ by means of Eq. (7-12). We find

$$E_{0,1}=\langle\phi_0|H'|\phi_0\rangle=-\frac{2}{\sqrt{\pi}}\int_{-\infty}^{\infty}xe^{-x^2}\,dx=0 \tag{7-27}$$

Next we evaluate the first-order correction term $\psi_{0,1}$ of the wave function. According to Eq. (7-13) we have to solve the equation

$$(H_o-\varepsilon_0)\psi_{0,1}=-H'\phi_0 \tag{7-28}$$

or, substituting Eqs. (7-24) and (7-26),

$$\left(-\frac{d^2}{dx^2}+x^2-1\right)\psi_{0,1}=\frac{2x}{(\pi)^{1/4}}e^{-x^2/2} \tag{7-29}$$

It follows from Eq. (7-25) that the function $xe^{-x^2/2}$ is an eigenfunction of $H_o$ belonging to the eigenvalue $\varepsilon_1$. We substitute, therefore,

$$\psi_{0,1} = axe^{-x^2/2} \tag{7-30}$$

into Eq. (7-29) and obtain

$$2axe^{-x^2/2} = 2x(\pi)^{-1/4}e^{-x^2/2} \tag{7-31}$$

Hence the solution is

$$\psi_{0,1} = (\pi)^{-1/4}xe^{-x^2/2} \tag{7-32}$$

and it is easily verified that this function is orthogonal to $\phi_0$.

The next energy correction $E_{0,2}$ is obtained by using Eq. (7-17):

$$E_{0,2} = \langle \phi_0 | H' | \psi_{0,1} \rangle = -\frac{2}{\sqrt{\pi}} \int_{-\infty}^{\infty} x^2 e^{-x^2} dx = -1 \tag{7-33}$$

We now proceed to evaluate $\psi_{0,2}$ from Eq. (7-18), which becomes, in this case,

$$\left( -\frac{d^2}{dx^2} + x^2 - 1 \right) \psi_{0,2} = (2x^2 - 1)\frac{e^{-x^2/2}}{\pi^{1/4}} \tag{7-34}$$

We note that the right side of this equation is proportional to $H_2(x)e^{-x^2/2}$, so that we substitute

$$\psi_{0,2} = \pi^{-1/4} b(2x^2 - 1)e^{-x^2/2} \tag{7-35}$$

into Eq. (7-34). We find that

$$4b = 1 \tag{7-36}$$

so that

$$\psi_{0,2} = \tfrac{1}{4}\pi^{-1/4}(2x^2 - 1)e^{-x^2/2} \tag{7-37}$$

The third-order energy is

$$E_{0,3} = \langle \phi_0 | H' | \psi_{0,2} \rangle = 0 \tag{7-38}$$

because $\phi_0$ and $\psi_{0,2}$ are both symmetric in $x$ and $H'$ is antisymmetric in $x$.

Summarizing the above results, we see that

$$E_0 = 1 - \mu^2 + \cdots$$

$$\psi_0 = \pi^{-1/4} e^{-x^2/2}\left[1 + \mu x + \mu^2\left(\tfrac{1}{2}x^2 - \tfrac{1}{4}\right) + \cdots\right] \tag{7-39}$$

These results can be verified, since it is possible to solve Eq. (7-24) exactly. If we write the equation as

$$\left[-\frac{d^2}{dx^2} + (x-\mu)^2 - \mu^2\right]\psi = E\psi \tag{7-40}$$

we can transform it to the Schrödinger equation for the harmonic oscillator

$$\left(-\frac{d^2}{dz^2} + z^2\right)\psi(z) = E'\psi(z) \tag{7-41}$$

if we substitute

$$z = x - \mu \qquad E' = E + \mu^2 \tag{7-42}$$

The energy and eigenfunction of the ground state are therefore

$$E_0 = 1 - \mu^2$$

$$\psi_0 = \pi^{-1/4} \exp\left[-\tfrac{1}{2}(x-\mu)^2\right] \tag{7-43}$$

which agrees with the results from our perturbation treatment.

## 7-4 Perturbation Expansions

Although the separation of the Hamiltonian according to Eq. (7-1) is always possible, we often find it convenient to use a different expansion instead:

$$H = H_0 + \lambda H^{(1)} + \lambda^2 H^{(2)} + \lambda^3 H^{(3)} + \cdots \tag{7-44}$$

Here each successive term is supposed to be an order of magnitude smaller than its predecessor, and $\lambda$ is again a scaling parameter. An example is an electron in a potential field $V(r)$ that moves in a homogeneous magnetic field $\mathbf{B}$. Here the Hamiltonian is

$$H = \frac{p^2}{2m} + V(r) - \frac{e}{2mc}(\mathbf{M}\cdot\mathbf{B}) + \frac{e^2}{8mc^2}\left[B^2 r^2 - (\mathbf{B}\cdot\mathbf{r})^2\right] \tag{7-45}$$

If we take the magnetic field as the scaling parameter, we see that the first two terms constitute $H_0$, the third term $\lambda H^{(1)}$, and the last term $\lambda^2 H^{(2)}$.

The perturbation equations now have a slightly different form from the expressions in Section 7-2. If we denote the eigenvalues and eigenfunctions of $H_0$ again by $\varepsilon_n$ and $\phi_n$, respectively, and expand the eigenvalues and eigenfunctions of $H$ as

$$E_n = \varepsilon_n + \sum_{k=1}^{\infty} \lambda^k E_{n,k} \tag{7-46}$$

$$\psi_n = \phi_n + \sum_{k=1}^{\infty} \lambda^k \psi_{n,k}$$

the analogue of Eq. (7-5) is

$$\left( H_0 + \sum_{k=1}^{\infty} \lambda^k H^{(k)} - \varepsilon_n - \sum_{k=1}^{\infty} \lambda^k E_{n,k} \right)\left( \phi_n + \sum_{k=1}^{\infty} \lambda^k \psi_{n,k} \right) = 0 \tag{7-47}$$

The first four perturbation equations are now

$$(H_0 - \varepsilon_n)\phi_n = 0 \tag{7-48a}$$

$$(H_0 - \varepsilon_n)\psi_{n,1} = -\left(H^{(1)} - E_{n,1}\right)\phi_n \tag{7-48b}$$

$$(H_0 - \varepsilon_n)\psi_{n,2} = -\left(H^{(1)} - E_{n,1}\right)\psi_{n,1} - \left(H^{(2)} - E_{n,2}\right)\phi_n \tag{7-48c}$$

$$(H_0 - \varepsilon_n)\psi_{n,3} = -\left(H^{(1)} - E_{n,1}\right)\psi_{n,2} - \left(H^{(2)} - E_{n,2}\right)\psi_{n,1} - \left(H^{(3)} - E_{n,3}\right)\phi_n \tag{7-48d}$$

If $H_0$ is Hermitian and if we take the $\psi_{n,k}$ to be orthogonal to $\phi_n$, we obtain

$$\begin{aligned} E_{n,1} &= \langle \phi_n | H^{(1)} | \phi_n \rangle \\ E_{n,2} &= \langle \phi_n | H^{(2)} | \phi_n \rangle + \langle \phi_n | H^{(1)} | \psi_{n,1} \rangle \\ E_{n,3} &= \langle \phi_n | H^{(3)} | \phi_n \rangle + \langle \phi_n | H^{(2)} | \psi_{n,1} \rangle + \langle \phi_n | H^{(1)} | \psi_{n,2} \rangle \end{aligned} \tag{7-49}$$

by multiplying Eq. (7-48) on the left by $\phi_n^*$ and subsequent integration.

In addition to the methods that we discussed in Section 7-3 there exists another procedure for determining the functions $\psi_{n,k}$. This approach is based on the expansion of the $\psi_{n,k}$ in terms of the complete set of eigenfunctions $\phi_k$. We will show how $\psi_{n,1}$ and $\psi_{n,2}$ are obtained in this way. Again, it is assumed that the state $n$ is nondegenerate.

We expand $\psi_{n,1}$ as

$$\psi_{n,1} = \sum_{k \neq n} a_k \phi_k \tag{7-50}$$

where we can exclude the term $a_n\phi_n$, since $\psi_{n,1}$ has to be orthogonal to $\phi_n$. It should be realized that this expansion should also be extended over the possible continuum states, as we discussed in Section 4-9. We assume that the continuum states are included in our summation sign.

Substitution of this expansion into Eq. (7-48b) gives

$$\sum_{k \neq n} a_k (H_0 - \varepsilon_n)\phi_k = -\left(H^{(1)} - E_{n,1}\right)\phi_n \tag{7-51}$$

Multiplication on the left by $\phi_m^*$ $(m \neq n)$ and subsequent integration yield

$$a_m(\varepsilon_m - \varepsilon_n) = -\left\langle \phi_m | H^{(1)} | \phi_n \right\rangle \tag{7-52}$$

which we abbreviate to

$$a_m = -\frac{H^{(1)}_{m,n}}{\varepsilon_m - \varepsilon_n} \tag{7-53}$$

Hence $\psi_{n,1}$ is obtained as

$$\psi_{n,1} = -\sum_{k \neq n} \frac{H^{(1)}_{k,n}}{\varepsilon_k - \varepsilon_n} \phi_k \tag{7-54}$$

By substituting this expression into Eq. (7-49), we derive an analytical expression for $E_{n,2}$:

$$E_{n,2} = H^{(2)}_{n,n} - \sum_{k \neq n} \frac{H^{(1)}_{n,k} H^{(1)}_{k,n}}{\varepsilon_k - \varepsilon_n} \tag{7-55}$$

containing only the known eigenvalues and eigenfunctions of $H_o$.

In the same way we can substitute the expansion

$$\psi_{n,2} = \sum_{k \neq n} b_k \phi_k \tag{7-56}$$

into Eq. (7-48c), and we find

$$(\varepsilon_k - \varepsilon_n) b_k = -\left\langle \phi_k | H^{(1)} | \psi_{n,1} \right\rangle + E_{n,1} \left\langle \phi_k | \psi_{n,1} \right\rangle - H^{(2)}_{k,n} \tag{7-57}$$

or

$$b_k = \frac{-1}{\varepsilon_k - \varepsilon_n} \left[ H^{(2)}_{k,n} + \frac{H^{(1)}_{n,n} H^{(1)}_{k,n}}{\varepsilon_k - \varepsilon_n} - \sum_{m \neq n} \frac{H^{(1)}_{k,m} H^{(1)}_{m,n}}{\varepsilon_m - \varepsilon_n} \right] \tag{7-58}$$

It can be shown that the energy perturbations $E_{n,2k}$ and $E_{n,2k+1}$ can be evaluated from the wave function perturbation $\psi_{n,k}$. For example, the energy perturbations $E_{n,2}$ and $E_{n,3}$ can be derived from the perturbation $\psi_{n,1}$. The energy perturbation $E_{n,2}$ is given by Eq. (7-49). The energy perturbation $E_{n,3}$ can be expressed in the form

$$E_{n,3}=\langle\phi_n|H^{(3)}|\phi_n\rangle+\langle\psi_{n,1}|H^{(2)}|\phi_n\rangle+\langle\phi_n|H^{(2)}|\psi_{n,1}\rangle$$
$$+\langle\psi_{n,1}|H^{(1)}-E_{n,1}|\psi_{n,1}\rangle \tag{7-59}$$

by repeated substitutions of the perturbation equations (7-48) into Eq. (7-48d).

There is a variety of expressions for the fourth-order energy perturbation $E_{n,4}$. We report only one of them for the situation where only the perturbation term $H'$ is nonzero:

$$E_{n,4}=-\langle\psi_{n,2}|H_o-\varepsilon_n|\psi_{n,2}\rangle-E_{n,2}\langle\psi_{n,1}|\psi_{n,1}\rangle \tag{7-60}$$

Again, this result can be derived by substituting the various perturbation equations into one another. By making use of other transformations we can express $E_{n,4}$ in terms of $\psi_{n,1}$, in terms of the unperturbed eigenvalues and eigenfunctions, and so on.

From the above results it can be seen that the correction terms $E_{n,k}$ and $\psi_{n,k}$ to the eigenvalues and eigenfunctions become increasingly more complicated as we proceed to higher orders. It is generally impractical to consider terms beyond the second-order correction to the energy and beyond the first order to the wave function. The basic assumption of perturbation theory is that $\lambda$ is so small that the various expansions in terms of $\lambda$ are rapidly convergent. If the convergence is so slow that we should consider higher orders, we should not have applied perturbation theory in the first place, but instead we should have looked for different approximate methods to solve our problem.

At first sight it seems that the perturbation treatment of the present section is easier to apply than the method we developed in Section 7-2. In our present description we obtain the $E_{n,k}$ and $\psi_{n,k}$ in closed form, expressed in a straightforward way in terms of a set of integrals that can all be evaluated in principle. In Section 7-2 we had to solve a set of differential equations. However, in practice there is little difference between the two approaches. We should realize that in Section 7-2 we needed to know only the ground-state wave function in order to evaluate the perturbation of the ground state. In the present section we must know the eigenfunctions of all states, and although we know them in principle, this is not quite the same as writing them all down, evaluating an infinite number of integrals, and adding up all the terms. In practice, only the method of Section 7-2 leads to exact perturbation results. The description in this section can be used for purely formal theoretical descriptions and also to obtain rough estimates of perturbation effects without much computational effort.

The latter estimates are obtained by utilizing an interesting summation theorem. If $\Omega$ and $\Lambda$ are two operators, $f$ and $g$ two functions, and $\phi_n$ a complete set of functions, we have

$$\sum_n \langle f|\Omega|\phi_n\rangle\langle\phi_n|\Lambda|g\rangle=\langle f|\Omega\Lambda|g\rangle \tag{7-61}$$

We can prove this by expanding the function $\Lambda g$ in terms of the $\phi_n$:

$$\Lambda g=\Sigma a_n\phi_n \tag{7-62}$$

$$a_n=\langle\phi_n|\Lambda|g\rangle$$

If we now operate on both sides of the first Eq. (7-62) by $\Omega$, we find

$$\Omega\Lambda g=\sum_n \Omega\phi_n\langle\phi_n|\Lambda|g\rangle \tag{7-63}$$

Multiplication on the left by $f^*$ and integration then lead to Eq. (7-61).

As an example, we consider the perturbation of a hydrogen atom in its ground state by a homogeneous electric field $F$. By using atomic units throughout, we write the Hamiltonian as

$$H=H_o+FH'$$

$$H_o=-\frac{1}{2}\Delta-\frac{1}{r} \tag{7-64}$$

$$H'=-z$$

We will show later, in Section 7-6, that

$$E_0=\varepsilon_0+E_{0,1}F+E_{0,2}F^2+\cdots$$

$$E_{0,1}=0 \tag{7-65}$$

$$E_{0,2}=-\frac{9}{4}$$

Let us now try to estimate these energy corrections from Eqs. (7-55) and (7-61). It follows easily that $E_{0,1}=0$, whereas $E_{0,2}$ is given by

$$E_{0,2}=-\sum_{k\neq 0}\frac{\langle\phi_0|H'|\phi_k\rangle\langle\phi_k|H'|\phi_0\rangle}{\varepsilon_k-\varepsilon_0} \tag{7-66}$$

where we have to sum over all excited states of the hydrogen atom. We note that every term of the infinite series is positive, and we replace the sum by

$$\sum_{k\neq 0}\frac{\langle\phi_0|H'|\phi_k\rangle\langle\phi_k|H'|\phi_0\rangle}{\varepsilon_k-\varepsilon_0}=\frac{1}{\Delta E}\sum_k\langle\phi_0|H'|\phi_k\rangle\langle\phi_k|H'|\phi_0\rangle \quad (7\text{-}67)$$

Here we call $\Delta E$ the effective average excitation energy. Although we do not know its exact value, we expect from our physical intuition that its value lies between the first excitation energy and the ionization energy of the hydrogen atom, that is,

$$\frac{3}{8}<\Delta E<\frac{1}{2} \quad (7\text{-}68)$$

The sum on the right side of Eq. (7-67) can now be evaluated by making use of the summation theorem (7-61):

$$\sum_k\langle\phi_0|H'|\phi_k\rangle\langle\phi_k|H'|\phi_0\rangle=\left\langle\phi_0|(H')^2|\phi_0\right\rangle=1 \quad (7\text{-}69)$$

We predict, therefore, that

$$-2>E_{0,2}>-\frac{8}{3} \quad (7\text{-}70)$$

which is in agreement with Eq. (7-65).

## 7-5 Perturbation Theory for a Degenerate State

It is easier to discuss the perturbations of degenerate states on the basis of the variational principle, which we discuss in Section 7-7, but we feel obliged to discuss the theory also from the point of view of perturbation theory. Again we start with a Hamiltonian $H$ that can be expanded as

$$H=H_0+\lambda H^{(1)}+\lambda^2H^{(2)}+\cdots \quad (7\text{-}71)$$

We denote the eigenvalues of $H_0$ by $\varepsilon_k$, and we assume that $\varepsilon_0$ is an $s$-fold degenerate eigenvalue with the orthonormal eigenfunctions $\phi_{0,1},\phi_{0,2},\phi_{0,3},\ldots,\phi_{0,s}$. Naturally any linear combination of the functions $\phi_{0,i}$ is also an eigenfunction of $H_0$ belonging to $\varepsilon_0$.

We now assume that the operator $H$ has one or more eigenvalues $E_0$ that differ from $\varepsilon_0$ by an amount that is of the order of magnitude of $\lambda$. Each of these eigenvalues can then be expanded as a power series in $\lambda$, and a typical term of this kind is

$$E_0=\varepsilon_0+\lambda E_0'+\lambda^2E_0''+\cdots \quad (7\text{-}72)$$

The corresponding eigenfunction can be written as

$$\psi_0 = \Psi_0 + \lambda\psi_0' + \lambda^2\psi_0'' + \cdots \tag{7-73}$$

where $\Psi_0$ is a linear combination of the $\phi_{0,i}$:

$$\Psi_0 = \sum_i c_i \phi_{0,i} \tag{7-74}$$

Our perturbation equations can be derived by considering the coefficients of successive powers of $\lambda$ in the equation

$$\left[(H_0 - \varepsilon_0) + \lambda\left(H^{(1)} - E_0'\right) + \lambda^2\left(H^{(2)} - E_0''\right) + \cdots\right]\left[\Psi_0 + \lambda\psi_0' + \lambda^2\psi_0'' + \cdots\right] = 0 \tag{7-75}$$

The perturbation equation of order zero is

$$(H_0 - \varepsilon_0)\Psi_0 = 0 \tag{7-76}$$

and is always satisfied. The first-order perturbation equation is

$$(H_0 - \varepsilon_0)\psi_0' + \left(H^{(1)} - E_0'\right)\Psi_0 = 0 \tag{7-77}$$

Let us first set out to determine $E_0'$. We multiply Eq. (7-77) on the left by $\phi_{0,j}^*$, and we integrate. Since $H_0$ is Hermitian, we obtain

$$\left\langle \phi_{0,j} | H^{(1)} | \Psi_0 \right\rangle - E_0' \left\langle \phi_{0,j} | \Psi_0 \right\rangle = 0 \tag{7-78}$$

If we now substitute Eq. (7-74) for $\Psi_0$, we obtain the following set of equations for the unknown coefficients $c_i$ and for $E_0'$:

$$\sum_i H_{j,i}^{(1)} c_i - E_0' c_j = 0 \qquad j = 1, 2, 3, \ldots, s \tag{7-79}$$

where

$$H_{j,i}^{(1)} = \left\langle \phi_{0,j} | H^{(1)} | \phi_{0,i} \right\rangle \tag{7-80}$$

We recognize this as the eigenvalue problem of the matrix $[H_{j,i}^{(1)}]$. Let the eigenvalues of this matrix be $E_{0,1}', E_{0,2}', \ldots, E_{0,s}'$ and the corresponding eigenvectors $\mathbf{c}^k$. Then the first-order approximation to the eigenvalues close to $\varepsilon_0$ is

$$E_{0,k} = \varepsilon_0 + \lambda E_{0,k}' \qquad k = 1, 2, \ldots, s \tag{7-81}$$

and the corresponding eigenfunctions, in zero-order approximation, are

$$\psi_{0,k} = \Psi_{0,k} = \sum_i c_i^k \phi_{0,i} \tag{7-82}$$

There are now two different situations to be considered. In the first case, none of the eigenvalues $E'_{0,k}$ is degenerate, and in the second, there are one or more degeneracies. We do not discuss the second case here since it will be dealt with in Section 7-7. Therefore, it is assumed that none of the eigenvalues $E'_{0,k}$ of the matrix $[H_{i,j}^{(1)}]$ is degenerate.

In order to determine the higher-order perturbation corrections, we employ now the expansions

$$E_{0,k} = \varepsilon_0 + \lambda E'_{0,k} + \lambda^2 E''_{0,k} + \cdots \tag{7-83}$$

$$\psi_{0,k} = \Psi_{0,k} + \lambda \psi'_{0,k} + \lambda^2 \psi''_{0,k} + \cdots$$

instead of Eqs. (7-72) and (7-73). After substituting these new expansions into the Schrödinger equation, we obtain, for the first-order perturbation equation,

$$(H_0 - \varepsilon_0)\psi'_{0,k} + (H^{(1)} - E'_{0,k})\Psi_{0,k} = 0 \tag{7-84}$$

For the unknown function $\psi'_{0,k}$ we substitute the expansion

$$\psi'_{0,k} = \sum_{n \neq 0} a_{k,n}\phi_n + \sum_{l \neq k} b_{k,l}\Psi_{0,l} \tag{7-85}$$

We have assumed that the functions $\phi_{0,k}$ form an orthonormal set, and consequently the $\Psi_{0,k}$ are also orthonormal. For convenience sake, we now assume that $\psi'_{0,k}$ is orthogonal to $\Psi_{0,k}$, and we omit, therefore, the term $l = k$ in the second summation of Eq. (7-85). The result of the substitution is

$$\sum_{n \neq 0} a_{k,n}(H_0 - \varepsilon_0)\phi_n + \sum_{l \neq k} b_{k,l}(H_0 - \varepsilon_0)\Psi_{0,l} + (H^{(1)} - E'_{0,k})\Psi_{0,k} = 0 \tag{7-86}$$

By multiplying on the left by $\phi_m^*$ and integrating, we find that

$$(\varepsilon_m - \varepsilon_0)a_{k,m} + \langle \phi_m | H^{(1)} | \Psi_{0,k} \rangle = 0 \tag{7-87}$$

or

$$a_{k,m} = -\frac{\langle \phi_m | H^{(1)} | \Psi_{0,k} \rangle}{\varepsilon_m - \varepsilon_0} \tag{7-88}$$

It is not possible to derive the values for the expansion coefficients $b_{k,l}$ from Eq. (7-84), and we proceed, therefore, to the next perturbation equation.

By substituting the expansions (7-83) into the Schrödinger equation, we find that the second perturbation equation for the state $(0, k)$ is

$$(H^{(0)} - \varepsilon_0)\psi''_{0,k} + (H^{(1)} - E'_{0,k})\psi'_{0,k} + (H^{(2)} - E''_{0,k})\Psi_{0,k} = 0 \tag{7-89}$$

First we multiply this equation on the left by $\Psi_{0,l}^*$, where $l \neq k$, and integrate to give

$$\left\langle \Psi_{0,l} | H^{(1)} - E'_{0,k} | \psi'_{0,k} \right\rangle + \left\langle \Psi_{0,l} | H^{(2)} | \Psi_{0,k} \right\rangle = 0 \tag{7-90}$$

Next we substitute the expansion (7-85) for $\psi'_{0,k}$ and obtain

$$\sum_{n \neq 0} \left\langle \Psi_{0,l} | H^{(1)} | \phi_n \right\rangle a_{k,n} + \sum_{j \neq k} \left\langle \Psi_{0,l} | H^{(1)} - E'_{0,k} | \Psi_{0,j} \right\rangle b_{k,j} + \left\langle \Psi_{0,l} | H^{(2)} | \Psi_{0,k} \right\rangle = 0 \tag{7-91}$$

Since

$$\left\langle \Psi_{0,l} | H^{(1)} | \Psi_{0,j} \right\rangle = E'_{0,l} \delta_{l,j} \tag{7-92}$$

we find that

$$(E'_{0,l} - E'_{0,k}) b_{k,l} = -\left\langle \Psi_{0,l} | H^{(2)} | \Psi_{0,k} \right\rangle - \sum_{n \neq 0} \left\langle \Psi_{0,l} | H^{(1)} | \phi_n \right\rangle a_{k,n} \tag{7-93}$$

or, using Eq. (7-88),

$$b_{k,l} = -\frac{\left\langle \Psi_{0,l} | H^{(2)} | \Psi_{0,k} \right\rangle}{E'_{0,l} - E'_{0,k}} + \sum_{n \neq 0} \frac{\left\langle \Psi_{0,l} | H^{(1)} | \phi_n \right\rangle \left\langle \phi_n | H^{(1)} | \Psi_{0,k} \right\rangle}{(\varepsilon_n - \varepsilon_0)(E'_{0,l} - E'_{0,k})} \tag{7-94}$$

The function $\psi'_{0,k}$ is now completely determined by Eqs. (7-85), (7-88), and (7-94).

The second-order energy correction $E''_{0,k}$ can be derived by multiplying Eq. (7-89) on the left by $\Psi_{0,k}^*$ and by subsequent integration:

$$E''_{0,k} = \left\langle \Psi_{0,k} | H^{(2)} | \Psi_{0,k} \right\rangle + \left\langle \Psi_{0,k} | H^{(1)} - E'_{0,k} | \psi'_{0,k} \right\rangle \tag{7-95}$$

If we replace $\psi'_{0,k}$ by the expansion (7-85), we find that

$$E''_{0,k} = \left\langle \Psi_{0,k} | H^{(2)} | \Psi_{0,k} \right\rangle + \sum_{n \neq 0} \left\langle \Psi_{0,k} | H^{(1)} | \phi_n \right\rangle a_{k,n} \tag{7-96}$$

or

$$E''_{0,k} = \left\langle \Psi_{0,k} | H^{(2)} | \Psi_{0,k} \right\rangle - \sum_{n \neq 0} \frac{\left\langle \Psi_{0,k} | H^{(1)} | \phi_n \right\rangle \left\langle \phi_n | H^{(1)} | \Psi_{0,k} \right\rangle}{\varepsilon_n - \varepsilon_0} \tag{7-97}$$

## 7-6 Perturbation Methods for the Hydrogen Atom

What we have discussed so far is known as the Rayleigh-Schrödinger perturbation theory. A similar approach is the Brillouin-Wigner perturbation theory, but we postpone its discussion until later since it is most conveniently derived from the variational principle. In fact, most of perturbation theory can be derived from the variational principle, and we discuss various variational perturbation methods at the end of this chapter. In the previous sections, we presented various formal solutions for the perturbation equations. It is necessary to know these formal theories, but if we seek to obtain numerical results for practical problems, then we realize very soon that our formal perturbation expressions are inadequate in many cases. Let us consider, for example, the problem of determining the effect of a homogeneous electric field on an atom or molecule. In general, we do not even know the eigenvalue and eigenfunction of the molecular ground state, let alone the eigenfunctions of the excited states. In order to deal with such cases, we must devise perturbation procedures involving approximate eigenfunctions.

Even in situations where we know the unperturbed eigenvalues and eigenfunctions, we encounter difficulties when using Rayleigh-Schrödinger perturbation theory. We have to evaluate and sum an infinite number of integrals in order to obtain the energy corrections. Often the continuum states contribute significantly and it becomes exceedingly difficult to evaluate, for example, the term $E_{0,2}$. In such cases, it seems to be more effective to try and solve the first-order perturbation equation directly. We consider the ground state 0 of the system. Then the perturbation equation (7-7) becomes

$$(H_0-\varepsilon_0)\psi_{0,1}+(H'-E_{0,1})\phi_0=0 \tag{7-98}$$

We can attempt to solve this equation by assuming that the perturbation function $\psi_{0,1}$ can be written as

$$\psi_{0,1}=F\phi_0 \tag{7-99}$$

where $F$ is a scalar function of the variables that occur in $H$. This approach was known in the theory of differential equations, and it was successfully applied by Dalgarno and various co-workers during the fifties. The most obvious situation for using this method is in perturbation treatments of the hydrogen atom, hence the title of this section.

If we substitute Eq. (7-99) into Eq. (7-98), we can write the first term as

$$(H_0-\varepsilon_0)F\phi_0=(H_0F-FH_0)\phi_0=[H_0, F]\phi_0 \tag{7-100}$$

We simplify the second term of Eq. (7-98) by introducing the reduced potential $V'$,

$$V'=H'-E_{0,1}=H'-\langle\phi_0|H'|\phi_0\rangle \tag{7-101}$$

The reduced potential $V'$ is easily derived from the unperturbed eigenfunction $\phi_0$. The perturbation equation (7-98) now takes the form

$$[H_0, F]\phi_0 + V'\phi_0 = 0 \tag{7-102}$$

In order to simplify this equation further, we note that the operator $H_0$ has the form

$$H_0 = -\frac{1}{2}\sum_i \nabla_i^2 + U \tag{7-103}$$

where the subscript $i$ can be equal to $x$ in one-dimensional problems, equal to $x$ and $y$ in two dimensions, and to $x$, $y$, and $z$ in three dimensions. Since $U$ and $F$ commute we have

$$[H_0, F] = -\frac{1}{2}\sum_i [\nabla_i^2, F] \tag{7-104}$$

For simplicity, we assume that we are dealing with a three-dimensional case. Substitution of Eq. (7-104) into (7-102) gives then

$$\phi_0(\Delta F) + 2(\nabla F \cdot \nabla \phi_0) = 2V'\phi_0 \tag{7-105}$$

Finally, we divide the equation by $\phi_0$ and we obtain

$$\Delta F + 2\nabla F \cdot \nabla(\log \phi_0) = 2V' \tag{7-106}$$

Here $\Delta$ is the Laplace operator,

$$\Delta = \nabla \cdot \nabla \tag{7-107}$$

In order to illustrate the method described above, we apply it to the Stark effect of the ground state of the hydrogen atom. Here the total Hamiltonian is

$$H = -\frac{1}{2}\Delta - \frac{1}{r} - \lambda z \tag{7-108}$$

if we use atomic units throughout and if we denote the electric field strength by $\lambda$. Consequently

$$H' = -z \tag{7-109}$$

We know that the ground-state wave function of the hydrogen atom is

$$\phi_0 = \pi^{-1/2} e^{-r} \tag{7-110}$$

It easily follows that

$$E_{0,1} = \langle \phi_0 | H' | \phi_0 \rangle = 0 \tag{7-111}$$

so that

$$V' = H' = -z \tag{7-112}$$

We also have

$$\nabla \log \phi_0 \cdot \nabla F = -\frac{\partial r}{\partial x}\frac{\partial F}{\partial x} - \frac{\partial r}{\partial y}\frac{\partial F}{\partial y} - \frac{\partial r}{\partial z}\frac{\partial F}{\partial z}$$

$$= -\frac{x}{r}\frac{\partial F}{\partial x} - \frac{y}{r}\frac{\partial F}{\partial y} - \frac{z}{r}\frac{\partial F}{\partial z} \tag{7-113}$$

We define the operators

$$\Omega = \frac{x}{r}\frac{\partial}{\partial x} + \frac{y}{r}\frac{\partial}{\partial y} + \frac{z}{r}\frac{\partial}{\partial z} \tag{7-114}$$

and

$$\Delta = \frac{\partial^2}{\partial x^2} + \frac{\partial^2}{\partial y^2} + \frac{\partial^2}{\partial z^2} \tag{7-115}$$

The perturbation equation (7-106) can then be written in the form

$$\Delta F - 2\Omega F = -2z \tag{7-116}$$

In order to solve the equation we note that

$$\Delta(z) = 0 \qquad \Delta(zr) = \frac{4z}{r}$$

$$\Omega(z) = \frac{z}{r} \qquad \Omega(zr) = 2z \tag{7-117}$$

Hence we substitute

$$F = az + bzr \tag{7-118}$$

into Eq. (7-116), and we find

$$(2a - 4b)\frac{z}{r} + 4bz = 2z \tag{7-119}$$

This equation is satisfied if $a=1$ and $b=\frac{1}{2}$, and the solution of Eq. (7-116) is

$$F = z(1 + r/2) \tag{7-120}$$

It follows from Eq. (7-17) that $E_{0,2}$ can be written as

$$E_{0,2} = \langle \phi_0 | H' | F\phi_0 \rangle \tag{7-121}$$

and we find

$$E_{0,2}=-\frac{9}{4} \tag{7-122}$$

This is the result that we used in Eq. (7-65).

If we consider a one-dimensional case, Eq. (7-105) can be integrated directly. If we multiply both sides of the equation by $\phi_0$, we can rewrite it as

$$\frac{d}{dx}\left(\phi_0^2\frac{dF}{dx}\right)=2\phi_0 V\phi_0 \tag{7-123}$$

Clearly the function $F(x)$ can now be obtained by integrating twice.

At this point we interrupt our presentation of the various perturbation theories to discuss the variational principle. We mentioned already that various aspects of perturbation theory, in particular perturbation theory involving approximate eigenfunctions, are more effectively derived by using variational techniques. On the other hand, we wished to present formal perturbation theory first in order to have a reference framework for comparing more approximate methods.

## 7-7 The Variational Principle

Before we discuss the variational principle, we wish to recall our description of matrix mechanics in Chapter 2 and our historical comments in Section 3-2. We mentioned there how it has been shown that matrix mechanics and the eigenvalue formalism of Schrödinger are formally equivalent. This equivalence can be proved also by means of a variational description. The mathematical formalism of these variational methods was already presented in the book by Courant and Hilbert in 1925. In addition to this important role in the theory of quantum mechanics, the variational principle is important also as a basis for various approximate methods for deriving eigenvalues and eigenfunctions. We first discuss the variational theorem itself and then we prove various inequalities that are closely related.

Let us first investigate what is meant by an infinitesimal variation of a function $f$ of a number of variables. As an example, we take the function

$$f(x)=(a_1+a_2x)\exp\left(-\frac{a_3x^2}{\left(a_4+a_5x^2\right)^{1/2}}\right) \tag{7-124}$$

If any of the parameters $a_k$ is varied by a small amount $\delta a_k$, we get a slightly different function $f(x)+\delta f(x)$, and we can write

$$\delta f=\sum_k \frac{\partial f}{\partial a_k}\delta a_k \tag{7-125}$$

However, there are many more ways of obtaining a slightly different function. For example, if we replace the term $a_3x^2$ by $a_3x^2+\delta a_6$, we also get a new function $f+\delta f$. In general, there are an infinite number of ways to get a different function $f+\delta f$. We can imagine that we expand the function $f(x)$ in terms of a complete set of functions $\phi_n$:

$$f(x)=\sum_n c_n\phi_n(x) \tag{7-126}$$

If we replace any of the coefficients $c_n$ by $c_n+\delta c_n$, we obtain a variation of the function $f(x)$ by an amount $\delta f$. Hence if we talk about all possible variations $\delta f$ of a function $f$, we should recognize that there are an infinite number of them.

We now consider a Hermitian operator $H$ and an arbitrary function $f$, and we construct the two integrals

$$E=\langle f|H|f\rangle \qquad S=\langle f|f\rangle \tag{7-127}$$

An arbitrary variation $\delta f$ in the function $f$ leads to a variation $\delta E$ in $E$ and a variation $\delta S$ in $S$:

$$\begin{aligned}\delta E&=\langle \delta f|H|f\rangle+\langle f|H|\delta f\rangle\\ \delta S&=\langle \delta f|f\rangle+\langle f|\delta f\rangle\end{aligned} \tag{7-128}$$

We now impose the condition that $\delta E=0$ for all possible variations $\delta f$ for which $\delta S=0$. This same condition can be formulated in a different way if we impose the condition that $\delta(E/S)=0$ for all possible variations $\delta f$. We write this condition as

$$\frac{E+\delta E}{S+\delta S}-\frac{E}{S}=0 \tag{7-129}$$

for all possible variations $\delta f$. By expanding we can write Eq. (7-129) as

$$\left(\frac{1}{S}\right)(\delta E-\lambda\,\delta S)=0$$

$$\lambda=\frac{E}{S} \tag{7-130}$$

By substituting Eq. (7-128) we obtain

$$\langle \delta f|H-\lambda|f\rangle+\langle f|H-\lambda|\delta f\rangle=0 \tag{7-131}$$

Since $H$ is Hermitian and $\lambda$ is real, we write Eq. (7-131) as

$$\langle \delta f|H-\lambda|f\rangle=0 \tag{7-132}$$

This equation can be satisfied for all possible variations $\delta f$ only if

$$(H-\lambda)f=0 \tag{7-133}$$

which means that $f$ is an eigenfunction of $H$ with eigenvalue $\lambda$. We find thus that $f$ is an eigenfunction of $H$ if $\delta(E/S)=0$ for all possible variations $\delta f$ of the function $f$.

The variation theorem above is in every respect equivalent to Heisenberg's matrix mechanics, which we discussed in Section 2-9. It also offers a matrix representation of the Schrödinger equation and it presents us with a way to demonstrate the equivalence between the Schrödinger equation and the matrix representation.

If we have a complete set of functions $\phi_n$, then any function $f$ can be expanded as

$$f=\Sigma_n c_n \phi_n \tag{7-134}$$

and any variation $\delta f$ in $f$ is obtained by varying the expansion coefficients $c_n$. The integrals $E$ and $S$ can now be reduced to

$$E=\langle f|H|f\rangle=\sum_n \sum_m c_n c_m^* H_{m,n}$$

$$S=\langle f|f\rangle=\sum_n \sum_m c_n c_m^* S_{m,n} \tag{7-135}$$

where

$$H_{m,n}=\langle \phi_m|H|\phi_n\rangle=H_{n,m}^*$$

$$S_{m,n}=\langle \phi_m|\phi_n\rangle=\delta_{m,n} \tag{7-136}$$

The variational problem is

$$\delta E-\lambda\delta S=\delta\sum_n \sum_m c_n c_m^*(H_{m,n}-\lambda\delta_{m,n})=0 \tag{7-137}$$

The variation is zero if the derivatives with respect to the parameter $c_n$ are zero. If we differentiate Eq. (7-137) with respect to the real and imaginary parts of $c_n$, we find that the variational problem is equivalent to the sets of equations

$$\sum_m (H_{n,m}-\lambda\delta_{n,m})c_m=0$$

$$\sum_m (H_{m,n}-\lambda\delta_{m,n})c_m^*=0 \tag{7-138}$$

The second set is the complex conjugate of the first set, and we can disregard it. The first set of equations represents the eigenvalue problem of the matrix $[H_{n,m}]$.

It follows that, in principle, the variation theorem above does not teach us anything that we did not know before. However, we can regard it as a starting point for deriving several useful approaches to solve the Schrödinger equation approximately. In practice, we can attempt to approximate some of the eigenfunctions of $H$ by the expansions

$$f_N(x) = \sum_{n=1}^{N} c_n \phi_n(x) \tag{7-139}$$

where $\phi_1, \phi_2, \ldots, \phi_N$ are the first $N$ functions of the complete set of functions $\phi_n$. We construct the integrals

$$E_N = \langle f_N | H | f_N \rangle \qquad S_N = \langle f_N | f_N \rangle \tag{7-140}$$

and we solve the variational problem

$$\delta E_N - \lambda \delta S_N = 0 \tag{7-141}$$

This supplies us with $N$ eigenvalues of the matrix $[H_{m,n}]$ of order $N$ and their corresponding eigenvectors. If $N$ tends to infinity, then in the limit we obtain the exact eigenvalues and eigenfunctions of $H$. The interesting question is how the eigenvalues and eigenfunctions of the truncated Eq. (7-141) compare with the true eigenvalues and eigenfunctions of the exact equation. The answer may be found in a number of theorems concerning upper and lower bounds of eigenvalues. These theorems are sometimes called variational theorems because they are to a certain extent related to the variational principle that we discussed above. We discuss the various inequalities in the following section.

## 7-8 Upper and Lower Bounds of Eigenvalues

One of the most useful theorems in quantum theory is the following. Let $H$ be a Hermitian operator with the eigenvalues $\varepsilon_n$ and the corresponding eigenfunctions $\phi_n$, and let $\varepsilon_1$ be the smallest of the eigenvalues. Then if $f$ is an arbitrary function, we have

$$\frac{\langle f | H | f \rangle}{\langle f | f \rangle} \geq \varepsilon_1 \tag{7-142}$$

The equality in Eq. (7-142) holds only when $f$ is identical with $\phi_1$. In order to prove this inequality, we imagine that $f$ is expanded in terms of the $\phi_n$:

$$f = \Sigma c_n \phi_n \tag{7-143}$$

It can then be derived that

$$\langle f|H-\varepsilon_1|f\rangle = \sum_n \sum_m c_n c_m^* \langle \phi_m|H-\varepsilon_1|\phi_n\rangle = \sum_n \sum_m c_m^* c_n (\varepsilon_n - \varepsilon_1)\delta_{m,n}$$

$$= \sum_n c_n c_n^* (\varepsilon_n - \varepsilon_1) \tag{7-144}$$

The result is positive or zero, since every term in the infinite series is nonnegative. Hence we have

$$\langle f|H-\varepsilon_1|f\rangle \geq 0 \tag{7-145}$$

or

$$\langle f|H|f\rangle \geq \varepsilon_1 \langle f|f\rangle \tag{7-146}$$

In practice, this theorem has proved to be very useful for deriving approximate eigenvalues and eigenfunctions for the ground states of various systems. It has been widely applied in quantum chemistry, and in reading various books on molecular orbital calculations, one gets the impression that it has almost replaced the Schrödinger equation as the basic equation of quantum theory. We assume that we have a Hermitian operator $H$ and that we have a rough idea of what its ground state eigenfunction $\phi_1$ looks like. We now construct a function $\Phi(x; s_1, s_2, \dots, s_N)$ that contains, in addition to the particle coordinates, which we symbolically denote by $x$, a number of arbitrary parameters $s_i$. The expectation value

$$E(s_1, s_2, \dots, s_N) = \frac{\langle \Phi|H|\Phi\rangle}{\langle \Phi|\Phi\rangle} \tag{7-147}$$

is then always larger than or equal to the lowest eigenvalue $\varepsilon_1$ of $H$. We now determine the set of values $s_i^o$ for which $E(s_1, s_2, \dots, s_N)$ has a minimum by solving the equations

$$\frac{\partial E(s_1, s_2, \dots, s_N)}{\partial s_i} = 0 \tag{7-148}$$

The energy $E_0 = E(s_1^o, s_2^o, \dots, s_N^o)$ and the function $\Phi_0 = \Phi(x; s_1^o, s_2^o, \dots, s_N^o)$ are then the best possible approximation to $\varepsilon_1$ and its eigenfunction $\phi_1$ that can be obtained from the function $\Phi$.

The essential feature of this procedure is the choice of the function $\Phi$, and this choice is a matter of physical or chemical intuition. In many cases this is a difficult decision to make, but sometimes the choice is obvious. As an example, we discuss the harmonic oscillator in a homogeneous electric field. This problem was treated in Section 7-3 from the viewpoint of perturbation theory.

According to Eq. (7-24) we can represent the Hamiltonian of a harmonic oscillator in an homogeneous electric field $F$ as

$$H=-\frac{d^2}{dx^2}+x^2-2\mu x \tag{7-149}$$

if we introduce the proper units. Since the ground-state wave function in the absence of an electric field is $\exp(-x^2/2)$, we take our variational function as

$$\Phi=(1+sx)e^{-x^2/2} \tag{7-150}$$

It is easily verified that

$$\langle\Phi|\Phi\rangle=\int_{-\infty}^{\infty}e^{-x^2}dx+s^2\int_{-\infty}^{\infty}x^2e^{-x^2}dx=\sqrt{\pi}\left(1+\frac{s^2}{2}\right) \tag{7-151}$$

From Eq. (7-25) we derive that

$$\left(-\frac{d^2}{dx^2}+x^2\right)e^{-x^2/2}=e^{-x^2/2}$$

$$\left(-\frac{d^2}{dx^2}+x^2\right)xe^{-x^2/2}=3xe^{-x^2/2} \tag{7-152}$$

Hence

$$\begin{aligned}\langle\Phi|H|\Phi\rangle&=\int_{-\infty}^{\infty}(1+sx)(1-2\mu x)e^{-x^2}dx+s\int_{-\infty}^{\infty}(1+sx)(3-2\mu x)xe^{-x^2}dx\\&=\int_{-\infty}^{\infty}e^{-x^2}dx+(3s^2-4\mu s)\int_{-\infty}^{\infty}x^2e^{-x^2}dx\\&=\frac{1}{2}\sqrt{\pi}\,(2+3s^2-4\mu s)\end{aligned} \tag{7-153}$$

We obtain, therefore,

$$E(s)=\frac{\langle\Phi|H|\Phi\rangle}{\langle\Phi|\Phi\rangle}=\frac{2+3s^2-4\mu s}{2+s^2} \tag{7-154}$$

This has a minimum for

$$s^o=\frac{1}{\mu}\left(\sqrt{1+2\mu^2}-1\right) \tag{7-155}$$

and the minimum is

$$E(s^o)=\frac{(3+6\mu^2)-(3+2\mu^2)\sqrt{1+2\mu^2}}{1+2\mu^2-\sqrt{1+2\mu^2}} \tag{7-156}$$

For small values of $\mu$ we can expand $E(s^o)$ as

$$E(s^o)\approx\frac{\mu^2-\dfrac{\mu^4}{2}}{\mu^2+\dfrac{\mu^4}{2}}\approx 1-\mu^2 \tag{7-157}$$

and $\Phi(s^o)$ as

$$\Phi(s^o)\simeq(1+\mu x)e^{-x^2/2} \tag{7-158}$$

These results are in agreement with Eq. (7-39).

From the inequality (7-142) we can also make some predictions about the results of a variational treatment with a limited set of functions as a basis. Let $H$ be a Hermitian operator with eigenvalues $\varepsilon_n$ and corresponding eigenfunctions $\phi_n$, and let $\psi_1, \psi_2, \ldots, \psi_n, \ldots,$ and so on be a complete set of functions. Now we construct the integrals

$$\begin{aligned} H_N &= \langle g_N|H|g_N\rangle \\ S_N &= \langle g_N|g_N\rangle \end{aligned} \tag{7-159}$$

where $g_N$ is the expansion

$$g_N=\sum_{n=1}^{N} c_n^N \psi_n \tag{7-160}$$

The solution of the variational problem

$$\delta(H_N-\lambda S_N)=0 \tag{7-161}$$

consists now of a set of $N$ eigenvalues $\lambda_1^N, \lambda_2^N, \ldots, \lambda_n^N$ with a corresponding set of eigenvectors $a_k^N$ from which we can construct the functions

$$f_k^N=\sum_{n=1}^{N} a_{k,n}^N \psi_n \tag{7-162}$$

We assume that the eigenvalues $\lambda_k$ and also the eigenvalues $\varepsilon_n$ are arranged in ascending order, so that

$$\lambda_1^N \leq \lambda_2^N \leq \lambda_3^N \leq \lambda_4^N \leq \cdots \leq \lambda_N^N$$

$$\varepsilon_1 \leq \varepsilon_2 \leq \varepsilon_2 \leq \varepsilon_4 < \cdots \tag{7-163}$$

It can be derived from Eq. (7-142) that

$$\lambda_1^N \geq \varepsilon_1 \tag{7-164}$$

Furthermore, we found in Section 7-7 that

$$\lim_{N \to \infty} \lambda_1^N = \varepsilon_1 \tag{7-165}$$

It has also been shown that for the other eigenvalues we have the inequalities

$$\lambda_2^N \geq \varepsilon_2, \lambda_3^N \geq \varepsilon_3, \ldots, \lambda_N^N \geq \varepsilon_N \tag{7-166}$$

The proof is rather complicated, and we do not attempt to reproduce it here.

It seems reasonable to conclude from the relations above that the difference between $\lambda_k^N$ and $\varepsilon_k$ becomes smaller with increasing $N$. However, it is quite another matter to predict exactly what the magnitudes of these differences are. This question has been studied extensively, but the results of these investigations cannot be expressed in any simple form.

Heretofore we have made predictions only about the upper bounds of the eigenvalues. There are also some theorems that are concerned with the lower bounds. In general, they are not very useful, and consequently they are not well known. As an illustration, we derive one of them. If the function $f$ is again given by Eq. (7-143), it follows readily that

$$\langle f|(H-\varepsilon_1)(H-\varepsilon_2)|f\rangle = \sum_n \sum_m c_n^* c_m \langle \phi_n|(H-\varepsilon_1)(H-\varepsilon_2)|\phi_m\rangle$$

$$= \sum_n (\varepsilon_n - \varepsilon_1)(\varepsilon_n - \varepsilon_2) c_n c_n^* \geq 0 \tag{7-167}$$

or

$$\langle f|H^2|f\rangle - (\varepsilon_1 + \varepsilon_2)\langle f|H|f\rangle + \varepsilon_1\varepsilon_2\langle f|f\rangle \geq 0 \tag{7-168}$$

We can write this as

$$\varepsilon_1[\langle f|H|f\rangle - \varepsilon_2\langle f|f\rangle] \leq [\langle f|H^2|f\rangle - \varepsilon_2\langle f|H|f\rangle] \tag{7-169}$$

Presumably $f$ is a reasonably good approximation to the ground-state eigenfunction of $H$, so that we can assume that

$$\langle f|H|f\rangle < \langle f|f\rangle \varepsilon_2 \tag{7-170}$$

We then find that

$$\varepsilon_1 \geq \frac{\varepsilon_2\langle f|H|f\rangle - \langle f|H^2|f\rangle}{\varepsilon_2\langle f|f\rangle - \langle f|H|f\rangle} \tag{7-171}$$

If we combine this with Eq. (7-142), we obtain

$$\frac{\varepsilon_2\langle f|H|f\rangle - \langle f|H^2|f\rangle}{\varepsilon_2\langle f|f\rangle - \langle f|H|f\rangle} \leq \varepsilon_1 \leq \frac{\langle f|H|f\rangle}{\langle f|f\rangle} \tag{7-172}$$

and we have obtained both an upper and a lower bound for $\varepsilon_1$. In practice, however, it is rather difficult to evaluate the lower bound, and predictions of lower bounds have not proved to be very useful.

## 7-9 Variational Procedures for Solving Perturbation Equations

We mentioned in Section 7-6 that the perturbation expansions of Section 7-5 are not always convenient for obtaining numerical results for atomic or molecular systems. More convenient expressions are obtained if the perturbation equations can be solved directly. For example, the second-order energy perturbation $E_{0,2}$ is evaluated by direct integration if we know the eigenfunction perturbation $\psi_{0,1}$, which is the solution of the equation

$$(H_0 - \varepsilon_0)\psi_{0,1} = (E_{0,1} - H')\phi_0 \tag{7-173}$$

In Section 7-6 we discussed one possible procedure for deriving the solution of the first-order perturbation equation above. We now derive an alternative, variational procedure for obtaining the function $\psi_{0,1}$

According to the inequality (7-142) of the previous section we have

$$\langle g - \psi_{0,1}|H_0 - \varepsilon_0|g - \psi_{0,1}\rangle \geq 0 \tag{7-174}$$

where $g$ is an arbitrary function and $\psi_{0,1}$ is a solution of Eq. (7-173). We can transform this expression into

$$\begin{aligned}
\langle g - \psi_{0,1}|H_0 - \varepsilon_0|g - \psi_{0,1}\rangle \\
&= \langle g|H_0 - \varepsilon_0|g\rangle - \langle g|H_0 - \varepsilon_0|\psi_{0,1}\rangle - \langle \psi_{0,1}|H_0 - \varepsilon_0|g\rangle \\
&\quad + \langle \psi_{0,1}|H_0 - \varepsilon_0|\psi_{0,1}\rangle \\
&= \langle g|H - \varepsilon_0|g\rangle + \langle g|H' - E_{0,1}|\phi_0\rangle + \langle \phi_0|H' - E_{0,1}|g\rangle \\
&\quad + \langle \psi_{0,1}|H_0 - \varepsilon_0|\psi_{0,1}\rangle
\end{aligned} \tag{7-175}$$

Let us now consider the second-order perturbation equation

$$(H_0-\varepsilon_0)\psi_{0,2}=(E_{0,1}-H')\psi_{0,1}+E_{0,2}\phi_0 \tag{7-176}$$

It is easily derived that

$$\begin{aligned}E_{0,2}&=-\langle\psi_{0,1}|H_0-\varepsilon_0|\psi_{0,1}\rangle\\&=-\sum_{n\neq 0}(\varepsilon_n-\varepsilon_0)^{-1}\langle\phi_0|H'|\phi_n\rangle\langle\phi_n|H'|\phi_0\rangle\end{aligned} \tag{7-177}$$

By substituting this into Eq. (7-175) we obtain

$$\begin{aligned}&\langle g|H_0-\varepsilon_0|g\rangle+\langle g|H'-E_{0,1}|\phi_0\rangle+\langle\phi_0|H'-E_{0,1}|g\rangle\\&\qquad=E_{0,2}+\langle g-\psi_{0,1}|H_0-\varepsilon_0|g-\psi_{0,1}\rangle\geq E_{0,2}\end{aligned} \tag{7-178}$$

Obviously, by minimizing the left side of Eq. (7-178) with respect to the variational function $g$, we obtain an approximate solution to the first-order perturbation equation (7-173). The minimum is always larger than the exact perturbation energy $E_{0,2}$ and, in the limit, approaches the exact value of $E_{0,2}$.

The advantage of the perturbation-variation procedure described above is that it can be generalized to situations where the exact ground-state eigenfunction $\phi_0$ is not known. Let us consider a situation where we only know an approximate ground-state eigenfunction $f_0$ of the operator $H_0$. We define the approximate ground-state energy $E_0$ as

$$E_0=\langle f_0|H_0|f_0\rangle \tag{7-179}$$

and we consider the equation

$$(H_0-E_0)f_0'=(\varepsilon_0'-H')f_0 \tag{7-180}$$

By analogy with Eqs. (7-175) and (7-178) we can derive that

$$\begin{aligned}&\langle g|H_0-E_0|g\rangle+\langle g|H'-\varepsilon_0'|f_0\rangle+\langle f_0|H'-\varepsilon_0'|g\rangle\\&\qquad=\varepsilon_0''+\langle g-f_0'|H_0-E_0|g-f_0'\rangle\geq\varepsilon_0''\end{aligned} \tag{7-181}$$

where $\varepsilon_0''$ is defined as

$$\varepsilon_0''=-\sum_{n\neq 0}(\varepsilon_n-E_0)^{-1}\langle f_0|H'-\varepsilon_0'|\phi_n\rangle\langle\phi_n|H'-\varepsilon_0'|f_0\rangle \tag{7-182}$$

It is advisable to impose the condition

$$\langle g|f_0\rangle=0 \tag{7-183}$$

on the variational function $g$. In that case, the inequality (7-181) reduces to

$$\langle g|H_0-E_0|g\rangle+\langle g|H'|f_0\rangle+\langle f_0|H'|g\rangle\geq\varepsilon_0'' \qquad (7\text{-}184)$$

By minimizing the left side of Eq. (7-184) with respect to the variational function $g$, subject to the condition (7-183), we obtain the approximate second-order energy perturbation. The accuracy of the result depends, of course, on the accuracy of the approximate eigenfunction $f_0$ and on the amount of effort we invest in minimizing the left side of Eq. (7-184). In many atomic and molecular problems, the above procedure offers a relatively convenient method for calculating second-order energy perturbations. At least the method is often more convenient than the various expansions that we described in Section 7-4.

If we have obtained an approximate solution $g_0$ to the perturbation equation (7-180) by minimizing the left side of Eq. (7-184), we can use this solution to evaluate the expectation value (or first-order energy perturbation):

$$E_{0,1}=\langle\phi_0|H'|\phi_0\rangle \qquad (7\text{-}185)$$

It can be shown that the expression

$$E_0'=\langle f_0|H'|f_0\rangle+\langle f_0|H_0-E_0|g_0\rangle+\langle g_0|H_0-E_0|f_0\rangle \qquad (7\text{-}186)$$

is a much better approximation to the exact expression $E_{0,1}$ than the first term on the right side of Eq. (7-186) by itself.

The variational procedures can also be used to derive approximate solutions to the second-order perturbation equation (7-176) and to deal with situations where the exact solutions $\psi_{0,1}$ and $\phi_0$ are not known. The various mathematical procedures are similar to the methods we described for solving the first-order perturbation equations. The advantages of these procedures are that the various variational problems that lead to the solutions of Eqs. (7-173) and (7-176) are similar to the variational methods that are used to derive the approximate ground-state eigenfunctions and that the mathematical techniques for solving these variational problems are fairly well known. The procedures seem to be particularly well suited for calculating molecular electric and magnetic susceptibilities and related quantities, and they have been used in many calculations of this nature.

## 7-10 Green Function Methods

Most of the results that were derived in Sections 7-4 and 7-5 can also be obtained by starting from the variation theorem of Section 7-7. If we follow that procedure, we obtain not only the conventional results but, in addition, we can derive some alternative perturbation expressions which present a different viewpoint and which lead to the solutions of certain problems that cannot be tackled with the conventional perturbation expansions. The perturbation expansions that we obtain in this section contain the energy eigenvalues of the exact Hamiltonian rather than the eigenvalues of the unperturbed Hamiltonian. The new results are denoted by the term "Brillouin-Wigner perturbation results" as opposed to the Rayleigh-Schrödinger expansions of Section 7-4.

We seek to derive the solutions of the eigenvalue problem

$$H\Psi = E\Psi \tag{7-187}$$

Here we assume that the Hamiltonian $H$ can be separated as

$$H(x) = H_0(x) + \lambda V(x) \tag{7-188}$$

The operators depend on a set of variables $q_1, q_2, \ldots, q_N$, which we denote symbolically by $x$. We denote the eigenvalues and eigenfunctions of the operator $H_0(x)$ by $\varepsilon_k$ and by $g_k(x)$, respectively:

$$H_0(x) g_k(x) = \varepsilon_k g_k(x) \tag{7-189}$$

According to the variational principle we can expand any eigenfunction $\Psi(x)$ of the operator $H(x)$ in terms of the complete set of functions $g_k(x)$:

$$\Psi(x) = \Sigma b_k g_k(x) \tag{7-190}$$

Substitution of this expansion into the Schrödinger equation (7-187) leads to the equations

$$\Sigma_k b_k H(x) g_k(x) = E \sum_k b_k g_k(x) \tag{7-191}$$

Rather than consider the general problem of Eq. (7-191), we consider a specific eigenvalue $E_0$ and its corresponding eigenfunction $\Psi_0(x)$. The eigenvalue $E_0$ should be close to a corresponding eigenvalue $\varepsilon_0$ of the Hamiltonian $H_0$, and the eigenfunction $\Psi_0(x)$ should be close to the corresponding eigenfunction $g_0(x)$, if we assume that $E_0$ and $\varepsilon_0$ are nondegenerate. We can now write the expansion (7-190) as

$$\Psi_0(x) = g_0(x) + \sum_{k \neq 0} a_k g_k(x) \tag{7-192}$$

The difference between $E_0$ and $\varepsilon_0$ is of the order of $\lambda$.

If we substitute the expansion (7-192) into the Schrödinger equation (7-187), we obtain

$$(H-E_0)\Psi_0(x)=(H_0+\lambda V-E_0)\left[g_0(x)+\sum_{k\neq 0} a_k g_k(x)\right]$$

$$=(\varepsilon_0+\lambda V-E_0)g_0(x)+\sum_{k\neq 0} a_k(\varepsilon_k+\lambda V-E_0)g_k(x)=0 \quad (7\text{-}193)$$

We multiply this equation on the left by $g_0^*(x)$ and integrate. This gives

$$\varepsilon_0-E_0+\lambda\langle g_0|V|g_0\rangle+\lambda\sum_{k\neq 0} a_k\langle g_0|V|g_k\rangle=0 \quad (7\text{-}194)$$

Similarly, multiplication on the left by $g_n^*(x)$ and subsequent integration gives

$$\lambda\langle g_n|V|g_0\rangle+a_n(\varepsilon_n-E_0)+\lambda\sum_{k\neq 0} a_k\langle g_n|V|g_k\rangle=0 \quad (7\text{-}195)$$

According to the variational principle, we obtain the exact eigenfunction $\Psi_0(x)$ if we vary all coefficients $a_k$. This means that we obtain the exact eigenfunction by solving the set of Eqs. (7-194) and (7-195) exactly.

First we consider Eq. (7-194). We can write the equation as

$$E_0=\varepsilon_0+\lambda\int g_0^*(x)V(x)\left[g_0(x)+\sum_{k\neq 0} a_k g_k(x)\right]dx \quad (7\text{-}196)$$

We substitute the expansion (7-192) and we obtain the simple expression

$$E_0=\varepsilon_0+\lambda\int g_0^*(x)V(x)\Psi_0(x)\,dx \quad (7\text{-}197)$$

Next we consider the set of Eqs. (7-195). We can rewrite the equation as

$$a_n=-\frac{\lambda\langle g_n|V|g_0\rangle}{\varepsilon_n-E_0}-\lambda\sum_{k\neq 0}\frac{\langle g_n|V|g_k\rangle a_k}{\varepsilon_n-E_0} \quad (7\text{-}198)$$

We now substitute this expression back into the expansion (7-192). This gives

$$\Psi_0(x)=g_0(x)+\sum_{n\neq 0} a_n g_n(x)$$

$$=g_0(x)-\lambda\sum_{n\neq 0}(\varepsilon_n-E_0)^{-1}\langle g_n|V|g_0\rangle g_n(x)$$

$$-\lambda\sum_{n\neq 0}\sum_{k\neq 0}(\varepsilon_n-E_0)^{-1}\langle g_n|V|g_k\rangle a_k g_n(x) \quad (7\text{-}199)$$

Again, we can combine the last two sums of this equation by using Eq. (7-192). The result is

$$\Psi_0(x)=g_0(x)-\lambda \sum_{n\neq 0} (\varepsilon_n - E_0)^{-1} g_n(x)\langle g_n|V|\Psi_0\rangle \tag{7-200}$$

or

$$\Psi_o(x)=g_0(x)-\lambda \sum_{n\neq 0} (\varepsilon_n - E_0)^{-1} g_n(x)\int g_n^*(x')V(x')\Psi_0(x')\,dx' \tag{7-201}$$

It is possible to transform this latter equation to an integral equation by introducing a function $G(x, x'; E)$, which is usually known as the Green function of the operator $H_0(x)$ and which is also called the resolvent by some people. The general Green function $G(x, x'; E)$ is defined as

$$G(x, x'; E)=\sum_n \frac{g_n(x)g_n^*(x')}{\varepsilon_n - E} \tag{7-202}$$

We prefer to deal with the reduced Green function $G_0(x, x'; E_0)$, which we define as

$$G_0(x, x'; E_0)=\sum_{n\neq 0} \frac{g_n(x)g_n^*(x')}{\varepsilon_n - E_0} \tag{7-203}$$

By making use of this Green function we can write Eq. (7-201) in the form of an integral equation

$$\Psi_0(x)=g_0(x)-\lambda\int G_0(x, x'; E_0)V(x')\Psi_0(x')\,dx' \tag{7-204}$$

We can substitute this result back into Eq. (7-197) for the energy. The result is

$$E_0=\varepsilon_0+\lambda\int g_0^*(x)V(x)g_0(x)\,dx$$

$$-\lambda^2\int\int g_0^*(x)V(x)G(x, x'; E_0)V(x')\Psi_0(x')\,dx\,dx' \tag{7-205}$$

It should be noted that Eqs. (7-204) and (7-205) are exact and that they are valid for all values of $\lambda$ or the perturbation $V$. The first equation lends itself to iterative solutions. By substituting the equation into itself we easily obtain successive perturbations to the eigenfunction. The only difference with the conventional Rayleigh-Schrödinger perturbation expansions is that the Green function contains the energy eigenvalue $E_0$ instead of the unperturbed eigenvalue $\varepsilon_0$. As we already mentioned, such expansions in terms of the exact eigenvalue $E_0$ are known as Brillouin-Wigner perturbation expansions. They are simpler in form than the Rayleigh-Schrödinger expansions, but they have the disadvantage that the exact value of $E_0$ is in many cases not available.

It may appear to the reader that the Green-function formalism described above is in every respect equivalent with the perturbation expansions of Section 7-4 because they both lead to exactly the same equations. However, there are differences. In conventional perturbation theory we calculate first the individual matrix elements $V_{n,m}$ and then we sum. In the Green-function formalism the order is reversed: Eq. (7-203) contains the infinite sum and we sum first and integrate later. In some cases the Green function may actually be calculated and expressed in closed form, and if that happens many perturbation results can be derived. It is well known that the Green function of a free particle can be expressed in closed form. This expression is often used in the theory of scattering processes. It was found a few years ago that the Green function of the hydrogen atom can be obtained in closed form also. The hydrogen atom Green function is obtained as a product of confluent hypergeometric functions. Consequently, some properties of the hydrogen atom can be calculated exactly by using the Green-function procedures, and these properties cannot be calculated by any other method. We have presented only a very brief outline of Green-function methods in perturbation theory. A more detailed discussion of recent developments falls outside the scope of this book.

## 7-11 Final Remarks

In this chapter we have tried to review the various approaches in perturbation theory that are being used today in practical applications. We have found it helpful to be familiar with all the different methods because in dealing with specific problems there is always one particular perturbation procedure that is more efficient than any of the others. For example, if we are interested in calculating various properties of the hydrogen atom, such as dynamic polarizabilities, two-photon absorption effects, and so on, the Green-function methods of Section 7-10 offer the most convenient approach. If we wish to calculate the corresponding properties for the helium atom or the hydrogen molecule, the variation-perturbation methods of Section 7-9 are the most effective. It is of course necessary to discuss the formal Rayleigh-Schrödinger theory in order to outline the problem and in order to define the various quantities that we are concerned with. Also, there are problems for which the formal perturbation expansions of Section 7-4 offer a convenient approach to the solution. However, a knowledge of formal perturbation theory only is not adequate for dealing with practical problems involving perturbations of atomic or molecular systems. Instead, it is necessary to know a variety of different approaches so that in any particular situation the most effective and efficient method can be selected.

## Problems

**1** Prove that the second-order energy perturbation $E_{0,2}$ of the lowest eigenstate of a Hamiltonian is always negative.

**2** We define the eigenvalues and the corresponding eigenfunctions of an unperturbed Hamiltonian $H_0$ by $\varepsilon_n$ and $\phi_n$, $H_0\phi_n = \varepsilon_n\phi_n$. We now consider the effect of a perturbation $\lambda V$ on these eigenvalues and eigenfunctions. Express the third-order energy perturbation $E_{0,3}$ of the nondegenerate groundstate 0 in terms of the unperturbed eigenvalues and eigenfunctions. Also, express $E_{0,3}$ in terms of the first-order eigenfunction perturbation $\psi_{0,1}$.

**3** Consider the situation defined in Problem 2 and express the fourth-order energy perturbation $E_{0,4}$ in terms of:

- **(a)** The unperturbed eigenvalues and eigenfunctions.
- **(b)** The first-order eigenfunction perturbation $\psi_{0,1}$ and the unperturbed eigenvalues and eigenfunctions.
- **(c)** The eigenfunction perturbations $\psi_{0,2}$ and $\psi_{0,1}$.

**4** Calculate the second-order energy perturbation of the ground state of the hydrogen atom due to a homogeneous electric field along the $z$ axis by means of the variational-perturbation method of Section 7-9.

- **(a)** Take the variational function $g$ for the first-order perturbed function as $g = az\exp(-r)$.
- **(b)** Take the variational function as $g = az\exp(-r/2)$.
- **(c)** Take the function $g$ as $g = (az + bzr)\exp(-r)$.

**5** Evaluate the first- and second-order energy perturbations of the ground state of the harmonic oscillator due to a perturbation $V = x^3$.

**6** Evaluate the effect of a homogeneous electric field along the $z$ axis on the fourfold degenerate $(2s, 2p)$ state of the hydrogen atom.

**7** Determine the best possible value of the ground-state energy of the hydrogen atom that can be obtained from the variational function

$$\psi = \exp(-\alpha r^2)$$

**8** Calculate the second-order energy correction due to a magnetic field of a hydrogen atom in its ground state. If we take the magnetic field **B** along the $z$ axis, the total Hamiltonian of the system is given by

$$H = -\frac{\hbar^2}{2m}\Delta - \frac{e}{2mc}BM_z + \frac{e^2}{8mc^2}B^2(x^2 + y^2)$$

**9** The interaction between a hydrogen atom and a point charge $e$ located at a large distance $R$ from the nucleus on the $z$ axis is represented by an interaction potential

$$V=\frac{e^2}{R}-\frac{e^2}{r'}=-\frac{e^2}{R}\sum_{n=1}^{\infty}\left(\frac{r}{R}\right)^n P_n(\cos\theta)$$

Treat $(1/R)$ as a perturbation parameter and evaluate the first nonzero energy perturbation of the hydrogen atom ground state.

**10** Derive the expressions for the third-order energy perturbation of a nondegenerate state according to the Rayleigh-Schrödinger theory and to the Brillouin-Wigner theory. Show that both expressions are consistent with one another.

**11** Derive the first-order eigenfunction perturbation and the second-order eigenvalue perturbation of the lowest eigenstate of a particle in a one-dimensional box due to a homogeneous electric field. The exact solution can be obtained by means of the procedure described in Section 7-6.

**12** Derive an approximate solution to Problem 11 by using the variation-perturbation method of Section 7-9. Take the variational function $g$ as $g=ax\phi_0$, where $a$ is an undetermined parameter and $\psi_0$ is the unperturbed ground-state eigenfunction.

## Recommended Reading

The variation principle was discussed as early as 1925 in the book by Courant and Hilbert. Chapter IV of the first volume contains a very thorough mathematical analysis of the calculus of variations. A more recent discussion of variational methods can be found in the text by Epstein. We list a few articles on perturbation theory. We found the review article by Hirschfelder, Byers Brown, and Epstein a very useful survey of the subject. We also list some of our own work where we discuss the procedures described in Sections 7-9 and 7-10.

R. Courant and D. Hilbert, *Methods of Mathematical Physics*, Vol. I, Interscience Publishers, Inc., New York, 1953; first published by Julius Springer, Berlin, 1925.

S. T. Epstein, *The Variation Method in Quantum Chemistry*, Academic Press, New York, 1974.

J. O. Hirschfelder, W. Byers Brown, and S. T. Epstein, *Recent Developments in Perturbation Theory*, Advances in Quantum Chemistry, Vol. I, Academic Press, New York, 1964.

H. F. Hameka and E. Nørby Svendsen, "Errors in the Calculation of First-Order and Second-Order Energy Perturbations," *Intl. J. Quant. Chem.*, **10**, 249 (1976).

H. F. Hameka, "On the Use of Green Functions in Atomic and Molecular Calculations I," *J. Chem. Phys.*, **47**, 2728 (1967); **48**, 4810 (1968).

CHAPTER EIGHT

# Time-Dependent Perturbation Theory

## 8-1 Introduction

In the previous chapter we discussed time-independent perturbation theory and in the present chapter we present time-dependent perturbation theory. The two theories have similar names, but they are quite different from each other. In the previous chapter we could outline the basic goal of perturbation theory in simple nonmathematical terms, but we cannot do that in the present case. Time-dependent perturbation theory is highly mathematical, and we have to make use of its mathematical formalisms in order to define our goals. In practical terms, it supplies the basis for the theoretical description of spectroscopic transitions in atoms or molecules. We will derive the mathematical description of the radiation field (both the classical and the quantum theory) in the next chapter (Chapter 9). The detailed theory of spectroscopic transitions will also be presented in Chapter 9. However, our treatment of time-dependent perturbation theory is certainly prejudiced toward spectroscopic problems, even though these problems are not discussed until the following chapter.

It should be noted that we considered only stationary states in the previous chapters, and we presented the general time-dependent Schrödinger equation in Section 3-10 [Eq. (3-168)]. If we make use of the operator formalism of Section 4-3, we may express the time-dependent Schrödinger equation of a system as

$$\mathcal{H}_{\text{op}}\Psi(x;t)=i\hbar\frac{\partial\Psi}{\partial t} \tag{8-1}$$

Here the symbol $x$ represents all spatial coordinates $q_1, q_2, \ldots, q_N$ of the system and $t$ stands for the time. We discussed in Section 3-10 that a stationary state of the system with energy $E$ is described by a function

$$\Psi(x;t)=\Phi(x)e^{-iEt/\hbar} \tag{8-2}$$

Naturally, there are stationary states only when the Hamiltonian is time independent. The wave function $\Phi(x)$ and the energy $E$ must satisfy the time-independent Schrödinger equation

$$\mathcal{H}_{\text{op}}(x)\Phi(x)=E\Phi(x) \tag{8-3}$$

We have seen that we obtain physically acceptable solutions only for certain specific energy values $E_1, E_2, \ldots, E_n$, which are known as eigenvalues of the operator $\mathcal{H}(x)$. The corresponding solutions are the eigenfunctions $\Phi_1(x), \Phi_2(x), \ldots, \Phi_n(x), \ldots,$ and so on. It follows that each stationary eigenstate $n$ is described by a time-dependent eigenfunction $\Psi_n$ which is given by

$$\Psi_n(x;t)=\Phi_n(x)\exp\left(\frac{-iE_n t}{\hbar}\right) \tag{8-4}$$

It is easily verified that each eigenfunction $\Psi_n(x;t)$ is a solution of the time-dependent Schrödinger equation (8-1). It represents a stationary state because it has a specific, well-defined energy, occurring in the time-dependent exponential part of the function.

Since each eigenfunction $\Psi_n(x;t)$ is a solution of the time-dependent Schrödinger equation (8-1), the general solution of (8-1) can be written as a linear combination of all functions $\Psi_n$:

$$\Psi(x;t)=\sum_n c_n\Psi_n(x;t)=\sum_n c_n\Phi_n(x)\exp\left(\frac{-iE_n t}{\hbar}\right) \tag{8-5}$$

It should be realized that in the preceding chapters we were interested in determining the stationary states of various systems and we discussed only the stationary states of these systems. However, a system does not have to be in a particular stationary state; in general, it can be distributed over a number of different stationary states. The most general case is represented by the general wave function $\Psi(x;t)$ which describes a superposition of various (or all) stationary states $n$. The coefficients $c_n$ are probability amplitudes that determine the probability of finding the system in a particular stationary state $n$. The probability $w_n$ of finding the system in a stationary state $n$ is given by

$$w_n = c_n c_n^* \tag{8-6}$$

if the system is described by the wave function $\Psi(x;t)$. We assume that the functions $\Phi_n$ form an orthonormal set. We then have

$$\langle\Psi(x;t)|\Psi(x;t)\rangle=\sum_n c_n c_n^*=1 \tag{8-7}$$

as our normalization condition.

Let us now consider a system that is described by a time-dependent Hamiltonian $\mathcal{H}(x;t)$ which can be separated as

$$\mathcal{H}(x;t)=\mathcal{H}_o(x)+\mathcal{H}'(x;t) \tag{8-8}$$

Here we assume that the time-dependent term $\mathcal{H}'$ is much smaller than the time-independent part $\mathcal{H}_o$. We also assume that we know the eigenvalues $E_n$ and eigenfunctions $\Phi_n$ of the operator $\mathcal{H}_o$:

$$\mathcal{H}_o(x)\Phi_n(x) = E_n\Phi_n(x) \tag{8-9}$$

We denote the time-dependent wave function of the operator $\mathcal{H}(x;t)$ by $\Psi(x;t)$. The functions $\Phi_n$ form a complete set, and at each given time $t$ we can expand the function $\Psi(x;t)$ in terms of the complete set of functions $\Phi_n(x)$:

$$\Psi(x;t) = \sum_n a_n(t)\Phi_n(x) \tag{8-10}$$

Even though this expansion is permissable from a mathematical point of view, it seems more suitable to expand the function $\Psi(x;t)$ in terms of the stationary state eigenfunctions of the operator $\mathcal{H}_o$, that is, as

$$\Psi(x;t) = \sum_n c_n(t)\exp\left(\frac{-iE_nt}{\hbar}\right)\Phi_n(x) \tag{8-11}$$

In the latter expansion we can still recognize the stationary states of the operator $\mathcal{H}_o(x)$ if the time dependence of the exponentials is not obliterated by the time dependence in the coefficients $c_n(t)$. In other words, if the condition

$$\frac{\partial c_n}{\partial t} \ll \frac{|E_n|}{\hbar} \tag{8-12}$$

is satisfied, we can identify the stationary states of the unperturbed Hamiltonian $\mathcal{H}_o(x)$ in the eigenfunction expansion of $\Psi(x;t)$. The same condition also describes the validity of the perturbation theory. As long as we can still identify the stationary states of the operator $\mathcal{H}_o(x)$, we can consider the effects of $\mathcal{H}'(x;t)$ as a small perturbation. On the other hand, if the time dependence of the coefficients $c_n(t)$ becomes comparable to the time dependence of the exponentials, we can no longer recognize the stationary states of $\mathcal{H}_o(x)$, and the representation (8-11) becomes meaningless.

An interesting situation occurs when the time dependence of the operator $\mathcal{H}'(x;t)$ of Eq. (8-8) has the form

$$\begin{aligned} \mathcal{H}'(x;t) &= 0 \qquad & t<0 \\ \mathcal{H}'(x;t) &= V(x) \qquad & t\geq 0 \end{aligned} \tag{8-13}$$

In other words, the time-independent perturbation $V(x)$ is suddently added to the system at time $t=0$. Obviously, the operator $\mathcal{H}_o + V$ also has a set of stationary states that are defined as

$$[\mathcal{H}_o(x) + V(x)]\psi_n(x) = \varepsilon_n\psi_n(x) \tag{8-14}$$

and the eigenfunctions of the system can be expanded also in terms of the eigenstates of $\mathcal{H}_o + V$:

$$\Psi(x;t) = \sum_n b_n(t) \exp\left(\frac{-i\varepsilon_n t}{\hbar}\right) \psi_n(x) \tag{8-15}$$

We may well ask which of the two expansions, Eq. (8-11) or (8-15), is the more appropriate one. Clearly, at times $t \leq 0$, the former expansion (8-11) is preferable, and at much later times $t \to \infty$, the latter expansions (8-15) become the more suitable. At intermediate times we must analyze the situation, and this is exactly what we discuss in the subsequent sections.

In the following section we derive the differential equations for the expansion coefficients, or probability amplitudes, $c_n(t)$. We shall see that it is not possible to solve these equations for the general case and that we have to define the situation that we are considering more precisely in order to derive solutions. Specifically, we consider situations where the system is in a particular stationary state of the operator $\mathcal{H}_o(x)$ until a certain time $t = t_o$ when the perturbation is switched on. We then have to differentiate between situations, depending on the specific form of the perturbation $\mathcal{H}'(x;t)$.

It is important to recall our discussions in Section 4-9 on eigenfunction expansions. We considered the case where the eigenvalue spectrum consists both of discrete and of continuum states and we analyzed the contribution from the continuum states to the eigenfunction expansions. We shall see that continuum states play an important role in time-dependent perturbation theory. In our expansions of Eqs. (8-10), (8-11), and so on, we did not write the continuum contributions separately, but it should be understood that the various summations should all be taken over both the discrete and the continuum part of the spectrum. In subsequent derivations we will have to analyze the continuum contributions in more detail and we will use some of the results of Section 4-9.

## 8-2 General Perturbation Equations

In this section we derive the general differential equations for the probability amplitudes of the system we described in the previous section. We may recall that this system is described by a Hamiltonian

$$\mathcal{H}(x;t) = \mathcal{H}_o(x) + \mathcal{H}'(x;t) \tag{8-16}$$

For simplicity, we assume first that the eigenvalue spectrum of the unperturbed Hamiltonian $\mathcal{H}_o(x)$ is discrete and that its eigenvalues and eigenfunctions are defined by Eq. (8-9). We can then expand the eigenfunction $\Psi(x;t)$ of $\mathcal{H}(x;t)$, according to Eq. (8-11). If we substitute this expansion into the time-dependent Schrödinger equation (8-1), we obtain

$$[\mathcal{H}_o(x)+\mathcal{H}'(x;t)]\Psi(t)=\sum_n c_n(t)[\mathcal{H}_o(x)+\mathcal{H}'(x;t)]\exp\left(\frac{-iE_nt}{\hbar}\right)\Phi_n(x)$$

$$=\sum_n c_n(t)\exp\left(\frac{-iE_nt}{\hbar}\right)\mathcal{H}'(x;t)\Phi_n(x)$$

$$+\sum_n c_n(t)\exp\left(\frac{-iE_nt}{\hbar}\right)E_n\Phi_n(x) \tag{8-17}$$

and

$$i\hbar\left(\frac{\partial}{\partial t}\right)\Psi(x;t)=i\hbar\sum_n\left(\frac{\partial}{\partial t}\right)\left[c_n(t)\exp\left(\frac{-iE_nt}{\hbar}\right)\Phi_n(x)\right]$$

$$=\sum_n c_n(t)\exp\left(\frac{-iE_nt}{\hbar}\right)E_n\Phi_n(x)$$

$$+(i\hbar)\sum_n\left(\frac{\partial c_n}{\partial t}\right)\exp\left(\frac{-iE_nt}{\hbar}\right)\Phi_n(x) \tag{8-18}$$

By equating Eqs. (8-17) and (8-18) we obtain

$$(i\hbar)\sum_n\left(\frac{\partial c_n}{\partial t}\right)\exp\left(\frac{-iE_nt}{\hbar}\right)\Phi_n(x)=\sum_n c_n(t)\exp\left(\frac{-iE_nt}{\hbar}\right)\mathcal{H}'(x;t)\Phi_n(x) \tag{8-19}$$

We note that the set of functions $\Phi_n(x)$ forms an orthonormal set, and we introduce the matrix elements

$$\mathcal{H}'_{m,n}(t)=\int\Phi_m^*(x)\mathcal{H}'(x;t)\Phi_n(x)\,dx \tag{8-20}$$

If we now multiply the Eqs. (8-19) on the left by $\Phi_k^*(x)$ and integrate, we obtain the following set of differential equations:

$$i\hbar\frac{\partial c_k}{\partial t}=\sum_n\mathcal{H}'_{k,n}(t)\exp\left[\frac{-i(E_n-E_k)t}{\hbar}\right]c_n(t) \tag{8-21}$$

We will generalize these equations so that they include the case where the eigenvalue spectrum of $\mathcal{H}_o$ consists of both discrete and continuum states. The continuum states are defined by

$$\mathcal{H}_o(x)\Phi(x;E)=E\Phi(x;E) \tag{8-22}$$

According to Section (4-9) the function $\Psi(x; t)$ can now be expanded as

$$\Psi(x;t)=\sum_n c_n(t)\Phi_n(x)\exp\left(\frac{-iE_nt}{\hbar}\right)$$

$$+\int c(E;t)\Phi(x;E)\exp\left(\frac{-iEt}{\hbar}\right)dE \qquad (8\text{-}23)$$

The wave functions $\Phi(x; E)$ satisfy the orthogonality relations

$$\int \Phi^*(x;E)\Phi(x;E')\,dx=\delta(E-E') \qquad (8\text{-}24)$$

according to Eq. (4-213). We also have

$$\int \Phi^*(x;E)\Phi_n(x)\,dx=0 \qquad (8\text{-}25)$$

We substitute the expansion (8-23) into the time-dependent Schrödinger equation and we follow the same procedure as in Eqs. (8-17) and (8-18). We just list the final results. The differential equations for the coefficients $c_k(t)$ and $c(E; t)$ are now

$$i\hbar\frac{\partial c_k}{\partial t}=\sum_n \mathcal{H}'_{k,n}(t)c_n(t)\exp\left[\frac{-i(E_n-E_k)t}{h}\right]$$

$$+\int \mathcal{H}_k^*(\varepsilon;t)c(\varepsilon;t)\exp\left[\frac{-i(\varepsilon-E_k)t}{\hbar}\right]d\varepsilon \qquad (8\text{-}26)$$

and

$$i\hbar\frac{\partial c(\varepsilon;t)}{\partial t}=\sum_n \mathcal{H}'_n(\varepsilon;t)c_n(t)\exp\left[\frac{-i(E_n-\varepsilon)t}{\hbar}\right]$$

$$+\int \mathcal{H}'(\varepsilon;E;t)c(E,t)\exp\left[\frac{-i(E-\varepsilon)t}{\hbar}\right]dE \qquad (8\text{-}27)$$

The various matrix elements are defined as

$$\mathcal{H}'(\varepsilon;E;t)=\int\Phi^*(x;\varepsilon)\mathcal{H}'(x;t)\Phi(x;E)\,dx$$

$$\mathcal{H}'_n(\varepsilon;t)=\int\Phi^*(x;\varepsilon)\mathcal{H}'(x;t)\Phi_n(x)\,dx$$

$$\mathcal{H}_k'^*(\varepsilon;t)=\int\Phi_k^*(x)\mathcal{H}'(x;t)\Phi(x;\varepsilon)\,dx \qquad (8\text{-}28)$$

The sets of equations (8-26) and (8-27) are quite general because we have not introduced any approximations up to this point. However, we cannot solve the general problem because we do not have sufficient information to arrive at a solution. The equations describe the time evolution of our system, and we must specify the state of the system at a particular time in order to predict the state of the system at subsequent times. There are various possibilities as far as the initial conditions for the coefficients $c(t)$ are concerned, but there is one particular case which is usually considered since it represents the most frequently occurring physical situation.

First we assume that the perturbation $\mathcal{H}'(x;t)$ is switched on at a particular time, which we take as $t=0$. Consequently, we have

$$\mathcal{H}'(x;t)=0 \qquad t<0 \tag{8-29}$$

We will consider various possibilities for the dependence of $\mathcal{H}'$ on $t$ for $t\geq 0$. In Section 8-3 we take $\mathcal{H}'$ time-independent for $t\geq 0$ and in Section 8-4 we consider time-dependence of the type $\exp(\pm i\omega t)$. It should be noted that even if $\mathcal{H}'$ is time-independent for $t\geq 0$, the total function $\mathcal{H}'(x;t)$ is still time dependent since it behaves as a step function at $t=0$. Of course, this step-function behavior of $\mathcal{H}'$ is usually an approximation and in reality $\mathcal{H}'$ is usually switched on more gradually. On the other hand, the step-function behavior is more easily treated mathematically and it is a generally accepted model.

Our subsequent derivations are limited to one situation, namely the case where the system is in a stationary state of the operator $\mathcal{H}_o(x)$ for times $t\leq 0$. We assume that this is a discrete state and we denote it by $o$. Consequently, we have

$$c_o(0)=1 \qquad c_k(0)=c(\varepsilon,0)=0 \tag{8-30}$$

We are interested only in the values of the probability amplitudes for $t\geq 0$. Their values for negative times have no physical relevance. However, in the last section of this chapter where we discuss more general perturbation methods, we make certain assumptions about the probability amplitudes for negative times. These assumptions serve mathematical convenience only.

It is possible to derive approximate solutions for the time-dependent perturbation equations (8-26) and (8-27) for the situation outlined above. For sufficiently small times we can assume that

$$c_o(t)\approx 1 \qquad c_k(t)\approx c(\varepsilon,t)\approx 0 \tag{8-31}$$

and we can substitute these approximate values into the right side of Eq. (8-26) and (8-27). The differential equations then take the form

$$\begin{aligned}\frac{\partial c_k}{\partial t} &= -\frac{i}{\hbar}\mathcal{H}'_{k,o}(t)\exp\left[\frac{-i(E_o-E_k)t}{\hbar}\right]\\ \frac{\partial c(\varepsilon,t)}{\partial t} &= -\frac{i}{\hbar}\mathcal{H}'_o(\varepsilon;t)\exp\left[\frac{-i(E_o-\varepsilon)t}{\hbar}\right]\end{aligned} \tag{8-32}$$

These equations can be integrated from $t=0$ to $t$ and we obtain

$$c_k(t) = -\frac{i}{\hbar}\int_0^t \mathcal{H}'_{k,o}(t)\exp\left[\frac{-i(E_o - E_k)t}{\hbar}\right]dt$$

$$c(\varepsilon, t) = -\frac{i}{\hbar}\int_0^t \mathcal{H}'_o(\varepsilon; t)\exp\left[\frac{-i(E_o - \varepsilon)t}{\hbar}\right]dt \tag{8-33}$$

It may seem that we have solved the problems of time-dependent perturbation theory, but that is far from true. It is relatively easy to calculate the integrals of Eq. (8-33) for discrete states, but the results that are obtained then are not relevant to the physical problems that we are interested in. It turns out that transitions between different stationary states have reasonably large probabilities only when the energy is conserved. It is not particularly interesting to calculate transition probabilities that are so small that they are not observable. Instead we are primarily interested in those transition probabilities that are large enough to be of physical interest. We shall find that these large transition probabilities occur in two different situations which we will discuss in Sections 8-3 and 8-4, respectively.

In Section 8-3 we take the perturbation $\mathcal{H}'$ to be time-independent, once it is switched on. We then consider a transition from a stationary state, embedded in a continuum of states, to the various continuum states, surrounding the state $o$. We may visualize this situation as a "gutter"-type potential field as we have sketched in Fig. 8-1. Here we have a two-dimensional potential field with a harmonic-oscillator-type potential in one dimension and a constant potential function in the other dimension. Each stationary state of the harmonic oscillator is surrounded by a continuum of states of the free particle. In this potential field we may have transitions between two different stationary states of the harmonic oscillator, and the energy difference may be either supplied or absorbed by transitions between different continuum states in the other dimension.

In Section 8-4 we take the perturbation $\mathcal{H}'$ as time dependent, and we take it as a superposition of exponential terms of the form $\exp(i\omega t)$. In that case we consider transitions between two discrete states of the operator $\mathcal{H}_o$. We shall see that these transitions are large if the perturbation $\mathcal{H}'$ contains terms of the

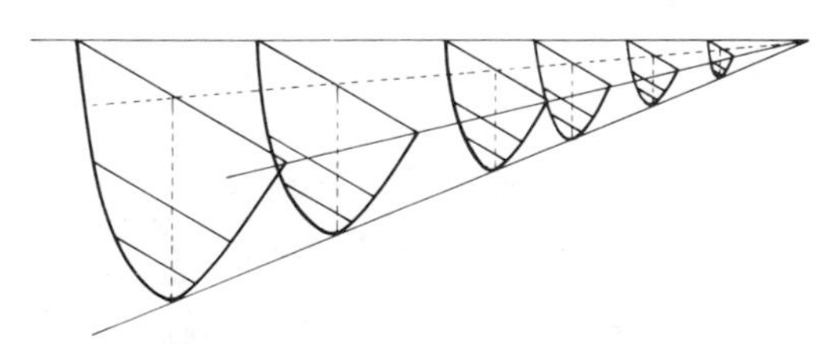

**Fig. 8-1** Sketch of a two-dimensional "gutter-type" potential function. The potential $V(x, y)$ behaves as a harmonic oscillator in the $Y$ direction and it is constant in the $X$ direction.

proper frequency. In simple words, the system selects the proper frequency terms out of the perturbation to give observable transition probabilities.

We see that the two cases outlined above, that is, the two cases that are of most interest from a physical point of view, require a more complete mathematical analysis than the perturbation equations (8-33). We shall take those perturbation equations as our starting points both in Section 8-3 and in Section 8-4.

## 8-3 Constant Perturbations

In this section we derive the approximate solution of the time-dependent perturbation equations for the case where the perturbation $\mathcal{H}'$ is time independent once it is switched on. The perturbation $\mathcal{H}'(x;t)$ is defined as

$$\begin{aligned} \mathcal{H}'(x;t)&=0 \qquad t<0 \\ \mathcal{H}'(x;t)&=V(x) \qquad t\geq 0 \end{aligned} \tag{8-34}$$

For sufficiently small times we can use Eqs. (8-33) and we substitute

$$\begin{aligned} \mathcal{H}'_{k,o}(t)&=\int \Phi_k^*(x)V(x)\Phi_o(x)\,dx=V_{ko} \\ \mathcal{H}'_o(\varepsilon;t)&=\int \Phi^*(x;\varepsilon)V(x)\Phi_o(x)\,dx=V_o(\varepsilon) \end{aligned} \tag{8-35}$$

since in both cases $t$ is larger than zero. The perturbation equations take the simple form

$$\begin{aligned} c_k(t)&=-\frac{iV_{ko}}{\hbar}\int_o^t \exp\left[\frac{-i(E_o-E_k)t}{\hbar}\right]dt \\ c(\varepsilon;t)&=-\frac{iV_o(\varepsilon)}{\hbar}\int_o^t \exp\left[\frac{-i(E_o-\varepsilon)t}{\hbar}\right]dt \end{aligned} \tag{8-36}$$

The two integrals are easily calculated and we obtain

$$\begin{aligned} c_k(t)&=V_{ko}\frac{\exp[i(E_k-E_o)t/\hbar]-1}{E_o-E_k} \\ c(\varepsilon;t)&=V_o(\varepsilon)\frac{\exp[i(\varepsilon-E_o)t/\hbar]-1}{E_o-\varepsilon} \end{aligned} \tag{8-37}$$

The probability of finding the system in the discrete state $k$ at a time $t$ is given by

$$c_k(t)c_k^*(t)=\frac{4V_{ko}V_{ko}^*\sin^2[(E_k-E_o)t/2\hbar]}{(E_k-E_o)^2} \tag{8-38}$$

Usually this expression is quite small. If the energy difference $E_k-E_o$ approaches zero, the expression (8-38) becomes significant, but we cannot expect this to happen if $k$ and $o$ are two different discrete eigenstates. It follows that the solution (8-38) is of little practical use because the probabilities for transitions between different discrete eigenstates are usually very small and they cannot be observed.

We try to identify situations where the probability for a transition is sufficiently large that it can be observed. For this purpose we consider transitions to continuum states. It follows from Eq. (8-37) that the probability of finding the system in any of the continuum states with energies between $\varepsilon$ and $\varepsilon+d\varepsilon$ is given by

$$c(\varepsilon;t)c^*(\varepsilon;t)\,d\varepsilon=\frac{4V_o(\varepsilon)V_o^*(\varepsilon)\sin^2[(\varepsilon-E_o)t/2\hbar]\,d\varepsilon}{(\varepsilon-E_o)^2} \tag{8-39}$$

Let us first study the behavior of the right side of Eq. (8-39) as a function of the variable $\varepsilon$. We can write this function as

$$f(x)=\frac{\sin^2\alpha x}{x^2} \tag{8-40}$$

For large values of the parameter $\alpha$ this function is a representation of the Dirac $\delta$ function that we introduced in Section 3-5. We have sketched the function in Fig. 8-2 and it can be seen that it has a very pronounced sharp maximum at the point $x=0$. It can be derived from a contour integration in the complex plane that

$$\int_{-\infty}^{\infty}\frac{\sin^2\alpha x}{x^2}\,dx=\alpha\pi \tag{8-41}$$

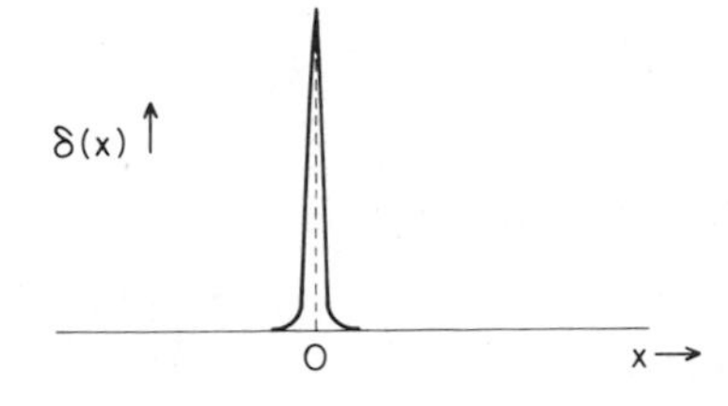


**Fig. 8-2** Sketch of a delta function $\delta(x)$ around the point $x=0$.

The width of the peak in Fig. 8-2 is inversely proportional to the parameter $\alpha$, and the function $f(x)$ becomes a representation of the $\delta$ function in the limit where $\alpha$ tends to infinity. In other words,

$$\lim_{\alpha\to\infty} \frac{\sin^2 \alpha(x-x_o)}{\alpha\pi(x-x_o)^2} = \delta(x-x_o) \tag{8-42}$$

We discuss various properties of the $\delta$ function in Section 8-5.

If we substitute these various results back into our perturbation equation (8-39), we obtain

$$c(\varepsilon;t)c^*(\varepsilon;t) = \left(\frac{2\pi t}{\hbar}\right) V_o(\varepsilon)V_o^*(\varepsilon)\delta(\varepsilon - E_o) \tag{8-43}$$

We predict a finite, observable transition only if the energy $\varepsilon$ of the final state is equal, or almost equal, to the initial energy $E_o$.

Since we have made several approximations, the validity of the above conclusions is subject to certain limitations. We mentioned in the beginning that our initial equations (8-36) are valid only for small times. It is clear that Eq. (8-43) cannot be valid for unlimited times because the probability amplitude is linearly proportional to $t$ according to the equation. We have assumed that the probability amplitude is much smaller than unity, and this implies that the condition

$$t \ll \frac{\hbar}{|V_o(\varepsilon)|} \tag{8-44}$$

should be satisfied. On the other hand, when we replaced the right side of Eq. (8-39) by a $\delta$ function, we assumed that

$$\frac{E_o t}{\hbar} \gg 1 \tag{8-45}$$

or

$$t \gg \frac{\hbar}{|E_o|} \tag{8-46}$$

The two Eq. (8-44) and (8-46) determine the time interval for which our perturbation expressions are valid.

The total probability $w(o\to\text{cont})$ for a transition from the discrete level $o$ into the continuum is derived by integrating Eq. (8-43):

$$\begin{aligned} w(o\to\text{cont}) &= \int c(\varepsilon;t)c^*(\varepsilon;t)\,d\varepsilon \\ &= \left(\frac{2\pi t}{\hbar}\right)\int V_o(\varepsilon)V_o^*(\varepsilon)\delta(\varepsilon-E_o)\,d\varepsilon \\ &= \left(\frac{2\pi t}{\hbar}\right) V_o(E_o)V_o^*(E_o) \end{aligned} \tag{8-47}$$

This is known as a time-proportional transition probability, but it can be seen that the linear time dependence is obtained only if we consider all continuum states. It follows from Eq. (8-43) that the transition probability has a sharp maximum for $\varepsilon = E_o$. The specific form of the maximum is not predicted correctly by the above perturbation theory due to the various approximations that we made. In Section 8-6 we discuss a more precise perturbation procedure which contains the correct shape of the maximum.

In Chapter 9 we will discuss the theoretical description of the radiation field and its interactions with atoms and molecules. We shall see that in the quantum description of the radiation field the interaction between matter and radiation is represented by a time-independent perturbation term. In that representation the spectroscopic transitions are described by the previously derived perturbation results, or by the more precise results of Section 8-6.

## 8-4 Periodic Perturbations

In the classical description of the radiation field spectroscopic transitions are due to a perturbation term that is periodic in time. In order to understand the theoretical description of such perturbations we first take the perturbation Hamiltonian $\mathcal{H}'(x;t)$ of Eq. (8-16) as

$$\mathcal{H}'(x;t) = V(x)e^{i\omega t} + V^*(x)e^{-i\omega t} \tag{8-48}$$

We should note that the perturbation of a molecule by a radiation field is actually described by a superposition of terms of the type of Eq. (8-48), but first we will study the effect of a single component.

We assume again that the system is in a stationary state $o$ of the Hamiltonian $\mathcal{H}_o$ until the time $t=0$, when we switch on the perturbation $\mathcal{H}'(x;t)$ of Eq. (8-48). This time we consider transitions to discrete states only, and we find from the first Eq. (8-33) that the probability amplitude $c_k(t)$ is given by

$$c_k(t) = -\frac{i}{\hbar}\int_o^t \mathcal{H}'_{k,o}(t)\exp(i\omega_k t)\,dt \tag{8-49}$$

Here we substituted

$$\hbar\omega_k = E_k - E_o \tag{8-50}$$

We introduce the abbreviations

$$V_{k,o} = \int \Phi_k^*(x)V(x)\Phi_o(x)\,dx$$

$$V'_{k,o} = \int \Phi_k^*(x)V^*(x)\Phi_o(x)\,dx \tag{8-51}$$

and we find that

$$\mathcal{H}'_{k,o}(t) = V_{k,o}e^{i\omega t} + V'_{k,o}e^{-i\omega t} \tag{8-52}$$

Substitution into Eq. (8-49) gives

$$c_k(t) = -\frac{i}{\hbar}\int_o^t \{V_{k,o}\exp[i(\omega+\omega_k)t] + V'_{k,o}\exp[i(-\omega+\omega_k)t]\}\,dt$$

$$= -\frac{V_{k,o}\{\exp[i(\omega_k+\omega)t]-1\}}{\hbar(\omega_k+\omega)} - \frac{V'_{k,o}\{\exp[i(\omega_k-\omega)t]-1\}}{\hbar(\omega_k-\omega)} \tag{8-53}$$

We should differentiate between the two situations where $E_k > E_o$ and where $E_k < E_o$. In the first situation $\omega_k$ is positive and we consider a transition to a state with higher energy, and in the second case $\omega_k$ is negative and we consider a transition to a state with lower energy. In general, the probability $c_k(t)$ is quite small and we will not be able to observe a transition. If the frequency $\omega$ becomes very close to either $\omega_k$ or $-\omega_k$, the probability may become large enough to be observable.

For convenience, we have considered the single perturbation (8-48) first, but in practical applications the perturbation $\mathcal{H}'$ is usually a superposition of terms of the type of Eq. (8-48). If we consider the interaction between a molecule and a radiation field, we have to represent the interaction as a superposition of different frequencies because it is not possible in practice to produce a beam of strictly monochromatic light. Therefore, in order to deal with situations of practical interest, we must take the perturbation as

$$\mathcal{H}'(x;t) = \int_{-\infty}^{\infty} V(x;\omega)e^{i\omega t}\,d\omega \qquad t \geq 0$$

$$\mathcal{H}'(x;t) = 0 \qquad t < 0 \tag{8-54}$$

We define the matrix elements as

$$\mathcal{H}'_{k,o}(t) = \int V_{k,o}(\omega)e^{i\omega t}\,d\omega$$

$$V_{k,o}(\omega) = \int \Phi_k^*(x)V(x;\omega)\Phi_o(x)\,dx \tag{8-55}$$

In general, the phase differences between the different components of the radiation field are randomly distributed. This means that the matrix elements satisfy the condition

$$V_{k,o}(\omega)V^*_{k,o}(\omega') = |V_{k,o}(\omega)|^2\delta(\omega-\omega') \tag{8-56}$$

Admittedly, the perturbation problem that we have described here is of a much greater mathematical complexity than the problem defined by Eq. (8-48). Unfortunately, we must deal with these mathematical complications because it is the latter problem that applies to realistic experimental situations and not the former.

We substitute the results of Eq. (8-55) into the perturbation equation (8-49) and we obtain a double integral:

$$c_k(t) = -\frac{i}{\hbar}\int_{-\infty}^{\infty} V_{k,o}(\omega)\, d\omega \int_o^t \exp[i(\omega+\omega_k)t]\, dt \tag{8-57}$$

Integration over $t$ gives

$$c_k(t) = -\int_{-\infty}^{\infty} \frac{V_{k,o}(\omega)\{\exp[i(\omega+\omega_k)t]-1\}}{\hbar(\omega+\omega_k)}\, d\omega \tag{8-58}$$

The probability amplitude of finding the system in the stationary state $k$ is given by

$$w_k(t) = c_k(t)c_k^*(t) \tag{8-59}$$

By making use of Eq. (8-56) we can write this as

$$w_k(t) = \int_{-\infty}^{\infty} \frac{4|V_{k,o}(\omega)|^2 \sin^2[\frac{1}{2}(\omega+\omega_k)t]}{\hbar^2(\omega+\omega_k)^2}\, d\omega \tag{8-60}$$

Part of the integrand can be replaced by the $\delta$ function, according to Eq. (8-42). This gives

$$\begin{aligned} w_k(t) &= \frac{2t}{\hbar^2}\int_{-\infty}^{\infty} |V_{k,o}(\omega)|^2 \delta(\omega+\omega_k)\, d\omega \\ &= \left(\frac{2t}{\hbar^2}\right)|V_{k,o}(-\omega_k)|^2 \end{aligned} \tag{8-61}$$

Again, the final result is a time-proportional transition probability.

In the above derivation we have made certain implicit assumptions about the functional behavior of the function $V_{k,o}(\omega)$. First, we have assumed that the dependence on $\omega$ of the matrix element is such that $V_{k,o}$ varies much slower than the $\delta$ function. This corresponds with the experimental situation where a beam of white light interacts with the system and contains the frequency $\omega_k$. We see that the system selects the proper frequency out of the radiation to give rise either to absorption or emission, depending on the sign of $\omega_k$.

If a monochromatic (or an almost monochromatic) beam of light interacts with the molecule, the situation becomes more complicated. Then we must compare the sharpness of the two peaks representing the $\delta$ function $\delta(\omega+\omega_k)$ and the intensity distribution of the light. Also, if the light does not contain the frequency $\omega_k$, the transition probability will be very small. The various questions that arise are related to the linewidths of the absorption or emission lines, and we cannot derive the correct answers to these questions from the above, approximate perturbation treatments. In order to deal with these problems we have to use the more general perturbation methods that we discuss in Section 8-6. Some of the necessary mathematics that we use in this discussion is presented separately in Section 8-5.

## 8-5 The Dirac $\delta$ and $\zeta$ Functions

The Dirac $\delta$ function was first introduced in Section 3-5, where we mentioned some of its properties. In the perturbation method of Section 8-6 we make use of the properties of the Dirac $\delta$ functions and also of the related Dirac $\zeta$ function. We present a more complete description of the two functions in this section.

The Dirac $\delta$ function $\delta(x-x_o)$ is defined by the relation

$$f(x_o)=\int_{-\infty}^{\infty} f(x)\delta(x-x_o)\,dx \tag{8-62}$$

where $f(x)$ is an arbitrary, continuous function of $x$. It follows from the definition also that

$$\int_{-\infty}^{\infty} \delta(x-x_o)\,dx=1 \tag{8-63}$$

We may visualize the $\delta$ function as being very small for all values of the variable $x$, except when $x$ is close to $x_o$ where it is very large. Hence the $\delta$ function has a sharp peak around the point $x_o$.

If we represent this general behavior by a specific mathematical function, we have a representation of the $\delta$ function. These representations are not unique. There are several different functions that give an adequate representation. We already encountered one such representation in Eq. (8-42):

$$\delta(x-x_o)=\lim_{\alpha\to\infty}\frac{\sin^2\alpha(x-x_o)}{\alpha\pi(x-x_o)^2} \tag{8-64}$$

Two alternative representations are

$$\delta(x-x_o)=\lim_{\alpha\to 0}\left(\frac{\alpha}{\pi}\right)^{1/2}\exp\left[-\alpha(x-x_o)^2\right]$$

$$\delta(x-x_o)=\frac{1}{\pi}\lim_{\alpha\to 0}\frac{\alpha}{(x-x_o)^2+\alpha^2} \tag{8-65}$$

We encountered a different type of representation in Eq. (3-49):

$$\delta(x-t)=\frac{1}{2\pi}\int_{-\infty}^{\infty}\exp\left[iu(x-t)\right]du \tag{8-66}$$

An equivalent expression is

$$\delta(x-t)=\frac{1}{\pi}\int_{0}^{\infty}\cos u(x-t)\,du \tag{8-67}$$

The latter two representations (8-66) and (8-67) cannot be easily visualized, but they are consistent with the definition (8-62).

The δ function is an even function of its variable and we have

$$\delta(x)=\delta(-x) \tag{8-68}$$

It follows from the definition that

$$x\delta(x)=0 \tag{8-69}$$

Finally, the three-dimensional δ function is defined as

$$\delta(\mathbf{r}-\mathbf{r}_o)=\delta(x-x_o)\delta(y-y_o)\delta(z-z_o) \tag{8-70}$$

The Dirac ζ function is to some extent related to the δ function. It is a complex function with a real and an imaginary part. The real part behaves as the function $(1/x)$ for all values of $x$, except for the value $x=0$, where it is equal to zero. The imaginary part of the ζ function is proportional to the δ function. We can write the definition as

$$\zeta(x)=\frac{\mathcal{P}}{x}-i\pi\delta(x) \tag{8-71}$$

Clearly,

$$\zeta(x)=\frac{1}{x}\qquad x\neq 0 \tag{8-72}$$

Again, we can write various different representations for the $\zeta$ function. One of our favorites is

$$\zeta(x) = \lim_{\rho\to 0} \frac{1}{x+i\rho} = \lim_{\rho\to 0} \frac{x-i\rho}{x^2+\rho^2} \tag{8-73}$$

This is consistent with the representation (8-65) for the $\delta$ function and with the definition (8-71) for the $\zeta$ function. Two alternate representations are

$$\zeta(x) = \lim_{N\to\infty} \frac{1-e^{iNx}}{x}$$

$$\zeta(x) = -i \lim_{N\to\infty} \int_0^N e^{ixt}\,dt \tag{8-74}$$

It is easily derived from the definition (8-71) that

$$\zeta(x) - \zeta^*(x) = -2\pi i\delta(x) \tag{8-75}$$

Furthermore, we always have

$$x\zeta(x) = 1 \tag{8-76}$$

This can be proved by using the representation (8-73) of the $\zeta$ function, but we must admit that it is not easy to prove the validity of Eq. (8-76) from the general definition (8-71).

We have two more properties of the $\zeta$ function that will be used in the subsequent section. By means of an integration in the complex plane along a suitable path of integration we can derive from the representation (8-73) that

$$\int_{-\infty}^{\infty} \zeta(x) e^{-ixt}\,dx = \begin{cases} -2\pi i & t>0 \\ 0 & t<0 \end{cases} \tag{8-77}$$

By means of a similar integration in the complex plane we can also prove that

$$\lim_{N\to\infty} \zeta(x) e^{ixNt} = \begin{cases} 0 & t>0 \\ -2\pi i\delta(x) & t<0 \end{cases} \tag{8-78}$$

## 8-6 General Perturbation Methods

The perturbation methods discussed previously have limited validity only. They can be used for limited ranges of the time $t$, they do not contain correct predictions about line shapes and so on, and they cannot be conveniently extended to higher orders. In this section we present a different approach which does not suffer from these limitations. On the other hand, we confine ourselves to a discussion of the situation defined by Eq. (8-34), where the perturbation is time independent once it is switched on. Also, we consider only discrete states in our equations. This is not really a limitation because continuum states can always be treated as a limiting case of a discrete spectrum, following the arguments of Section 4-9. It may be helpful to note that the perturbation methods that we describe in this section are particularly suitable for describing interactions between radiation and matter in the quantum theory of the radiation field.

The perturbation equations for the situation described above are given by Eq. (8-21). Since the perturbation is time independent once it is switched on, we can represent it by Eq. (8-34). The matrix elements are defined as

$$V_{k,n} = \int \Phi_k^*(x) V(x) \Phi_n(x)\, dx \tag{8-79}$$

The perturbation equations (8-21) take the form

$$\frac{\partial c_k}{\partial t} = -\frac{i}{\hbar} \sum_n V_{k,n} c_n(t) \exp\left[\frac{i(E_k - E_n)t}{\hbar}\right] \tag{8-80}$$

We assume again that the system is in a stationary state $o$ when the perturbation is switched on at the time $t=0$. We have, therefore,

$$c_o(0) = 1 \qquad c_n(0) = 0 \qquad n \neq o \tag{8-81}$$

The perturbation equations (80) are valid for positive times $t \geq 0$ only.

In our present perturbation method we express the various probability amplitudes $c_k(t)$ as Fourier integrals:

$$c_k(t) = -\frac{1}{2\pi i} \int_{-\infty}^{\infty} G_k(E) \exp\left[\frac{i(E_k - E)t}{\hbar}\right] dE \tag{8-82}$$

Obviously, the probability amplitude $c_k(t)$ and its Fourier transform $G_k(E)$ are closely related to one another, and it is easy to derive one of the two quantities once we know the other. Our purpose is to derive a set of perturbation equations for the Fourier transforms $G(E)$. This procedure bears a resemblance to the Green-function formalism which we discussed in Section 7-10.

We must make some mathematical adjustments in order to use the Fourier transform method. From a physical point of view we have observed that our perturbation equations (8-80) are valid only for $t \geq 0$ and that the behavior of the coefficients $c_k(t)$ for $t<0$ is of no importance. However, in our procedure we must define our coefficients for all values of $t$, both positive and negative, and we must define a set of equations that are valid for all times. From a physical point of view we may choose anything we want for the coefficients at negative times and we make a choice that is mathematically convenient, that is,

$$c_o(t) = c_k(t) = 0 \qquad t<0 \tag{8-83}$$

Our perturbation equations take a slightly different form, namely

$$\frac{dc_k}{dt} = \delta_{k,o}\delta(t) - \frac{i}{\hbar}\sum_n V_{k,n}c_n(t)\exp\left[\frac{i(E_k - E_n)t}{\hbar}\right] \tag{8-84}$$

These equations are identical with Eq. (8-80) for $k \neq o$. In the equation where $k=o$ we have added a term $\delta(t)$ which accounts for the discontinuous jump from zero to unity at the point $t=0$.

We represent the $\delta$ function $\delta(t)$ according to Eq. (8-66),

$$\delta(t) = \frac{1}{2\pi\hbar}\int_{-\infty}^{\infty}\exp\left[\frac{i(E_o - E)t}{\hbar}\right]dE \tag{8-85}$$

and we substitute this representation, together with the Fourier transforms, into the perturbation equations (8-84). We consider the equations for $c_o(t)$ and for $c_k(t)$ separately. In the first case we have

$$\begin{aligned}\int_{-\infty}^{\infty}(E - E_o)G_o(E)\exp\left[\frac{i(E_o - E)t}{\hbar}\right]dE &= \int_{-\infty}^{\infty}\exp\left[\frac{i(E_o - E)t}{\hbar}\right]dE \\ &+ \sum_n V_{o,n}\int_{-\infty}^{\infty}G_n(E)\exp\left[\frac{i(E_o - E)t}{\hbar}\right]dE\end{aligned} \tag{8-86}$$

and in the second case, $k \neq o$, we have

$$\int_{-\infty}^{\infty}(E - E_k)G_k(E)\exp\left[\frac{i(E_k - E)t}{\hbar}\right]dE = \sum_n V_{k,n}\int_{-\infty}^{\infty}G_n(E)\exp\left[\frac{i(E_k - E)t}{\hbar}\right]dE \tag{8-87}$$

Obviously, the equations for the Fourier transform functions $G_n(E)$ are

$$(E - E_o)G_o(E) = 1 + \sum_n V_{o,n}G_n(E) \tag{8-88}$$

for $G_o(E)$ and

$$(E - E_k)G_k(E) = \sum_n V_{k,n}G_n(E) \tag{8-89}$$

for $k \neq o$.

In order to solve the above equations we express all $G_k(E)$ in terms of $G_o(E)$ and of the new set of unknowns $W_k(E)$ by making the substitution

$$G_k(E) = W_k(E)G_o(E) \tag{8-90}$$

The first equation (8-88) then becomes

$$(E-E_o)G_o(E) = 1 + G_o(E)\left[V_{o,o} + \sum_{n\neq o} V_{o,n}W_n(E)\right] \tag{8-91}$$

This gives an expression for $G_o(E)$ in terms of $W_n(E)$:

$$G_o(E) = \left[(E-E_o) - V_{o,o} - \sum_{n\neq o} V_{o,n}W_n(E)\right]^{-1} \tag{8-92}$$

The other equations (8-89) for $k\neq o$ become

$$(E-E_k)W_k(E) = V_{k,o} + \sum_{n\neq o} V_{k,n}W_n(E) \tag{8-93}$$

Again, Eq. (8-93) bears a resemblance to thc Green-function equations of Section 7-10. It may seem at first that we can solve by means of iterative methods once we have divided by $E-E_k$. However, we have to be careful in doing so. If we consider the equation

$$xf(x) = g(x) \tag{8-94}$$

then it follows from Eq. (8-69) that we can write

$$xf(x) = g(x) + \lambda x\delta(x)g(x) \tag{8-95}$$

where $\lambda$ is an arbitrary parameter. If we divide the latter equation by $x$, we obtain

$$f(x) = x^{-1}g(x) + \lambda\delta(x)g(x) \tag{8-96}$$

It appears that the solution of Eq. (8-94), obtained by dividing the equation by $x$, is not unique because every value of $\lambda$ gives a legitimate solution. It can be shown now that the proper solutions of Eq. (8-93), which are consistent with our initial values for $c_k(t)$ and which are consistent with the physical requirements, are obtained by using the Dirac $\zeta$ function which we introduced in the previous section. We take the functions $W_k(E)$ as

$$W_k(E) = U_k(E)\zeta(E-E_k) \tag{8-97}$$

and we substitute this into Eq. (8-93). The result is

$$U_k(E)=V_{k,o}+\sum_{n\neq o} V_{k,n}U_n(E)\zeta(E-E_n) \tag{8-98}$$

Also, if we substitute the transformations (8-97) into Eq. (8-92), we obtain

$$G_o(E)=\left[(E-E_o)-V_{o,o}-\sum_{n\neq o} V_{o,n}U_n(E)\zeta(E-E_n)\right]^{-1} \tag{8-99}$$

The two Eq. (8-98) and (8-99) are our new purturbation equations.

In many situations the perturbation equations (8-98) can be solved by iterative procedures. If the perturbation is small and the infinite sum is smaller than the first term, then we obtain in successive approximations

$$\begin{aligned} &U_k(E)\approx V_{k,o} \\ &U_k(E)\approx V_{k,o}+\sum_{n\neq o} V_{k,n}V_{n,o}\zeta(E-E_n) \quad \text{and so on} \end{aligned} \tag{8-100}$$

These results are consistent with Section 8-3. The difference is that the higher-order terms are more easily derived by means of the present procedure. The specific form of the solutions $U_n(E)$ depends, of course, on the system under consideration and on the nature of the perturbation. We will discuss some specific applications in Chapter 9.

It should be noted that we have not introduced any approximations in the present section and that our results should have general validity. It is particularly useful that the results are correct for all values of the time and that we can determine the probability amplitudes $c_o(t)$ and $c_k(t)$ for various limiting cases.

We first consider $c_o(t)$, which is determined by the function $G_o(E)$ of Eq. (8-99). In order to determine $G_o(E)$ we introduce the "damping constant" $\Gamma(E)$ by means of

$$\hbar\Gamma(E)=2iV_{o,o}+2i\sum_{n\neq o} V_{o,n}U_n(E)\zeta(E-E_n) \tag{8-101}$$

This "constant" is actually a function of $E$, but in many situations it is a slowly varying function of $E$, and in determining $c_o(t)$ we may treat it as a constant. If we substitute Eq. (8-101) into Eq. (8-99), we obtain

$$G_o(E)=\left[(E-E_o)+\left(\frac{i\hbar}{2}\right)\Gamma(E)\right]^{-1} \tag{8-102}$$

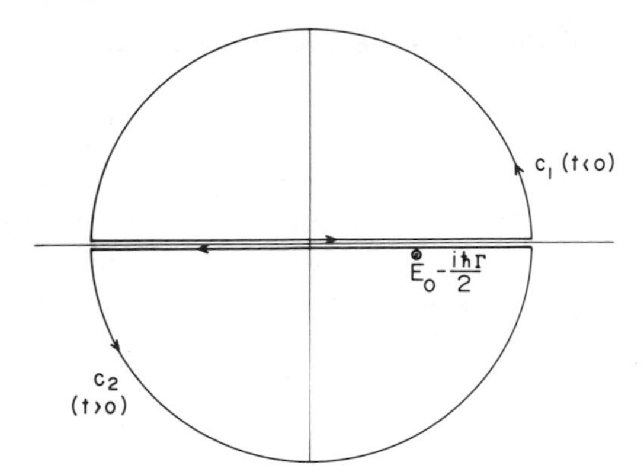


**Fig. 8-3** Integration contours that are used for the evaluation of the probability amplitude $c_0(t)$ of Eq. (8-103).

and by substituting this result into Eq. (8-82), we find

$$c_o(t) = -\frac{1}{2\pi i}\int_{-\infty}^{\infty} \frac{\exp[i(E_o - E)t/\hbar]}{E - E_o + (i\hbar/2)\Gamma(E)}\, dE \tag{8-103}$$

The integral can be evaluated by means of an integration in the complex plane along the paths that we have sketched in Fig. 8-3, that is, along a large half-circle either above or below the real axis and along the real axis. The integrand has a singularity that is determined by the equation

$$E - E_o + \left(\frac{i\hbar}{2}\right)\Gamma(E) = 0 \tag{8-104}$$

We denote the position of the singularity by

$$E = E_o - \left(\frac{i\hbar}{2}\right)[\Gamma_r + i\Gamma_i] \tag{8-105}$$

where $\Gamma_r$ is the real part and $\Gamma_i$ is the imaginary part of the damping constant at the singularity. It is important to note that the singularity is located slightly below the real axis. We evaluate $c_o(t)$ by integrating along either one of the contours that we have sketched in Fig. 8-3. Depending on the value of $t$, we select the one contour for $t<0$ and the other contour for $t>0$. We find that

$$c_o(t) = 0 \qquad t<0$$

$$c_o(t) = \exp\left[-\left(\frac{\Gamma_r t}{2}\right)\right] \cdot \exp\left[-\left(\frac{i\Gamma_i t}{2}\right)\right] \qquad t>0 \tag{8-106}$$

The probability of finding the system in state $o$ is given by

$$c_o(t)c_o^*(t)=\exp(-\Gamma_r t) \qquad (t>0) \tag{8-107}$$

It follows that the parameter $\Gamma_r$ is the average lifetime of the system in the state $o$ and that the probability of finding the system in the state $o$ decays exponentially. The imaginary part $\Gamma_i$ of $\Gamma$ has a different physical meaning; it represents a change in the position of the energy level $o$ due to the perturbation. Such "level shifts" due to the interaction between an atom and a radiation field have in fact been observed.

In order to determine the other probability amplitudes $c_k(t)$ $(k\neq o)$ we note that

$$G_k(E)=\frac{U_k(E)\zeta(E-E_k)}{E-E_o+(i\hbar/2)\Gamma(E)} \tag{8-108}$$

Substitution into Eq. (8-82) gives

$$c_k(t)=-\frac{1}{2\pi i}\int_{-\infty}^{\infty}\frac{U_k(E)\zeta(E-E_k)\exp[i(E_k-E)t/\hbar]}{E-E_o+(i\hbar/2)\Gamma(E)}\,dE \tag{8-109}$$

The results of the integration for the two limiting cases $t\rightarrow\infty$ and $t\rightarrow-\infty$ can be derived by making use of the properties of the $\zeta$ function, in particular, Eq. (8-78). We find that

$$\lim_{t\rightarrow-\infty} c_k(t)=0 \tag{8-110}$$

and

$$\lim_{t\rightarrow\infty} c_k(t)=\frac{U_k(E_k)}{E_k-E_o+(i\hbar/2)\Gamma(E_k)} \tag{8-111}$$

The latter result is quite useful and it cannot be obtained from the conventional perturbation methods described in Section 8-3.

Finally, we wish to mention two additional relations that can be derived from the properties of the Dirac $\zeta$ functions. We do not give the derivations since they are fairly complicated.

In many cases the damping constant $\Gamma_r$ is much smaller than the frequencies $\omega_k=E_k/\hbar$. In such cases we can consider values of $t$ in the interval

$$|\omega_k|^{-1}<t<|\Gamma_r|^{-1} \tag{8-112}$$

For those values of $t$ we can again represent the probability of finding the system in the state $k$ by a time-proportional transition probability $w_k$:

$$c_k(t)c_k^*(t)=w_k t \tag{8-113}$$

This probability is given by

$$w_k = \left(\frac{2\pi}{\hbar}\right) U_k(E_o) U_k^*(E_o) \delta(E_o - E_k) \tag{8-114}$$

It can also be shown that the damping constant $\Gamma_r$, which represents the lifetime of the system in the state $o$, is the sum of all transition probabilities:

$$\Gamma_r = \sum_k w_k \tag{8-115}$$

Again, we do not present the proofs of the two relations (8-113) and (8-115) since they are rather complicated, but we felt that it might be helpful to at least mention them.

The perturbation procedures discussed above are convenient for describing higher-order effects and for dealing with many situations that are not covered by the conventional perturbation methods of Sections 8-3 and 8-4.

## Problems

**1** Consider a system that has two and only two stationary eigenstates 1 and 2 with energy eigenvalues $E_1$ and $E_2$ and eigenfunctions $\Phi_1$ and $\Phi_2$. We assume that the system is in eigenstate 1 until time $t=0$ when we switch on a time-independent perturbation $V$. The matrix elements of $V$ are $V_{11} = V_{22} = 0$ and $V_{12} = V_{21} = \Delta$. Calculate the probability amplitudes $c_1(t)$ and $c_2(t)$ of the system for $t > 0$.

**2** Consider a harmonic oscillator with Hamiltonian $H_o$ which is in its ground state $n=0$ until time $t=0$ when we switch on a time-independent perturbation $V(x) = \hbar\omega x/\sqrt{\alpha}$, where $\omega$ and $\alpha$ are defined in Section 5-7. Determine the various probability amplitudes $c_n(t)$ and, in particular, their limit for $t \to \infty$ by using the procedure of Section 8-6.

**3** Problem 2 can also be solved by expanding its wave function $\Psi(x;t)$ in terms of the stationary states of the Hamiltonian $H = H_o + V$ since the eigenvalues and eigenfunctions of $H$ are easily derived. Give the expressions for the probability amplitudes $b_n(t)$ for the new expansion in terms of the stationary states of the operator $H$.

## Recommended Reading

Time-dependent perturbation theory is discussed in most quantum mechanics textbooks. We quote Kramers' book as well as a text we published previously. The perturbation procedures of Section 8-6 are discussed in the book by Heitler, where a reference is made to a paper by Heitler and Ma. We presume that the perturbation method was first presented in this paper, but we have never been able to read it. We also quote some early work by Wigner and Weisskopf on the theory of the natural line width because it is relevant to the material in this chapter.

H. A. Kramers, *Quantum Mechanics*, North-Holland Publishing Company, Amsterdam, Netherland, 1958.

H. F. Hameka, *Advanced Quantum Chemistry*, Addison-Wesley Publishing Co., Reading, MA., 1965.

W. Heitler, *The Quantum Theory of Radiation*, Oxford University Press, London, 1957.

W. Heitler and S. T. Ma, *Proc. Roy. Irish Ac.*, **52**, 109 (1949).

V. Weisskopf and E. Wigner, *Zeits. f. Physik*, **63**, 54 (1930); **65**, 18 (1930).

CHAPTER NINE

# Interactions Between Radiation and Matter

## 9-1 Introduction

In the present chapter we discuss spectroscopic transitions in atoms or molecules, that is, transitions between different stationary states in atoms or molecules that are caused by radiation fields.

It may seem at first that this is not a complicated problem. All we have to do is take the perturbation term representing the interaction between molecule and radiation field and substitute it into the perturbation expressions of the previous chapter. However, a thorough discussion of radiation-induced transitions requires much more than that. First we must present the mathematical description of the radiation field itself. This involves the definition and the solution of the Maxwell equations. Next we must discuss the interactions between matter and radiation.

We will show that there are two different, alternative mathematical descriptions of the radiation field. The first one is based on the conventional solution of the Maxwell equations and it is known as the semiclassical description. The second one relates the mathematical equations to the quantum mechanical description of the harmonic oscillator and it is known as the quantum description of the radiation field.

These mathematical models are hard to visualize, and sometimes it is difficult to understand how they relate to the observable physics. It may be helpful to recall our discussion in Section 5-11 of the rigid rotor. We separated the coordinates of the center of gravity and we obtained the Schrödinger equation (5-245) as a representative equation for the motion of the rigid rotor. This Schrödinger equation is identical with the equation representing the motion of a particle on the surface of a sphere. Since the two equations of motion are the same, their solutions are identical also. This means that we can represent the motion of a rigid rotor by the motion of a particle on a sphere, and vice versa, since the mathematical descriptions of the two models are the same and their conclusions should be consistent with experimental observations.

In judging the validity of a theoretical model we should consider (1) the situations that it applies to, (2) the applicability of its mathematical conclusions to the prediction of experimental observables, (3) the agreement between its conclusions and the available experimental information, (4) its logical consistency, and (5) its resemblance to reality. A theoretical model may be quite acceptable even if it does not describe all aspects of a given area as long as it is well understood which areas it applies to and as long as it is not inconsistent with experimental facts. It is possible to have different theoretical models for the same phenomenon as long as the different models all satisfy the above five conditions.

In the theory of the radiation field it is customary to use different theoretical models. According to the above arguments this is quite permissible. Each of the models is satisfactory for dealing with some aspects of the situation, but it becomes less suitable if we want to include additional features. For instance, the quantum description of the radiation field is quite adequate for questions concerning energy, energy transfer between field and molecules, transition probabilities, and so on. The model is much less suitable for questions concerning coherence effects and various other problems.

We also discuss the relations between the theoretical results and the experimental parameters that describe the emission and absorption of radiation. We take the latter as the Einstein coefficients of absorption and emission. They were first introduced by Einstein in 1917 in order to describe the two different types of emission, stimulated and spontaneous, and in order to relate the various coefficients to Planck's radiation theory.

We wish to discuss a variety of topics in addition to the theory of transition probabilities, namely the Einstein coefficients, the Maxwell equations, and the quantum theory of the radiation field. We decided, therefore, to dedicate a separate chapter to the discussion of these various theories.

## 9-2 The Einstein Coefficients of Absorption and Emission

During the years 1916 and 1917 Einstein published an analysis of the various mechanisms in emission and absorption of radiation by molecules. At the time, the purpose of this analysis was the derivation of Planck's radiation expression, which we discussed in Section 1-4. However, Einstein's work made an important contribution to the theoretical understanding of spectroscopic transitions. In fact, Einstein's papers supply the theoretical basis for the recent invention of the laser.

It can be seen from our treatment of perturbation in the previous chapter and from general considerations that a radiation field can cause transitions from one stationary state to another in an atomic or molecular system. If the molecular system is in its lowest eigenstate $o$ and if we switch on an appropriate radiation field at a time $t=0$, the radiation field can cause absorption transitions to the various higher stationary states of the molecule. We expect that we may observe an absorption of radiation while the molecule changes from its ground state $o$ to a specific excited state $k$.

In order to give a quantitative expression for the absorption process, Einstein considered a container filled with isotropic radiation with energy density $\rho$ and with a certain number of atoms. The energy per unit volume of the part of the radiation with frequencies between $\nu$ and $\nu + d\nu$ is given by

$$dE = \rho(\nu)\, d\nu \tag{9-1}$$

Furthermore, there are $N_o$ molecules per unit volume in the ground state $o$ and $N_k$ molecules per unit volume in the stationary state $k$. The probability per unit time of an upward transition from state $o$ to state $k$ due to the perturbation of the radiation field can then be written as

$$P_{\text{abs}} = B_{o\to k}\rho(\nu_k) \tag{9-2}$$

Here $B_{o\to k}$ is known as the Einstein coefficient of absorption.

It is easily seen that there is an equivalent mechanism for a downward transition from state $k$ to state $o$ due to the perturbation of the radiation field. If we follow the perturbation procedures of the previous chapter and if we assume that the system is initially in the state $k$, until we switch on the perturbation, it can be seen that the probability of a transition from $k$ to $o$ is exactly the same as for a transition from $o$ to $k$. We can write the probability of this radiation-induced downward transition as

$$P'_{em} = B_{k\to o}\rho(\nu_k) \tag{9-3}$$

We know from experience that there is a second mechanism for downward transitions. If we excite an atom or molecule to an excited state $k$, for example, by means of an electric discharge, the molecule will drop back to its ground state spontaneously, even in the absence of a radiation field. Einstein denoted this type of process by spontaneous emission. The corresponding transition probability per unit time is given by

$$P''_{em} = A_{k\to o} \tag{9-4}$$

Here $A_{k\to o}$ is known as the Einstein coefficient of spontaneous emission. The other emission process, represented by Eq. (9-3), became known as stimulated emission, and $B_{k\to o}$ is the Einstein coefficient of stimulated emission.

It can be seen that the total number of upward transitions per unit time is given by

$$N_o B_{o\to k}\rho(\nu_k) \tag{9-5}$$

and the total number of downward transitions per unit time is

$$N_k[B_{k\to o}\rho(\nu_k) + A_{k\to o}] \tag{9-6}$$

We have thus

$$\frac{dN_o}{dt} = -\frac{dN_k}{dt} = N_k[B_{k\to o}\rho(\nu_k) + A_{k\to o}] - N_o B_{o\to k}\rho(\nu_k) \qquad (9\text{-}7)$$

In the equilibrium condition we have a steady state, and Eq. (9-7) should be equal to zero:

$$\frac{N_o}{N_k} = \frac{B_{k\to o}\rho(\nu_k) + A_{k\to o}}{B_{o\to k}\rho(\nu_k)} \qquad (9\text{-}8)$$

We know that in the equilibrium condition the ratio between $N_k$ and $N_o$ should be given by the Maxwell distribution function:

$$\frac{N_o}{N_k} = \exp\left(\frac{h\nu_k}{kT}\right) \qquad (9\text{-}9)$$

where

$$h\nu_k = E_k - E_o \qquad (9\text{-}10)$$

By equating Eqs. (9-8) and (9-9) we obtain

$$\rho(\nu_k) = \frac{A_{k\to o}}{B_{o\to k}\exp(h\nu_k/kT) - B_{k\to o}} \qquad (9\text{-}11)$$

It may be recalled from Eqs. (1-50) and (1-76) that Planck's radiation formula is

$$\rho(\nu) = \frac{8\pi\nu^2}{c^3}\,\frac{h\nu}{\exp(h\nu/kT) - 1} \qquad (9\text{-}12)$$

The two results (9-11) and (9-12) are identical if we assume that the Einstein coefficients obey the relations

$$B_{o\to k} = B_{k\to o} \qquad (9\text{-}13)$$

and

$$A_{k\to o} = \left(\frac{8\pi h\nu^3}{c^3}\right)B_{k\to o} \qquad (9\text{-}14)$$

We indicated already that the first relation (9-13) is consistent with the perturbation treatment since the absorption and induced emission should have the same transition probabilities. The second relation (9-14) gives the ratio between spontaneous and induced emission.

The above results are useful for two different reasons. First, in some theoretical models it is relatively easy to calculate one of the Einstein coefficients, but it is awkward to calculate the coefficient of spontaneous emission. If that happens we can derive its value from Eq. (9-14). Second, the Einstein coefficients are generally used to represent the magnitudes of the various emission and absorption processes, and they supply the connection between the theoretical perturbation results and the experimental quantities. The problem is that in the subsequent perturbation treatments we introduce a beam of linearly polarized light, whereas in Einstein's analysis the radiation is supposed to be isotropic and three-dimensional. We must therefore relate these two different situations to one another.

In the following sections we derive the theoretical description of the radiation field by solving the Maxwell equations. This will be preceded by a brief review of vector analysis.

## 9-3 Vector Analysis

We discussed some properties of vectors and vector fields in Sections 1-2 and 1-4, but in order to deal with the Maxwell equations in the following sections we need to know some additional results of vector analysis. We consider only three-dimensional vectors.

A vector $\mathbf{v}$ is determined by its three components $v_x$, $v_y$, and $v_z$. If each component is a continuous and single-valued function of the Cartesian coordinates $x$, $y$, and $z$, we have a vector field. The components may also depend on the time $t$, in addition to $x$, $y$, and $z$. A scalar field is determined by a single function $\phi(x, y, z; t)$ of the Cartesian coordinates $x$, $y$, and $z$ and of the time $t$.

In Section 1-2 we introduced the symbol $\nabla$, which stands for the three operators $(\partial/\partial x)$, $(\partial/\partial y)$, and $(\partial/\partial z)$. The vector equation

$$\mathbf{u} = \nabla \phi(x, y, z; t) \tag{9-15}$$

stands for the three equations

$$u_x = \frac{\partial \phi}{\partial x}$$

$$u_y = \frac{\partial \phi}{\partial y}$$

$$u_z = \frac{\partial \phi}{\partial z} \tag{9-16}$$

If the vector $\mathbf{u}$ is defined by Eq. (9-15), it is called the gradient of the function $\phi$.

From the vector field $\mathbf{v}$ we derive two quantities that are widely used in the mathematical description of the radiation field. The first one is a scalar. It is called the divergence of $\mathbf{v}$ or div $\mathbf{v}$ and it is defined as

$$\operatorname{div} \mathbf{v} = \frac{\partial v_x}{\partial x} + \frac{\partial v_y}{\partial y} + \frac{\partial v_z}{\partial z} = \nabla \cdot \mathbf{v} \tag{9-17}$$

It can be written as the dot product of $\nabla$ and of the vector $\mathbf{v}$. The second quantity is a vector. It is called curl $\mathbf{v}$ and it is defined by making use of the cross product:

$$\operatorname{curl} \mathbf{v} = \nabla \times \mathbf{v} \tag{9-18}$$

According to the definition (1-86) of the cross product its components are given by

$$\begin{aligned}
(\operatorname{curl} \mathbf{v})_x &= \frac{\partial v_z}{\partial y} - \frac{\partial v_y}{\partial z} \\
(\operatorname{curl} \mathbf{v})_y &= \frac{\partial v_x}{\partial z} - \frac{\partial v_z}{\partial x} \\
(\operatorname{curl} \mathbf{v})_z &= \frac{\partial v_y}{\partial x} - \frac{\partial v_x}{\partial y}
\end{aligned} \tag{9-19}$$

The following two vector properties are used in solving the Maxwell equations. First, if the vector $\mathbf{v}$ can be expressed as the gradient of a scalar field $\phi$, its curl is $\mathbf{0}$. In other words,

$$\operatorname{curl} \operatorname{grad} \phi = \nabla \times (\nabla \phi) = \mathbf{0} \tag{9-20}$$

This is easily verified by substituting Eq. (9-16) into Eq. (9-19). In a similar fashion it can be shown that

$$\operatorname{div} \operatorname{curl} \mathbf{v} = \nabla \cdot (\nabla \times \mathbf{v}) = 0 \tag{9-21}$$

by substituting Eq. (9-19) into Eq. (9-17).

We mention two additional relations, namely

$$\begin{aligned}
\operatorname{curl} \operatorname{curl} \mathbf{v} &= \operatorname{grad} \operatorname{div} \mathbf{v} - \Delta \mathbf{v} \\
\Delta &= \frac{\partial^2}{\partial x^2} + \frac{\partial^2}{\partial y^2} + \frac{\partial^2}{\partial z^2}
\end{aligned} \tag{9-22}$$

and

$$\operatorname{div} (\mathbf{u} \times \mathbf{v}) = \mathbf{v} \cdot \operatorname{curl} \mathbf{u} - \mathbf{u} \cdot \operatorname{curl} \mathbf{v} \tag{9-23}$$

They can both be proved by substituting the various definitions.

As we shall see, the above vector properties will prove to be useful in solving the Maxwell equations in the following section.

## 9-4 The Maxwell Equations

An electromagnetic field is determined by the electric and magnetic field strengths. The electric field strengths are represented by a vector field $\mathbf{E}(x, y, z; t)$, which depends on the Cartesian coordinates $x$, $y$, and $z$ and on the time $t$. We can measure the value of $\mathbf{E}$ at a given point $P$ and at a given time $t_o$ by placing a small test charge $\delta q$ at $P$ at the time $t_o$. The force $\mathbf{F}$ that the charge experiences is given by

$$\mathbf{F}=\mathbf{E}\,\delta q \tag{9-24}$$

The vector $\mathbf{E}$ is then defined as the electric field strength. In a similar fashion we can define the magnetic field strengths, which are represented by a vector field $\mathbf{H}(x, y, z; t)$.

The field strengths $\mathbf{E}$ and $\mathbf{H}$ and their time dependence are determined by the electric charges and currents. The electric charges are described by the charge density function $\rho(x, y, z; t)$. By definition the charge $\delta q$ in a small volume element $d\tau$ surrounding the point $(x_o, y_o, z_o)$ at a time $t_o$ is given by

$$\delta q=\rho(x_o, y_o, z_o; t_o)\,d\tau \tag{9-25}$$

In addition to the continuous function $\rho$ we can also have point charges at various points in space. If the velocity of the charge at a given point is given by the vector $\mathbf{v}(x, y, z; t)$, the current density vector $\mathbf{i}$ is defined as

$$\mathbf{i}(x, y, z; t)=\mathbf{v}(x, y, z; t)\rho(x, y, z; t) \tag{9-26}$$

The Maxwell equations are the fundamental equations that are the basis for the theory of electromagnetic fields. They are a set of four differential equations that describe the relations between the electric and magnetic field strengths $\mathbf{E}$ and $\mathbf{H}$ and the charge density and current density $\rho$ and $\mathbf{i}$. Two of the equations just contain the electric and magnetic field vectors $\mathbf{E}$ and $\mathbf{H}$:

$$\begin{aligned} &\operatorname{div}\mathbf{H}=0 \\ &\operatorname{curl}\mathbf{E}+\frac{1}{c}\frac{\partial\mathbf{H}}{\partial t}=\mathbf{0} \end{aligned} \tag{9-27}$$

The other two equations contain the current density $\mathbf{i}$ and the charge density $\rho$ as well:

$$\begin{aligned} &\operatorname{div}\mathbf{E}=4\pi\rho \\ &\operatorname{curl}\mathbf{H}-\frac{1}{c}\frac{\partial\mathbf{E}}{\partial t}=\frac{4\pi}{c}\mathbf{i} \end{aligned} \tag{9-28}$$

The set of equations was first proposed by Maxwell in 1865. From the two Eqs. (9-28) it can be derived that

$$\operatorname{div}\mathbf{i}+\frac{\partial\rho}{\partial t}=0 \tag{9-29}$$

which represents the conservation of charge.

The Maxwell equations are most conveniently solved by introducing a set of potential functions, namely the scalar potential function $\phi$ and the vector potential $\mathbf{A}$. The potential functions are defined in such a way that the Maxwell equations (9-27) are automatically satisfied.

It follows from Eq. (9-21) that the first Maxwell equation (9-27) is automatically satisfied if we represent the magnetic field strength $\mathbf{H}$ as the curl of another vector field $\mathbf{A}$:

$$\mathbf{H}=\operatorname{curl}\mathbf{A} \tag{9-30}$$

If we substitute this into the second Eq. (9-27), we obtain

$$\operatorname{curl}\left(\mathbf{E}+\frac{1}{c}\frac{\partial\mathbf{A}}{\partial t}\right)=\mathbf{0} \tag{9-31}$$

It follows from Eq. (9-20) that this equation is automatically satisfied if we represent the left side as

$$\mathbf{E}+\frac{1}{c}\frac{\partial\mathbf{A}}{\partial t}=-\operatorname{grad}\phi \tag{9-32}$$

where $\phi$ is a scalar field. By means of Eqs. (9-30) and (9-32) we have expressed the electric and magnetic field vectors $\mathbf{E}$ and $\mathbf{H}$ in terms of a scalar and a vector potential, $\phi$ and $\mathbf{A}$, respectively:

$$\begin{aligned}\mathbf{E}&=-\operatorname{grad}\phi-\frac{1}{c}\frac{\partial\mathbf{A}}{\partial t}\\ \mathbf{H}&=\operatorname{curl}\mathbf{A}\end{aligned} \tag{9-33}$$

By substituting Eqs. (9-33) into the second set of Maxwell equations (9-28) we obtain the differential equations for the scalar potential $\phi$ and the vector potential $\mathbf{A}$:

$$\begin{aligned}-\operatorname{div}\operatorname{grad}\phi-\frac{1}{c}\operatorname{div}\frac{\partial\mathbf{A}}{\partial t}&=4\pi\rho\\ \operatorname{curl}\operatorname{curl}\mathbf{A}+\frac{1}{c}\operatorname{grad}\frac{\partial\phi}{\partial t}+\frac{1}{c^2}\frac{\partial^2\mathbf{A}}{\partial t^2}&=\frac{4\pi}{c}\mathbf{i}\end{aligned} \tag{9-34}$$

These equations can be transformed further by making use of the vector relation (9-22):

$$\Delta\phi + \frac{1}{c}\operatorname{div}\frac{\partial \mathbf{A}}{\partial t} = -4\pi\rho$$

$$\frac{1}{c^2}\frac{\partial^2 \mathbf{A}}{\partial t^2} - \Delta \mathbf{A} + \operatorname{grad}\left(\operatorname{div}\mathbf{A} + \frac{1}{c}\frac{\partial \phi}{\partial t}\right) = \frac{4\pi}{c}\mathbf{i} \tag{9-35}$$

These are the basic differential equations that represent $\phi$ and $\mathbf{A}$ as functions of $\rho$ and $\mathbf{i}$.

Before making any attempts to solve the differential equations (9-35) we should discuss the gauge of the potentials. We pointed out in the beginning of this section that the field strengths $\mathbf{E}$ and $\mathbf{H}$ are physical observables. They should, therefore, be real and single-valued functions of the variables $x$, $y$, $z$, and $t$. No such restriction applies to the potentials $\phi$ and $\mathbf{A}$. The potentials are mathematical quantities that we introduced in order to help us solve the equations, but their magnitudes cannot be measured directly. We should realize that a certain ambiguity in the definition of the potentials is quite permissible as long as this ambiguity has no effect on the definition of the electric and magnetic field vectors $\mathbf{E}$ and $\mathbf{H}$.

We imagine that our system is represented by a specific set of potentials $\phi_o$ and $\mathbf{A}_o$. Let us now consider a vector $\mathbf{A}$, which we define as

$$\mathbf{A} = \mathbf{A}_o - \operatorname{grad}\chi \tag{9-36}$$

where $\chi$ is a function of $x$, $y$, $z$, and $t$. It follows from Eq. (9-20) that

$$\operatorname{curl}\mathbf{A} = \operatorname{curl}\mathbf{A}_o - \operatorname{curl}\operatorname{grad}\chi = \operatorname{curl}\mathbf{A}_o \tag{9-37}$$

According to Eq. (9-37) the magnetic field vector $\mathbf{H}$ derived from $\mathbf{A}$ is the same as thc field vector derived from $\mathbf{A}_o$. We also define a function $\phi$ as

$$\phi = \phi_o + \frac{1}{c}\frac{\partial \chi}{\partial t} \tag{9-38}$$

We have

$$-\operatorname{grad}\phi - \frac{1}{c}\frac{\partial \mathbf{A}}{\partial t} = -\operatorname{grad}\phi_o - \frac{1}{c}\frac{\partial \mathbf{A}_o}{\partial t} \tag{9-39}$$

The electric field vector $\mathbf{E}$ derived from $\phi$ and $\mathbf{A}$ is equal to the field vector derived from $\phi_o$ and $\mathbf{A}_o$. We see that we can transform the set of potentials $\phi_o$ and $\mathbf{A}_o$ according to Eqs. (9-36) and (9-38) without changing the physical observables of the system. Such a potential transformation is known as a gauge transformation.

The ambiguity in the definition of the potentials can be used to help solve the Maxwell equations. For instance, it is possible to choose a gauge so that

$$\operatorname{div}\mathbf{A}+\frac{1}{c}\frac{\partial\phi}{\partial t}=0 \tag{9-40}$$

This is known as the Lorentz gauge. If we substitute Eq. (9-40) into the Maxwell equations (9-35), they become much simpler, namely

$$\left(\frac{1}{c^2}\frac{\partial^2}{\partial t^2}-\Delta\right)\phi=4\pi\rho \qquad \left(\frac{1}{c^2}\frac{\partial^2}{\partial t^2}-\Delta\right)\mathbf{A}=\frac{4\pi}{c}\mathbf{i} \tag{9-41}$$

A different choice of gauge is the Coulomb gauge, which is defined by the condition

$$\operatorname{div}\mathbf{A}=0 \tag{9-42}$$

If we substitute this constraint into the Maxwell equations (9-35), they take the form

$$\begin{gathered}\Delta\phi=-4\pi\rho\\ \left(\frac{1}{c^2}\frac{\partial^2}{\partial t^2}-\Delta\right)\mathbf{A}+\frac{1}{c}\operatorname{grad}\frac{\partial\phi}{\partial t}=\frac{4\pi}{c}\mathbf{i}\end{gathered} \tag{9-43}$$

In Section 9-4 we discuss the solution of the Maxwell equations for the radiation field.

Finally, we wish to report the form of the Hamiltonian of a particle in an electromagnetic field. We consider an electron with mass $m$ and charge $(-e)$, moving in an electromagnetic field described by the potentials $\phi$ and $\mathbf{A}$. It can be deduced from relativity theory that the energy of the electron is given by

$$E=\frac{1}{2m}\left(\mathbf{p}+\frac{e}{c}\mathbf{A}\right)^2-e\phi \tag{9-44}$$

Obviously, the corresponding Hamiltonian is

$$\mathcal{H}=\frac{1}{2m}\left(\frac{\hbar}{i}\nabla+\frac{e}{c}\mathbf{A}\right)^2-e\phi \tag{9-45}$$

The Hamiltonian contains the potentials and not the field strengths, and we should consider the question of how a gauge transformation affects the solutions of the corresponding Schrödinger equation.

We consider a stationary state of the Hamiltonian (9-45) and we assume that for a given choice of potentials $A_o$ and $\phi_o$ the Hamiltonian $\mathcal{H}_o$ has an eigenvalue $E_n$ and a corresponding eigenfunction $\Phi_n$:

$$\left[\frac{1}{2m}\left(\frac{\hbar}{i}\nabla+\frac{e}{c}\mathbf{A}_o\right)^2-e\phi_o\right]\Phi_n=E_n\Phi_n \tag{9-46}$$

A gauge transformation requires a simultaneous transformation of the potentials, according to

$$\mathbf{A}'=\mathbf{A}-\nabla\chi$$

$$\phi'=\phi+\frac{1}{c}\frac{\partial\chi}{\partial t} \tag{9-47}$$

and of the eigenfunction $\Phi_n$, according to

$$\Phi_n'=\Phi_n\exp\left(\frac{ie\chi}{\hbar c}\right) \tag{9-48}$$

Since we deal with a stationary state we restrict $\chi$ to being time independent. We have

$$\left(\frac{\hbar}{i}\nabla+\frac{e}{c}\mathbf{A}'\right)\Phi_n'=\left(\frac{\hbar}{i}\nabla+\frac{e}{c}\mathbf{A}-\frac{e}{c}(\nabla\chi)\right)\Phi_n\exp\left(\frac{ie\chi}{\hbar c}\right)$$

$$=\exp\left(\frac{ie\chi}{\hbar c}\right)\left(\frac{\hbar}{i}\nabla+\frac{e}{c}\mathbf{A}\right)\Phi_n \tag{9-49}$$

Consequently,

$$\mathcal{H}'\Phi_n'=\left[\frac{1}{2m}\left(\frac{\hbar}{i}\nabla+\frac{e}{c}\mathbf{A}'\right)^2-e\phi'\right]\Phi_n\exp\left(\frac{ie\chi}{\hbar c}\right)$$

$$=\mathrm{cxp}\left(\frac{ie\chi}{\hbar c}\right)\left[\frac{1}{2m}\left(\frac{\hbar}{i}\nabla+\frac{e}{c}\mathbf{A}\right)^2-e\phi\right]\Phi_n$$

$$=\exp\left(\frac{ie\chi}{\hbar c}\right)\mathcal{H}\Phi_n=E_n\Phi_n' \tag{9-50}$$

It follows that the function $\Phi_n'$ of Eq. (9-48) is an eigenfunction of the Hamiltonian $\mathcal{H}'$ containing the new potentials $\mathbf{A}'$ and $\phi'$. The eigenvalue $E_n$ is not affected by the gauge transformation. We see that a gauge transformation in quantum mechanics involves a simultaneous transformation of the potentials and the eigenfunction as described by Eqs. (9-47) and (9-48). This transformation does not affect any quantities that are physically observable.

The above argument is easily extended to apply also to the time-dependent Schrödinger equation, but we do not present this generalization here.

## 9-5 The Radiation Field

A radiation field can be defined as the solution of the Maxwell equations in the absence of any charges or currents. In other words, the mathematical representation of the radiation field is obtained from the Maxwell equations (9-35) by substituting $\rho=0$ and $\mathbf{i}=\mathbf{0}$. In order to solve the equations we choose a gauge where

$$\operatorname{div}\mathbf{A}=0 \tag{9-51}$$

The first Eq. (9-35) is then solved by

$$\phi=0 \tag{9-52}$$

and the second equation takes the form

$$\left(\frac{1}{c^2}\frac{\partial^2}{\partial t^2}-\Delta\right)\mathbf{A}=\mathbf{0} \tag{9-53}$$

It can be seen that the radiation field can be represented by a vector potential $\mathbf{A}$, since the scalar potential $\phi$ may be taken equal to zero.

In order to find a solution of the differential equation (9-53) we substitute

$$\mathbf{A}(\mathbf{r};t)=q_\lambda(t)\mathbf{A}_\lambda(x,y,z) \tag{9-54}$$

into the equation. This gives

$$\mathbf{A}_\lambda(x,y,z)\frac{\partial^2 q_\lambda}{\partial t^2}-c^2 q_\lambda \Delta\mathbf{A}_\lambda(x,y,z)=\mathbf{0} \tag{9-55}$$

We can separate this into two equations, an equation for $q_\lambda(t)$ and an equation for $\mathbf{A}_\lambda(x,y,z)$:

$$\frac{d^2 q_\lambda}{dt^2}+\omega_\lambda^2 q_\lambda=0$$

$$\Delta\mathbf{A}_\lambda+\frac{\omega_\lambda^2}{c^2}\mathbf{A}_\lambda=\mathbf{0} \tag{9-56}$$

In addition we must consider the restraint (9-51), which reduces to

$$\operatorname{div}\mathbf{A}_\lambda=0 \tag{9-57}$$

First we consider the equations for $\mathbf{A}_\lambda$. The differential equation (9-56) has an infinite number of solutions. Each solution can be written as

$$\mathbf{A}_\lambda=\mathbf{v}_\lambda \exp[i(\mathbf{k}_\lambda\cdot\mathbf{r})] \tag{9-58}$$

In order to satisfy Eq. (9-56) the vector $\mathbf{k}_\lambda$ must satisfy the condition

$$k_\lambda^2 = \frac{\omega_\lambda^2}{c^2} \tag{9-59}$$

The condition (9-57) is satisfied if

$$(\mathbf{k}_\lambda \cdot \mathbf{v}_\lambda) = 0 \tag{9-60}$$

In order to label the various solutions it is helpful to assume that the radiation field is enclosed in a large cubic box of length $L$. We then impose periodic boundary conditions on the solutions

$$\mathbf{A}_\lambda\left(\frac{-L}{2}, y, z\right) = \mathbf{A}_\lambda\left(\frac{L}{2}, y, z\right)$$

$$\mathbf{A}_\lambda\left(x, \frac{-L}{2}, z\right) = \mathbf{A}_\lambda\left(x, \frac{L}{2}, z\right)$$

$$\mathbf{A}_\lambda\left(x, y, \frac{-L}{2}\right) = \mathbf{A}_\lambda\left(x, y, \frac{L}{2}\right) \tag{9-61}$$

This means that the vector $\mathbf{k}_\lambda$ can have the values

$$\mathbf{k}_\lambda = \left(\frac{2\pi n_{\lambda x}}{L}, \frac{2\pi n_{\lambda y}}{L}, \frac{2\pi n_{\lambda z}}{L}\right) \tag{9-62}$$

where $n_{\lambda x}$, $n_{\lambda y}$, and $n_{\lambda z}$ are integers.

For each value of the vector $\mathbf{k}_\lambda$ there are two different solutions, because we still have to choose the direction of the vector $\mathbf{v}_\lambda$. We know that $\mathbf{v}_\lambda$ must be perpendicular to $\mathbf{k}_\lambda$ and we can choose two independent perpendicular directions for $\mathbf{v}_\lambda$. For instance, if $\mathbf{k}_\lambda$ is given by

$$\mathbf{k}_\lambda = (k_\lambda \sin\theta\cos\phi, k_\lambda \sin\theta\sin\phi, k_\lambda\cos\theta) \tag{9-63}$$

then the two possible choices for $\mathbf{v}_\lambda$ are

$$\mathbf{v}_{\lambda,1} = (v_\lambda\cos\theta\cos\phi, v_\lambda\cos\theta\sin\phi, -v_\lambda\sin\theta)$$

$$\mathbf{v}_{\lambda,2} = (v_\lambda\sin\phi, -v_\lambda\cos\phi, 0) \tag{9-64}$$

Because of the boundary conditions (9-61), it is easily proved that different solutions $\mathbf{A}_\lambda$ and $\mathbf{A}_\mu$ are orthogonal:

$$\int \mathbf{A}_\lambda^* \cdot \mathbf{A}_\mu \, d\tau = 0 \tag{9-65}$$

We normalize the solutions by imposing the condition

$$\int \mathbf{A}_\lambda^* \cdot \mathbf{A}_\lambda \, d\tau = 4\pi c^2 \tag{9-66}$$

In the classical description we solve the differential equation (9-56) for the time-dependent coefficients $q_\lambda(t)$. It is obvious that the equation has two solutions for each value $\omega_\lambda$, namely

$$q_\lambda(t) = q_\lambda^o \exp(\pm i\omega_\lambda t) \tag{9-67}$$

Each particular solution of the Maxwell equations can now be expressed as a product of the vector $\mathbf{A}_\lambda$ of Eq. (9-58) and one of the time-dependent coefficients, $q_\lambda(t)$, of Eq. (9-67). The general solution of the Maxwell equations (9-53) can then be written as

$$\mathbf{A} = \sum_\lambda (q_\lambda + q_\lambda^*)\mathbf{A}_\lambda \tag{9-68}$$

We have selected this specific superposition of all particular solutions because it is a real function. It can be seen from Eq. (9-58) that the complex conjugate of $\mathbf{A}_\lambda$ is obtained by replacing $\mathbf{k}_\lambda$ by $-\mathbf{k}_\lambda$. The sum of both contributions gives a real vector.

Now that we have obtained the general solution for the vector potential $\mathbf{A}$ we proceed to derive expressions for the electric and magnetic field strengths $\mathbf{E}$ and $\mathbf{H}$ and for the energy $U$ of the radiation field. It follows from Eq. (9-33) that

$$\mathbf{H} = \operatorname{curl} \mathbf{A} = \sum_\lambda (q_\lambda + q_\lambda^*) \operatorname{curl} \mathbf{A}_\lambda \tag{9-69}$$

and that

$$\mathbf{E} = -\frac{1}{c}\frac{\partial \mathbf{A}}{\partial t} = -\frac{1}{c}\sum_\lambda \left( \frac{\partial q_\lambda}{\partial t} + \frac{\partial q_\lambda^*}{\partial t} \right)\mathbf{A}_\lambda \tag{9-70}$$

By substituting Eq. (9-67) we obtain

$$\mathbf{E} = \sum_\lambda \left( \frac{-i\omega_\lambda}{c} \right)(q_\lambda - q_\lambda^*)\mathbf{A}_\lambda \tag{9-71}$$

According to the theory of electromagnetic fields the energy density of the field is given by

$$\frac{1}{8\pi}(E^2 + H^2) \tag{9-72}$$

and the total energy of the radiation field can be written as

$$U = \frac{1}{8\pi} \int (\mathbf{E} \cdot \mathbf{E}^* + \mathbf{H} \cdot \mathbf{H}^*) \, d\tau \tag{9-73}$$

Here the integration should be performed over the volume of the cubic box of length $L$. The field strengths $\mathbf{E}$ and $\mathbf{H}$ are real and we can represent $U$ in the form of Eq. (9-73).

We evaluate the energy by substituting the results (9-69) and (9-70) for the field strengths into Eq. (9-73):

$$\begin{aligned} U = & \frac{1}{8\pi} \sum_{\lambda} \sum_{\mu} \frac{\omega_\lambda \omega_\mu}{c^2} (q_\lambda - q_\lambda^*)(q_\mu^* - q_\mu) \int (\mathbf{A}_\lambda \cdot \mathbf{A}_\mu^*) \, d\tau \\ & + \frac{1}{8\pi} \sum_{\lambda} \sum_{\mu} (q_\lambda + q_\lambda^*)(q_\mu + q_\mu^*) \int \operatorname{curl} \mathbf{A}_\lambda \cdot \operatorname{curl} \mathbf{A}_\mu^* \, d\tau \end{aligned} \tag{9-74}$$

The first sum is easily evaluated by using Eqs. (9-65) and (9-66). The second sum can be determined by an integration by parts. We have

$$\begin{aligned} \int (\operatorname{curl} \mathbf{A}_\lambda \cdot \operatorname{curl} \mathbf{A}_\mu^*) \, d\tau &= \int (\mathbf{A}_\mu^* \cdot \operatorname{curl} \operatorname{curl} \mathbf{A}_\lambda) \, d\tau \\ &= -\int \mathbf{A}_\mu^* \cdot \Delta \mathbf{A}_\lambda \, d\tau = \left( \frac{\omega_\lambda^2}{c^2} \right) \int \mathbf{A}_\mu^* \cdot \mathbf{A}_\lambda \, d\tau \end{aligned} \tag{9-75}$$

Here we have used the vector relation (9-22) and the differential equations (9-51) and (9-56). The final integral of Eq. (9-75) can again be calculated by using Eqs. (9-65) and (9-66). By substituting all these results into Eq. (9-74) we obtain

$$\begin{aligned} U &= \sum_{\lambda} \left( \frac{\omega_\lambda^2}{8\pi c^2} \right) (4\pi c^2) [(q_\lambda - q_\lambda^*)(q_\lambda^* - q_\lambda) + (q_\lambda + q_\lambda^*)(q_\lambda + q_\lambda^*)] \\ &= \sum_{\lambda} \omega_\lambda^2 (q_\lambda q_\lambda^* + q_\lambda^* q_\lambda) \end{aligned} \tag{9-76}$$

Finally, we wish to report the classical expression for the interaction between an electron in a potential field and a radiation field. This expression is obtained by substituting Eq. (9-68) for the vector potential $\mathbf{A}$ of the radiation field into Eq. (9-45). We should realize that the last term in Eq. (9-45) is usually written as $V(\mathbf{r})$, the potential energy of the electron. The result is

$$\mathcal{H} = \frac{1}{2m} \left( \frac{\hbar}{i} \nabla + \frac{e}{c} \sum_{\lambda} (q_\lambda + q_\lambda^*) \mathbf{A}_\lambda \right)^2 + V(\mathbf{r}) \tag{9-77}$$

We can separate this into

$$\mathcal{H} = \mathcal{H}_o + \mathcal{H}_{\text{int}} \tag{9-78}$$

where

$$\mathcal{H}_o = \frac{1}{2m}\left(\frac{\hbar}{i}\nabla\right)^2 + V(\mathbf{r}) \tag{9-79}$$

represents the electronic energy. The remaining terms

$$\mathcal{H}_{\text{int}} = \frac{e\hbar}{mci}\sum_{\lambda}(q_\lambda + q_\lambda^*)(\mathbf{A}_\lambda \cdot \nabla) + \frac{e^2}{2mc^2}\sum_{\lambda}\sum_{\mu}(q_\lambda + q_\lambda^*)(q_\mu + q_\mu^*)(\mathbf{A}_\lambda \cdot \mathbf{A}_\mu) \tag{9-80}$$

represent the interaction between the electron and the radiation field. Usually the first term of Eq. (9-80) is much larger than the second term, and in most calculations of radiation-induced transitions it is customary to neglect the second term.

## 9-6 The Quantum Theory of the Radiation Field

In the previous section we presented the classical mathematical description of the radiation field. There is an alternative theoretical representation of the radiation field, which is known as the quantum model. Here the radiation field is treated as a superposition of an infinite number of harmonic oscillators. Each harmonic oscillator is described by a set of Hamiltonian equations, and by representing each harmonic oscillator by its quantum description it is possible to construct a quantum theory of the radiation field.

We take the set of vectors $\mathbf{A}_\lambda$ that is defined by Eq. (9-58) as the basis for our theoretical description of the radiation field. We recall that the most general representation of the radiation field is given by Eq. (9-68), namely

$$\mathbf{A} = \sum_{\lambda}(q_\lambda + q_\lambda^*)\mathbf{A}_\lambda \tag{9-81}$$

The radiation field is then described by the set of coefficients $q_\lambda$, since the vectors $\mathbf{A}_\lambda$ are exactly defined. We recall that the coefficients $q_\lambda$ are described by the first of the differential equations (9-56), namely

$$\frac{d^2 q_\lambda}{dt^2} + \omega_\lambda^2 q_\lambda = 0 \tag{9-82}$$

We set out to derive a set of Hamiltonian equations of motion for the coefficients $q_\lambda$, consistent with the Hamiltonian formalism which we discussed in Section 1-2. We introduce a new set of canonical variables and their conjugate momenta by defining

$$Q_\lambda = q_\lambda(t) + q_\lambda^*(t)$$

$$P_\lambda = \frac{\partial Q_\lambda}{\partial t} \tag{9-83}$$

We note that

$$\mathbf{H} = \sum_\lambda Q_\lambda \operatorname{curl} \mathbf{A}_\lambda$$

$$\mathbf{E} = \left(\frac{-1}{c}\right) \sum_\lambda P_\lambda \mathbf{A}_\lambda \tag{9-84}$$

By substituting this into Eq. (9-73) for the energy of the field we can derive the analogue of Eq. (9-76):

$$U = \tfrac{1}{2} \sum_\lambda (P_\lambda^2 + \omega^2 Q_\lambda^2) \tag{9-85}$$

Clearly, the radiation field can be represented by a Hamiltonian $H_f$, which we write as

$$H_f = \sum_\lambda H_\lambda(P_\lambda, Q_\lambda)$$

$$H_\lambda = \tfrac{1}{2}(P_\lambda^2 + \omega^2 Q_\lambda^2) \tag{9-86}$$

The Hamiltonian equations

$$\frac{\partial H_\lambda}{\partial Q_\lambda} = -\frac{\partial P_\lambda}{\partial t} \qquad \frac{\partial H_\lambda}{\partial P_\lambda} = \frac{\partial Q_\lambda}{\partial t} \tag{9-87}$$

are then identical with the differential equations (9-82) and with the definition (9-83) of the momenta $P_\lambda$.

We have learned in the previous chapters that a harmonic oscillator can be described either by means of classical mechanics or by means of quantum mechanics. We are now dealing with a radiation field that is represented by the Hamiltonian (9-86). If we solve the set of Hamiltonian equations (9-87) classically and substitute the results into Eq. (9-81), we obtain the classical description of the radiation field that we presented in the previous section. On the other hand, if we take the quantum mechanical description of the harmonic oscillators of the Hamiltonian (9-86), we move in a very different theoretical direction. Let us investigate where this procedure leads us.

We discussed the quantum theory of the harmonic oscillator in Section 5-7. We have shown that a harmonic oscillator has a discrete eigenvalue spectrum, that the stationary states are denoted by a quantum number $n=0,1,2,3,\ldots$, and so on, and that the energy eigenvalues are given by Eq. (5-105). In the case of the radiation field we have an infinite number of harmonic oscillators. One particular harmonic oscillator, $\lambda$, is described by a Hamiltonian $H_\lambda$ and its stationary state is described by a quantum number $n_\lambda$. Its energy is then

$$E_\lambda=(n_\lambda+\tfrac{1}{2})\hbar\omega_\lambda \tag{9-88}$$

The radiation field is thus described by a set of quantum numbers $(n_1, n_2, n_3, \ldots, n_\lambda, \ldots)$ and its energy is given by

$$E_f=\sum_\lambda (n_\lambda+\tfrac{1}{2})\hbar\omega_\lambda \tag{9-89}$$

According to quantum theory the field is also described by a wave function $\Psi$, which is a product of the wave functions of the individual oscillators

$$\Psi=\prod_\lambda \psi_\lambda(n_\lambda, Q_\lambda) \tag{9-90}$$

It may be somewhat difficult to give a physical interpretation of this wave function, but we shall see that this is not really essential. We have the alternative of using the matrix representation of the harmonic oscillators and this offers a much more convenient approach to the theory of the radiation field.

We have to make some adjustments in the above description of the radiation field because it contains some features that cause it to be unacceptable. We note that we have an infinite number of harmonic oscillators. Consequently, the energy $E_f$ of Eq. (9-89) is infinite because the sum of the zero-point energies of the oscillators is infinite.

This is not a serious difficulty since it can be dealt with by changing the zero point of the energies, which is accomplished by subtracting a constant term from the Hamiltonian $H_\lambda$. We redefine the Hamiltonians $H_\lambda$ of Eq. (9-86) as

$$H_\lambda=\tfrac{1}{2}(P_\lambda^2+\omega^2Q_\lambda^2-\hbar\omega_\lambda) \tag{9-91}$$

Its eigenvalues are

$$E_\lambda=n_\lambda\hbar\omega_\lambda \tag{9-92}$$

and the energy of the radiation field is now

$$E_f=\sum_\lambda E_\lambda=\sum_\lambda n_\lambda\hbar\omega_\lambda \tag{9-93}$$

It can be seen that the energy of the radiation field is zero if all quantum numbers $n_\lambda$ are zero and that it has a finite value when some of the numbers $n_\lambda$ are different from zero.

If we just consider the energy, the quantum description of the radiation field is consistent in every respect with the quantum theories of Planck and Einstein which we discussed in Section 1-4. Each radiation mode $\lambda$ is quantized, and the energy of the mode is a multiple of the quantized energy parcels or energy quanta that are known as photons. We have $n_1$ photons of mode 1, $n_2$ photons of mode 2, $n_\lambda$ photons of species $\lambda$, and so on. If we compare our present treatment with the discussion of Section 1-4, it can be seen that now we have quantized the radiation field itself rather than the wall of the black-body container or the interactions between wall and radiation as in the old theories. Since the radiation field can be considered as a superposition of an infinite number of harmonic oscillators, its quantum description becomes the quantum theory of the harmonic oscillator. The only adjustment that is necessary is the correction of the Hamiltonian, so that the zero-point energies of the harmonic oscillators are eliminated.

Again, if we just consider the energy of the radiation field, we can look upon it either as an assembly of particles, the photons, or as a superposition of modes of radiation. Both approaches give the same energy. In the theory of spectroscopic transitions we are primarily interested in the energies of the radiation field and of the molecules and in the transfer of energy between the radiation field and the molecules. Again, it makes little difference whether we visualize the radiation field from a particle or from a wave viewpoint.

Let us now proceed to consider the matrix representation of the harmonic oscillators. In Section 5-7 we derived the eigenvalues and eigenfunctions of a harmonic oscillator which was represented by a Hamiltonian $\mathcal{H}_o$:

$$\mathcal{H}_o = \frac{p^2}{2m} + \frac{kx^2}{2} \tag{9-94}$$

It can be seen that this Hamiltonian becomes identical with $H_\lambda$ of Eq. (9-91) if we take

$$m = 1 \qquad k = \omega^2 \tag{9-95}$$

and if we change the zero point of the energy. In Section 5-7 we introduced the unit of length $\alpha$ in Eq. (5-95). In the present case we can derive from Eq. (9-95) that $\alpha$ is now given by

$$\alpha = \frac{\hbar}{\omega} \tag{9-96}$$

By substituting the above values into Eq. (5-117) we find that the matrix representation of the coordinate $Q_\lambda$ with respect to the eigenfunctions

$\psi_\lambda(n_\lambda, Q_\lambda)$ is given by

$$\langle n_\lambda | Q_\lambda | n_\lambda + 1 \rangle = \left( \frac{\hbar}{2\omega_\lambda} \right)^{1/2} (n_\lambda + 1)^{1/2}$$

$$\langle n_\lambda | Q_\lambda | n_\lambda - 1 \rangle = \left( \frac{\hbar}{2\omega_\lambda} \right)^{1/2} (n_\lambda)^{1/2}$$

$$\langle n | Q_\lambda | m \rangle = 0 \qquad \text{if } m \neq n \pm 1 \tag{9-97}$$

In the same way, by substituting the values of Eqs. (9-95) and (9-96) into the Eqs. (5-113) and (5-120) we find that the matrix representation of the momentum $P_\lambda$ with respect to the eigenfunctions $\psi(n_\lambda, Q_\lambda)$ is given by

$$\langle n_\lambda | P_\lambda | n_\lambda + 1 \rangle = -i \left( \frac{\hbar \omega_\lambda}{2} \right) (n_\lambda + 1)^{1/2}$$

$$\langle n_\lambda | P_\lambda | n_\lambda - 1 \rangle = i \left( \frac{\hbar \omega_\lambda}{2} \right) (n_\lambda)^{1/2}$$

$$\langle n | P_\lambda | m \rangle = 0 \qquad \text{if } m \neq n \pm 1 \tag{9-98}$$

It may be recalled from Section 2-9 that $Q_\lambda$ and $P_\lambda$ always obey the relation

$$P_\lambda Q_\lambda - Q_\lambda P_\lambda = \left( \frac{\hbar}{i} \right) \tag{9-99}$$

This means that the matrix elements of the above operator are given by

$$\langle n | P_\lambda Q_\lambda - Q_\lambda P_\lambda | m \rangle = \left( \frac{\hbar}{i} \right) \delta_{n,m} \tag{9-100}$$

It is advantageous to introduce a new set of variables, namely

$$q_\lambda = \frac{1}{2} \left[ Q_\lambda + \frac{iP_\lambda}{\omega_\lambda} \right]$$

$$q_\lambda^* = \frac{1}{2} \left[ Q_\lambda - \frac{iP_\lambda}{\omega_\lambda} \right] \tag{9-101}$$

We construct the matrix elements of the new variables by substitution into

Eqs. (9-97) and (9-98). We have

$$\langle n_\lambda|q_\lambda|n_\lambda+1\rangle=\left(\frac{\hbar}{2\omega_\lambda}\right)^{1/2}(n_\lambda+1)^{1/2}$$

$$\langle n_\lambda|q_\lambda^*|n_\lambda+1\rangle=0$$

$$\langle n_\lambda|q_\lambda|n_\lambda-1\rangle=0$$

$$\langle n_\lambda|q_\lambda^*|n_\lambda-1\rangle=\left(\frac{\hbar}{2\omega_\lambda}\right)^{1/2}(n_\lambda)^{1/2} \tag{9-102}$$

and all other matrix elements are zero. We can also write Eqs. (9-102) in the form

$$\langle n_\lambda-1|q_\lambda|n_\lambda\rangle=\left(\frac{\hbar}{2\omega_\lambda}\right)^{1/2}(n_\lambda)^{1/2}$$

$$\langle n_\lambda+1|q_\lambda^*|n_\lambda\rangle=\left(\frac{\hbar}{2\omega_\lambda}\right)^{1/2}(n_\lambda+1)^{1/2} \tag{9-103}$$

with all other matrix elements zero. Since the eigenfunctions $\psi_\lambda(n_\lambda, q_\lambda)$ form an orthonormal set, we can write Eq. (9-103) in operator form as

$$q_\lambda\psi_\lambda(n_\lambda)=\left(\frac{\hbar}{2\omega_\lambda}\right)^{1/2}(n_\lambda)^{1/2}\psi_\lambda(n_\lambda-1)$$

$$q_\lambda^*\psi_\lambda(n_\lambda)=\left(\frac{\hbar}{2\omega_\lambda}\right)^{1/2}(n_\lambda+1)^{1/2}\psi_\lambda(n_\lambda+1) \tag{9-104}$$

It is also useful to give the commutator relations between $q_\lambda$ and $q_\lambda^*$. Since we have

$$Q_\lambda=q_\lambda+q_\lambda^*$$

$$iP_\lambda=\omega_\lambda(q_\lambda-q_\lambda^*) \tag{9-105}$$

it is easily derived from Eq. (9-99) that

$$P_\lambda Q_\lambda-Q_\lambda P_\lambda=\left(\frac{2\omega_\lambda}{i}\right)(q_\lambda q_\lambda^*-q_\lambda^* q_\lambda)=\left(\frac{\hbar}{i}\right) \tag{9-106}$$

A similar result can be derived directly from the operator relations (9-104):

$$(q_\lambda q_\lambda^*-q_\lambda^* q_\lambda)\psi_\lambda(n_\lambda)=\left(\frac{\hbar}{2\omega_\lambda}\right)\psi_\lambda(n_\lambda) \tag{9-107}$$

Finally, we express the Hamiltonian $H_\lambda$ of Eq. (9-91) in terms of $q_\lambda$ and $q_\lambda^*$. By substituting Eqs. (9-105) it is easily verified that

$$H_\lambda = 2\omega_\lambda^2 q_\lambda^* q_\lambda \tag{9-108}$$

It is interesting to note that this result is consistent with the classical expression (9-76) for the energy of the radiation field. The slight difference is due to the correction for the zero-point energy of the harmonic oscillator.

It follows from Eq. (9-104) that the effect of $q_\lambda^*$, considered as an operator, is a change from the function $\psi_\lambda(n_\lambda)$ to the function $\psi_\lambda(n_\lambda+1)$. Consequently, $q_\lambda^*$ changes the stationary state $n_\lambda$ to the stationary state $n_\lambda+1$, and we might say that the effect of $q_\lambda^*$ is the creation of an additional photon to the radiation field. The operator $q_\lambda^*$ is known as a creation operator. The operator $q_\lambda$ has exactly the opposite effect; it changes the wave function $\psi_\lambda(n_\lambda)$ to the wave function $\psi_\lambda(n_\lambda-1)$, and it has the effect of removing a photon from the radiation field. The operator $q_\lambda$ is therefore known as an annihilation operator.

The two operators $q_\lambda$ and $q_\lambda^*$ are sometimes referred to as ladder operators; $q_\lambda^*$ is called the "step-up" operator since it increases the quantum number $n_\lambda$ by one, and $q_\lambda$ is called the "step-down" operator since it decreases the quantum number by one. It should be noted that the operators $M_1$ and $M_{-1}$, which we introduced in Section 6-3 in order to describe the angular momentum, are very similar to $q_\lambda^*$ and $q_\lambda$. We showed in Eq. (6-63) that $M_1$ has the effect of increasing the quantum number $m$ by unity and it may be called the step-up operator. Similarly, it follows from Eq. (6-64) that $M_{-1}$ corresponds to the step-down operator since it lowers the quantum number $m$ by unity.

By using the concept of ladder operators it is easily seen that the functions $\psi_\lambda(n_\lambda)$ must be eigenfunctions of both operators $q_\lambda q_\lambda^*$ and $q_\lambda^* q_\lambda$ because we return to the original position of the ladder if we first go up the ladder by one step and then go down by one step, or vice versa. In both cases the operators $q_\lambda q_\lambda^*$ or $q_\lambda^* q_\lambda$ maintain the original number of photons.

Any operator that is linear in $q_\lambda^*$ will cause the creation of a photon, which represents an emission process because it adds a photon to the radiation field. Similarly, an operator that is linear in $q_\lambda$ represents an absorption process because it causes the removal of a photon from the field. We may recall Eq. (9-80) which represents the interaction between a one-electron system and a radiation field. This expression was derived in relation to the classical description of the radiation field, but it is equally valid in the quantum theory of the field since it contains the quantities $q_\lambda$ and $q_\lambda^*$. In the classical representation $q_\lambda$ and $q_\lambda^*$ refer to the time-dependent coefficients that determine the form of the radiation field. In the quantum representation $q_\lambda$ and $q_\lambda^*$ are the creation and annihilation operators; the term in $q_\lambda$ describes an absorption process and the term in $q_\lambda^*$ describes an emission process. The double sum in Eq. (9-80), containing terms that are quadratic in $q_\lambda$ and $q_\lambda^*$, causes higher-order optical processes, and it may be neglected if we are interested only in conventional absorption or emission of radiation.

It should be noted that the interaction Hamiltonian may have the same form in the classical and in the quantum theories of the radiation field, but it is really quite different in the two models. In the classical theory the $q_\lambda$ are time-dependent coefficients and the interaction energy is periodic in the time. In the quantum theory the $q_\lambda$ and $q_\lambda^*$ are time-independent operators and the interaction Hamiltonian is independent of the time once the interaction is switched on. In the following two sections we discuss the theory of radiation-induced transition probabilities, first in the classical model of the radiation field and next with the quantum description of the field. We shall see that the two theories exhibit some fundamental differences. The differences are not just a matter of mathematical details but they involve the basic framework of the theory. The quantum theory gives an easy and logical description of both stimulated and spontaneous emission, whereas the classical theory does not offer a convenient explanation of spontaneous emission. We feel that the quantum model gives a more complete and consistent description of phenomena, but it may be interesting to present both the classical and the quantum theories.

## 9-7 The Semiclassical Theory of Radiation Absorption

In this section we use the classical description of the radiation field for the derivation of the interaction between a molecule and a radiation field. This theory is known as semiclassical because one part of the system, the molecule, is described by quantum theory and the other part, the radiation field, is described by classical theory.

The interaction between a one-electron system and a radiation field is given by Eq. (9-45), but we wish to generalize the expression to a many-electron system so that it has more general validity. The Hamiltonian of a many-electron system in the presence of an electromagnetic field is given by

$$\mathcal{H}=\frac{1}{2m}\sum_j\left(\mathbf{p}_j+\frac{e}{c}\mathbf{A}_j\right)^2-\sum_j e\phi_j+V \tag{9-109}$$

Here $\mathbf{A}_j$ and $\phi_j$ are the vector and scalar potentials, respectively, acting on the $j$th electron. In a radiation field we can take $\phi$ equal to zero and we have

$$\mathcal{H}=\frac{1}{2m}\sum_j p_j^2+V+\frac{e}{mc}\sum_j(\mathbf{A}_j\cdot\mathbf{p}_j)+\frac{e^2}{2mc^2}\sum_j A_j^2 \tag{9-110}$$

We neglect the quadratic term in $A$ and we have

$$\mathcal{H}=\mathcal{H}_o+\mathcal{H}'$$

$$\mathcal{H}'=\frac{e}{mc}\sum_j(\mathbf{A}_j\cdot\mathbf{p}_j) \tag{9-111}$$

The latter term $\mathcal{H}'$ represents the interaction between the molecule and the radiation field.

We take the radiation field as a plane polarized beam of light. One of its components can be represented as

$$\mathbf{A}_\lambda = A_o(\omega)\mathbf{u}\{\exp[i(\mathbf{k}_\lambda \cdot \mathbf{r}) - \omega_\lambda t] + \exp[-i(\mathbf{k}_\lambda \cdot \mathbf{r}) + \omega_\lambda t]\} \tag{9-112}$$

where $\mathbf{k}_\lambda$ denotes the direction of propagation and the unit vector $\mathbf{u}$ denotes the direction of polarization of the light.

We should point out that the Einstein coefficients are defined in Section 9-2 for isotropic nonpolarized radiation and that we derive the interactions by means of a perturbation treatment involving a linearly polarized beam of light. The reason for this inconsistency should be obvious; it is much easier to apply perturbation theory to a linearly polarized beam of light than to the most general case. Nevertheless, we must be careful to make sure that our perturbation results can be related to the Einstein coefficients.

The perturbation term representing the interaction between the molecule and the radiation field is obtained by substituting Eq. (9-112) into Eq. (9-111). We write it as

$$\mathcal{H}'(\omega) = \left\{ \frac{eA_o(\omega)}{mc} \right\} \sum_j (\mathbf{u} \cdot \mathbf{p}_j)[\gamma_j \exp(-\omega_\lambda t) + \gamma_j^* \exp(+\omega_\lambda t)] \tag{9-113}$$

We consider now the situation that we analyzed in Section 8-4. We assume that the molecule is in a stationary state $o$ until the time $t=0$ when we switch on the radiation field. Due to the perturbation there are, in principle, transitions to the other stationary states of the molecule. We consider the transitions to a particular state $k$. We assume that $E_k$ is larger than $E_o$ so that

$$\hbar\omega_k = E_k - E_o \tag{9-114}$$

is a positive quantity.

The interaction term of Eq. (9-113) refers to a monochromatic beam of light. We wish to generalize to a beam of light with a continuous frequency distribution. The corresponding perturbation term is

$$\mathcal{H}_{\text{int}} = \int \mathcal{H}'(\omega)\, d\omega \tag{9-115}$$

The phase constants of the waves are incorporated in the coefficients $A_o(\omega)$, and the matrix elements of the perturbation satisfy the condition (8-56).

We make use of the results of Section 8-4 and we see that the probability for a transition from state $o$ to state $k$ is given by Eq. (8-61). This equation contains the matrix element $V_{k,o}$ of the perturbation with respect to the eigenfunctions $\Phi_k$ and $\Phi_o$ of the states $k$ and $o$, respectively. In our present case this matrix element reduces to

$$V_{k,o}(-\omega)=\left(\frac{eA_o(\omega)}{mc}\right)\left\langle \Phi_k\left|\sum_j \gamma_j(\mathbf{u}\cdot\mathbf{p}_j)\right|\Phi_o\right\rangle \qquad (9\text{-}116)$$

According to Eq. (8-61) the transition probability $w_{o\to k}$ is then given by

$$w_{o\to k}=\left(\frac{2t}{\hbar^2}\right)\int |V_{k,o}(-\omega)|^2\delta(\omega-\omega_k)\,d\omega \qquad (9\text{-}117)$$

We have adapted Eq. (8-61) to our present situation.

First we transform the matrix elements of Eq. (9-116) by making use of what is known as the dipole transformation. We note that the commutator between the molecular Hamiltonian $\mathcal{H}_o$ and one of the electron coordinates, $x_j$, is given by

$$\mathcal{H}_o x_j - x_j\mathcal{H}_o=\left(\frac{-\hbar^2}{2m}\right)\left[\frac{\partial^2}{\partial x_j^2}, x_j\right]$$

$$=\left(\frac{-\hbar^2}{m}\right)\left(\frac{\partial}{\partial x_j}\right)=\left(\frac{\hbar}{im}\right)p_{jx} \qquad (9\text{-}118)$$

Hence

$$p_{jx}=\left(\frac{im}{\hbar}\right)[\mathcal{H}_o, x_j] \qquad (9\text{-}119)$$

We also note that the quantities $\gamma_j$ that occur in Eq. (9-116) have the form

$$\gamma_j=\exp[i(\mathbf{k}_\lambda\cdot\mathbf{r}_j)] \qquad (9\text{-}120)$$

Here $k_\lambda$ is the inverse of the wave length of the light which is of the order of 5000 Å. The coordinate $\mathbf{r}_j$ varies over the dimensions of the molecule, but this dimension is usually a few Å only. We can expand $\gamma_j$ as a power series:

$$\gamma_j=1+i(\mathbf{k}_\lambda\cdot\mathbf{r}_j)+\cdots \qquad (9\text{-}121)$$

In most cases we can take $\gamma_j$ equal to unity. This is known as the dipole approximation.

We transform the matrix element $V_{o,k}$ by using the dipole approximation and by substituting the commutator relation (9-119). We have then

$$\langle\Phi_k|p_{jx}|\Phi_o\rangle=\left(\frac{im}{\hbar}\right)\langle\Phi_k|[\mathcal{H}_o, x_j]|\Phi_o\rangle$$

$$=\left(\frac{im}{\hbar}\right)(E_k-E_o)\langle\Phi_k|x_j|\Phi_o\rangle$$

$$=im\omega_k\langle\Phi_k|x_j|\Phi_o\rangle \qquad (9\text{-}122)$$

We introduce the dipole moment operator

$$\boldsymbol{\mu}_{\text{op}} = e \sum \mathbf{r}_j \tag{9-123}$$

and its matrix element

$$\boldsymbol{\mu}_{k,o} = \langle \Phi_k | \boldsymbol{\mu}_{\text{op}} | \Phi_o \rangle \tag{9-124}$$

which is known as the transition moment between $k$ and $o$. We can then write $V_{k,o}$ as

$$V_{k,o} = \left( \frac{i\omega_k A_o(\omega)}{c} \right) (\boldsymbol{\mu}_{k,o} \cdot \mathbf{u}) \tag{9-125}$$

By substituting this into Eq. (9-117) we obtain

$$w_{o\to k} = \left( \frac{2t\omega_k^2}{\hbar^2 c^2} \right) \int A_o^2(\omega)(\boldsymbol{\mu}_{k,o} \cdot \mathbf{u})(\boldsymbol{\mu}_{k,o}^* \cdot \mathbf{u}) \delta(\omega - \omega_k)\, d\omega \tag{9-126}$$

We define $\theta$ as the angle between the polarization direction $\mathbf{u}$ of the light and the transition moment. We can then write $w_{o\to k}$ as

$$w_{o\to k} = \left( \frac{2t\omega_k^2}{\hbar^2 c^2} \right) A_o^2(\omega_k)(\boldsymbol{\mu}_{k,o} \cdot \boldsymbol{\mu}_{o,k}) \cos^2\theta \tag{9-127}$$

We like to relate the above transition probability, which was derived for a plane-polarized, plane radiation wave, to the Einstein coefficient $B_{o\to k}$, which was defined for isotropic radiation with energy density $\rho(\nu)$. It can be shown that by averaging over all directions of propagation and polarization and over all phases of the radiation field vector potential we obtain the isotropic transition probability by substituting

$$2\omega_k^2 \cos^2\theta A_o^2(\omega_k) = \left( \frac{2\pi}{3} \right) c^2 \rho(\nu_k) \tag{9-128}$$

By combining the two Eqs. (9-127) and (9-128) we find that the Einstein coefficient $B_{o\to k}$ is given by

$$B_{o\to k} = \frac{2\pi}{3\hbar^2} (\boldsymbol{\mu}_{o,k} \cdot \boldsymbol{\mu}_{k,o}) \tag{9-129}$$

The Einstein coefficients of emission can now be derived from Eqs. (9-13) and (9-14). The coefficient $B_{k\to o}$ of induced emission is equal to $B_{o\to k}$, and the coefficient of spontaneous emission is given by

$$A_{k\to o} = \frac{32\pi^3 \nu_{ok}^3}{3\hbar c^3} (\boldsymbol{\mu}_{o,k} \cdot \boldsymbol{\mu}_{k,o}) \tag{9-130}$$

We have made a number of approximations in deriving the result of Eq. (9-129). The most obvious one is taking the function $\gamma$ of Eq. (9-121) equal to unity, but we have also neglected a number of contributions that we did not even mention. For instance we did not consider the interactions between the radiation field and the nuclear motion in the molecule and the interactions between the electromagnetic field and the electron spins that are discussed in the next chapter. In fact there exist many higher-order effects that may contribute to the transition probability. We define a spectroscopic transition as allowed if its transition moment, defined by Eq. (9-124), is different from zero and we define the transition as forbidden if the transition moment is zero. A forbidden transition may be observed due to higher-order effects, and the experimental study of forbidden transitions may supply us with useful information about the system.

In the following section we present the theory of radiation-induced transitions by using the quantum theory of the radiation field. This alternate theory may give us a broader understanding of the nature of emission and absorption of radiation.

## 9-8 Emission and Absorption of Radiation in the Quantized Field Theory

In the previous section we presented the theory of emission and absorption of radiation according to the classical theory of the radiation field. Here we discuss the same problem by following the quantum theory of the radiation field. We believe that the quantum model of the radiation field leads to a more logical and comprehensive theory of emission and absorption. All mechanisms (absorption, stimulated, and spontaneous emission) are directly derived from the theory. Also, the mathematical derivation of higher-order and nonlinear effects from the quantum model is much simpler than the corresponding derivations according to the classical model. One aspect of the quantum theory is more cumbersome, namely the precise quantitative definitions of the various results. We avoid these difficulties by limiting ourselves to a more qualitative discussion.

In the quantum description the radiation field is represented by a Hamiltonian $H_f$, which is an infinite sum of a set of Hamiltonians $H_\lambda$:

$$H_f = \sum_{\lambda} H_\lambda \tag{9-131}$$

The individual Hamiltonians $H_\lambda$ are defined by Eq. (9-108), and each $H_\lambda$ refers to a particular mode of oscillation or to a particular harmonic oscillator component of the radiation field. The state of the radiation field is described by an infinite set of quantum numbers $n_\lambda$.

We consider also a molecule (or a set of molecules) represented by a Hamiltonian $H_m$. The latter Hamiltonian has a set of eigenvalues $\varepsilon_k$ and corresponding eigenfunctions $\Phi_k$.

In the quantum description of radiation processes we consider first the combined system of molecule and radiation field, assuming that there is no interaction between the two systems. This system is represented by a time-independent Hamiltonian $H_o$, which is the sum of the Hamiltonians of the molecule and the field:

$$H_o = H_m + H_f \tag{9-132}$$

We assume that this system is in a stationary state of the operator $H_o$ until the time $t=0$ when we introduce the interaction between the molecule and the field. The interaction is represented by the Hamiltonian $H_{\text{int}}$, which we derive from Eq. (9-80) by generalizing to a many-electron system:

$$H_{\text{int}} = \left(\frac{e}{mc}\right) \sum_{\lambda} \sum_{j} (q_\lambda + q_\lambda^*)(\mathbf{A}_{\lambda j} \cdot \mathbf{p}_j) \tag{9-133}$$

We have neglected the second term of Eq. (9-80) since we consider only one-photon processes. We should note here that $q_\lambda$ and $q_\lambda^*$ are time-independent annihilation and creation operators and that $H_{\text{int}}$ is a time-independent perturbation, once it has been switched on at $t=0$.

Our present approach to the problem is quite different from the one described in the previous section. In the semiclassical description we considered the molecule only until the time $t=0$ when we switched on the radiation field. The effect of the radiation field on the molecule was described by a time-dependent interaction term, and we considered transitions to different stationary states of the molecule. In the present theory we consider the stationary states of the combined system of molecule and radiation field until the time $t=0$ when we switch on the interaction between the two subsystems. This interaction is now represented by a time-independent interaction Hamiltonian.

The stationary states of the Hamiltonian $H_o$ of Eq. (9-132) are described by the quantum number $k$ of the molecular stationary states and by the infinite set of quantum numbers $(n_1, n_2, n_3, \ldots, n_\lambda, \ldots)$ which describe the state of the radiation field. We assume now that at the time $t=0$ the combined system of molecule and field is in a stationary state which we denote by $(o;0)$; this means that the molecule is in its stationary state $o$ and that the radiation field is in a given stationary state described by a known set of quantum numbers $(n_\lambda)$. We take the state 0 as our reference state of the radiation field, and we consider various states that differ by one, two, or more photons from the state 0. In our notation the state $(k;\lambda)$ describes a stationary state $k$ of the molecule and a state of the radiation field that has an additional photon of species $\lambda$ as compared to 0; in other words, the quantum number $n_\lambda$ of 0 is changed to $n_\lambda+1$ and all other quantum numbers are unchanged. Similarly, the state $(k;\bar{\lambda})$ has a photon of species $\lambda$ less than 0, and the quantum number $n_\lambda$ of 0 has been changed to $n_\lambda - 1$. The states $(o;\lambda\mu)$ and $(k;\lambda\mu)$ have two photons $\lambda$ and $\mu$, respectively, in addition to 0. The state $(k;\lambda\bar{\mu})$ has one more photon of species $\lambda$ and one less photon of species $\mu$ than the state $(o;0)$.

The energies of these various states are given by

$$E(k;\lambda)-E(o;0)=\varepsilon_k-\varepsilon_o+\hbar\omega_\lambda$$

$$E(k;\bar{\lambda})-E(o;0)=\varepsilon_k-\varepsilon_o-\hbar\omega_\lambda$$

$$E(k,\lambda\mu)-E(o;0)=\varepsilon_k-\varepsilon_o+\hbar\omega_\lambda+\hbar\omega_\mu \qquad \text{and so on} \tag{9-134}$$

where $\varepsilon_k$ and $\varepsilon_o$ are the energy eigenfunctions of the molecule.

In order to calculate the various transition probabilities we use the perturbation treatment of Section 8-6. The relevant perturbation equation is Eq. (8-98), and we see that the transition probabilities are determined by the matrix elements of the interaction Hamiltonian with respect to the initial state $(o;0)$ and the other states:

$$H(K|o;0)=\langle K|H_{\text{int}}|o;0\rangle \tag{9-135}$$

Since $H_{\text{int}}$ is linear in the operators $q_\lambda$ and $q_\lambda^*$, the only nonzero matrix elements occur when the state $K$ differs by one photon only from the state $(o;0)$; in other words, the matrix elements are nonzero only if the state $K$ is either $(k;\lambda)$ or $(k;\bar{\lambda})$. We can use Eq. (9-103) and we find

$$H(k;\lambda|o;0)=\left(\frac{e}{mc}\right)\left(\frac{\hbar}{2\omega_\lambda}\right)^{1/2}(n_\lambda+1)^{1/2}\left\langle\Phi_k|\sum_j(\mathbf{A}_{\lambda j}\cdot\mathbf{p}_j)|\Phi_o\right\rangle$$

$$H(k;\bar{\lambda}|o;0)=\left(\frac{e}{mc}\right)\left(\frac{\hbar}{2\omega_\lambda}\right)^{1/2}(n_\lambda)^{1/2}\left\langle\Phi_k|\sum_j(\mathbf{A}_{\lambda j}\cdot\mathbf{p}_j)|\Phi_o\right\rangle \tag{9-136}$$

Here $\Phi_k$ and $\Phi_o$ are the molecular eigenfunctions.

The specific form of $\mathbf{A}_\lambda$ is given by Eq. (9-58), together with the normalization condition (9-66). It may be recalled that we normalized the vector potential with respect to a cubic box of length $L$. We may take the length $L$ equal to unity without loss of generality. We then have

$$\mathbf{v}_\lambda=(4\pi c^2)^{1/2}\mathbf{e}_\lambda$$

$$\mathbf{A}_\lambda=(4\pi c^2)^{1/2}\mathbf{e}_\lambda\exp\left[i(\mathbf{k}_\lambda\cdot\mathbf{r})\right] \tag{9-137}$$

where $\mathbf{e}_\lambda$ is a unit vector denoting the polarization direction of the photon.

In evaluating the matrix elements we introduce again the dipole approximation where the exponential in Eq. (9-137) for $\mathbf{A}_\lambda$ is taken equal to unity. The result is

$$\left\langle\Phi_k|\sum_j(\mathbf{A}_{\lambda j}\cdot\mathbf{p}_j)|\Phi_o\right\rangle=(4\pi c^2)^{1/2}\left\langle\Phi_k|\sum_j(\mathbf{e}_\lambda\cdot\mathbf{p}_j)|\Phi_o\right\rangle \tag{9-138}$$

Next we introduce the dipole transformation (9-122) and we obtain

$$\left\langle \Phi_k | \sum_j (\mathbf{A}_{\lambda j} \cdot \mathbf{p}_j) | \Phi_o \right\rangle = (4\pi c^2)^{1/2} \left( \frac{im\omega_k}{e} \right) (\mathbf{e}_\lambda \cdot \boldsymbol{\mu}_{k,o}) \tag{9-139}$$

with

$$\hbar\omega_k = \varepsilon_k - \varepsilon_o \tag{9-140}$$

By substituting all these results into Eq. (9-136) we find

$$H(k;\lambda|o;0) = (i\omega_k)\left(\frac{2\pi\hbar}{\omega_\lambda}\right)^{1/2} (n_\lambda + 1)^{1/2} (\mathbf{e}_\lambda \cdot \boldsymbol{\mu}_{k,o})$$

$$H(k;\bar{\lambda}|o;0) = (i\omega_k)\left(\frac{2\pi\hbar}{\omega_\lambda}\right)^{1/2} (n_\lambda)^{1/2} (\mathbf{e}_\lambda \cdot \boldsymbol{\mu}_{k,o}) \tag{9-141}$$

The corresponding transition probabilities can now be derived from Eq. (8-114). We denote them by $w(k;\lambda)$ and by $w(k;\bar{\lambda})$, respectively. We have

$$w(k;\lambda) = \left(\frac{2\pi}{\hbar}\right) H(k;\lambda|o;0) H^*(k;\lambda|o;0) \delta(E_{k;\lambda} - E_{o;0})$$

$$w(k;\bar{\lambda}) = \left(\frac{2\pi}{\hbar}\right) H(k;\bar{\lambda}|o;0) H^*(k;\bar{\lambda}|o;0) (E_{k;\bar{\lambda}} - E_{o;0}) \tag{9-142}$$

We should differentiate between the two cases where $\varepsilon_k > \varepsilon_o$ and where we have absorption of light, and the other case where $\varepsilon_k < \varepsilon_o$ and where we have emission of light. For convenience we introduce the symbol

$$\varepsilon_\lambda = \hbar\omega_\lambda \tag{9-143}$$

Substitution of Eq. (9-141) into Eq. (9-142) gives then

$$w(k;\lambda) = \left(\frac{4\pi^2\omega_k^2}{\omega_\lambda}\right)(n_\lambda + 1)(\mathbf{e}_\lambda \cdot \boldsymbol{\mu}_{k,o})(\mathbf{e}_\lambda \cdot \boldsymbol{\mu}_{k,o}^*)\delta(\varepsilon_k - \varepsilon_o + \varepsilon_\lambda)$$

$$w(k;\bar{\lambda}) = \left(\frac{4\pi^2\omega_k^2}{\omega_\lambda}\right)(n_\lambda)(\mathbf{e}_\lambda \cdot \boldsymbol{\mu}_{k,o})(\mathbf{e}_\lambda \cdot \boldsymbol{\mu}_{k,o}^*)\delta(\varepsilon_k - \varepsilon_o - \varepsilon_\lambda) \tag{9-144}$$

First we consider the case where $\varepsilon_k - \varepsilon_o$ is negative; this corresponds to emission of light. It follows from Eq. (9-144) that $w(k;\bar{\lambda})$ is always zero because of the $\delta$ function. The probability $w(k;\lambda)$ describes the probability of a transition from state $o$ to state $k$ with the simultaneous emission of a photon of species $\lambda$. The total probability of a transition from state $o$ to state $k$ is obtained by summing over all possible photon species $\lambda$:

$$w(o \to k) = \sum_{\lambda} w(k; \lambda) \tag{9-145}$$

We evaluate $w(o \to k)$ from Eqs. (9-144) and (9-145) by replacing the summation by an integration. It follows from Eq. (9-62) that there are $(2\pi)^3 \cdot 4\pi k_\lambda^2 \cdot dk_\lambda$ values of $k_\lambda$ between $k_\lambda$ and $k_\lambda + dk_\lambda$. Since there are two possible polarization directions for each value $k_\lambda$, we have

$$\sum_{\lambda} \to \int 64\pi^4 k_\lambda^2 \, dk_\lambda \tag{9-146}$$

By substituting Eq. (9-59) we obtain

$$\sum_{\lambda} \to \left( \frac{64\pi^4}{c^3} \right) \int \omega_\lambda^2 \, d\omega_\lambda \tag{9-147}$$

Finally we must replace the $\delta$ function in Eq. (9-144) so that it is compatible with the integration over $\omega_\lambda$. It follows from the definition of the $\delta$ function in Section 8-5 that

$$\delta(\varepsilon_k - \varepsilon_o + \varepsilon_\lambda) = \left( \frac{1}{\hbar} \right) \delta(\omega_k + \omega_\lambda)$$

$$\hbar \omega_\lambda = \varepsilon_k - \varepsilon_o \tag{9-148}$$

By substituting these various results into Eqs. (9-144) and (9-145) we obtain

$$w(o \to k) = \frac{256\omega_k^2 \pi^6}{\hbar c^3} \int_0^\infty (n_\lambda + 1)(\boldsymbol{\mu}_{k,o} \cdot \boldsymbol{\mu}_{k,o}^*) \cos^2\theta \, \omega_\lambda \delta(\omega_k + \omega_\lambda) \, d\omega_\lambda \tag{9-149}$$

The transition probability is the sum of two contributions. One contribution contains the quantum number $n_\lambda$ and is related to the energy density of the radiation field at $\omega_\lambda$; this contribution represents the stimulated emission term. The second contribution is independent of $n_\lambda$ and it must represent the spontaneous emission. We evaluate the spontaneous transition probability $w'(o \to k)$, and it is given by

$$w'(o \to k) = \left( \frac{256\omega_k^2 \pi^6}{\hbar c^3} \right) (\boldsymbol{\mu}_{o,k} \cdot \boldsymbol{\mu}_{k,o}) (\cos^2\theta)_{\text{av}} \int_0^\infty \omega_\lambda \delta(\omega_\lambda + \omega_k) \, d\omega_\lambda \tag{9-150}$$

Since

$$(\cos^2\theta)_{\text{av}}=\tfrac{1}{3} \tag{9-151}$$

we obtain

$$w'(o\to k)=\frac{256\pi^6(-\omega_k)^3}{3\hbar c^3}(\boldsymbol{\mu}_{o,k}\cdot\boldsymbol{\mu}_{k,o}) \tag{9-152}$$

It can be verified that this is identical with Eq. (9-130) for the Einstein coefficient of spontaneous emission, which was derived indirectly from the semiclassical theory of transition probabilities.

The Einstein coefficient of stimulated emission can also be derived from Eq. (9-144), and it is given by

$$w''(o\to k)=\left(\frac{256\omega_k^2\pi^6}{3\hbar c^3}\right)(\boldsymbol{\mu}_{o,k}\cdot\boldsymbol{\mu}_{k,o})\int_0^\infty n_\lambda\omega_\lambda\delta(\omega_\lambda+\omega_k)\,d\omega_\lambda \tag{9-153}$$

Its exact expression can be derived by relating the quantum numbers $n_\lambda$ to the energy density of the radiation field, but we do not present this derivation here. The Einstein coefficient of absorption is given by the same expression as the Einstein coefficient of stimulated emission, and it is easily seen that the two coefficients are equal to one another.

We have shown that the quantum theory of transition probabilities predicts all three mechanisms in a straightforward fashion. In fact, the absorption, stimulated, and spontaneous emission mechanisms are all contained in Eqs. (9-144). We only derived the exact expression for the spontaneous emission, but it can be seen that the other two can be derived also from Eq. (9-144), although we do not present those derivations here.

It is relatively easy to derive the line shape of an absorption or emission line by combining the above results with the various expressions of Section 8-6. We consider an absorption line from the molecular ground state $o$ to an excited state $m$. We derive the line shape of the absorption line by calculating the probability amplitudes $c(m,\bar{\lambda})$ for the stationary state where the time $t$ tends to infinity.

We derived in Eq. (8-111) that $c(m,\bar{\lambda})$ is given by

$$c(m,\bar{\lambda})=\frac{U(m,\bar{\lambda})}{\varepsilon_m-\varepsilon_o-\hbar\omega_\lambda+(i\hbar/2)\Gamma} \tag{9-154}$$

The quantity $U(m,\bar{\lambda})$ is obtained from Eqs. (8-98) and (8-100). In the present case

$$U(m,\bar{\lambda})=H(m;\bar{\lambda}|o;0)=(i\omega_m)\left(\frac{2\pi\hbar}{\omega_\lambda}\right)^{1/2}(n_\lambda)^{1/2}(\mathbf{e}_\lambda\cdot\boldsymbol{\mu}_{m,o}) \tag{9-155}$$

The probability of an upward transition from the molecular ground state $o$ to an excited state $m$ with the simultaneous absorption of a photon of species $\lambda$ with frequency $\omega_\lambda$ is given by

$$w(o,m;\bar{\lambda}) = c(m,\bar{\lambda})c^*(m,\bar{\lambda}) \tag{9-156}$$

It follows now from Eqs. (9-154), (9-155), and (9-156) that

$$w(o,m;\bar{\lambda}) = \frac{2\pi\omega_m^2}{\hbar\omega_\lambda}(\boldsymbol{\mu}_{o,m}\cdot\boldsymbol{\mu}_{m,o})\cos^2\theta_\lambda \frac{n_\lambda}{(\omega_\lambda - \omega_m)^2 + (\Gamma_r^2/4)} \tag{9-157}$$

The function

$$f(x) = \frac{\alpha}{(x-x_o)^2 + \alpha^2} \tag{9-158}$$

is called a Lorentzian function. It has a maximum at the point $x = x_o$. We have seen in Section 8-5 that it has a sharp maximum when $\alpha$ becomes smaller and that it is a representation of the $\delta$ function in the limit when $\alpha$ tends to zero.

In Eq. (9-157) the real part $\Gamma_r$ of the damping constant $\Gamma$ is usually small and the Lorentzian part of Eq. (9-157) represents a very sharp maximum of $\omega_\lambda$ around the point $\omega_m$. Usually $n_\lambda$ varies much more slowly as a function of $\omega_\lambda$ than the Lorentzian. In that case the line shape can be described as

$$I(\omega_\lambda) = \frac{I_o}{(\omega_\lambda - \omega_m)^2 + (\Gamma_r/2)^2} \tag{9-159}$$

If $n_\lambda$ varies more rapidly with $\omega_\lambda$ than the Lorentzian, the line shape is determined by the behavior of $n_\lambda(\omega_\lambda)$.

## 9-9 Final Remarks

In the preceding sections we have attempted to present a survey of the various theoretical descriptions of the radiation field and of the interactions between a radiation field and a molecule. We felt that it might be instructive to present both the semiclassical and the quantum theories of transition probabilities. In our opinion, the quantum representation of the radiation field, in combination with the general perturbation theory of Section 8-6, offers the most comprehensive and logical theoretical description of optical effects. This approach can be used to deal with scattering, optical activity, and the various multiple-photon effects that have been discovered since the use of high-intensity lasers. However, in the present book we limit ourselves to a discussion of simple absorption or emission of light only.

## Problems

**1** Calculate the transition moment between a $2p_z$ and a $1s$ state of the hydrogen atom. Evaluate the Einstein coefficient of spontaneous emission for the transition from the $2p_z$ state to the $1s$ state and the lifetime of a H atom in its $2p_z$ state.

**2** Derive the relation between the quantum numbers $n_\lambda$ that describe the state of the radiation field in its quantum representation and the energy density $\rho(\nu)$ of the radiation field. Use this relation to calculate the Einstein coefficient of stimulated emission from Eq. (9-153).

**3** Consider an electron, moving in a harmonic oscillator potential such that the energy difference between the state $n=0$ and $n=1$ is exactly 1 eV. Evaluate the transition moment and the Einstein coefficient of spontaneous emission for a transition from the state $n=1$ to the state $n=0$.

**4** Calculate the transition moments between a hydrogen atom $2p_z$ state and the $3s$ and the various $3d$ states. Which transition has a larger probability: from $2p_z$ to $3s$ or from $2p_z$ to $3d$?

## Recommended Reading

One of the classical presentations of the quantization of the radiation field is the book by Dirac. A more modern treatment can be found in the text by Bogoliubov and Shirkov. We again quote the book by Heitler because we enjoyed reading it, and we cite our own book because of its more extensive discussion of multiple-photon processes.

P. A. M. Dirac, *The Principles of Quantum Mechanics*, Oxford University Press, London, 1947.

N. N. Bogoliubov and D. V. Shirkov, *Introduction to the Theory of Quantized Fields*, Interscience Publishers, Inc., New York, 1959.

W. Heitler, *The Quantum Theory of Radiation*, Oxford University Press, London, 1957.

H. F. Hameka, *Advanced Quantum Chemistry*, Addison-Wesley Publishing Co., Inc., Reading, MA, 1965.

CHAPTER TEN

# Electron Spin and the Helium Atom

## 10-1 Introduction

Up to this point we have considered one-electron systems only. There is a good reason for this limitation. We must discuss some additional aspects of quantum theory, in particular, the exclusion principle and the electron spin, in order to present the quantum theory of many-electron systems. The exclusion principle was proposed by Wolfgang Pauli in 1925 and the electron spin was introduced by George E. Uhlenbeck and Samuel Goudsmit, also in 1925. We will give a brief description of the various experimental discoveries that preceded these theoretical developments.

We discussed in Section 1-6 how the experimental information on the hydrogen atom spectrum helped Bohr in formulating his quantum theory. Clearly, there was much more experimental information available on larger atoms and molecules. Among the abundance of experimental data, certain experimental discoveries were particularly relevant to quantum theory. In our opinion these experiments were (1) the discovery of the Zeeman effect by Pieter Zeeman in 1896, (2) the work by Stern and Gerlach on the splitting of a molecular beam in a magnetic field, (3) the emission of X-rays by atoms, and (4) the doublet structure of the spectra of alkali atoms. We discuss these experiments separately.

Zeeman felt that an atomic spectral line might be affected by the presence of a magnetic field and he decided to investigate the effect of a magnetic field on the *D* lines of the sodium atom. We know now that the sodium *D* lines, located at 5890.0 Å and at 5895.9 Å, represent the 3*s* to 3*p* transition of the outer electron of the sodium atom. Zeeman found a distinct widening of the sodium *D* lines when the magnetic field was turned on. He could not resolve the lines, but in a later experiment on one of the *Cd* lines he observed a splitting of the spectral line in the presence of a magnetic field. If the magnetic field is parallel to the direction of observation, the line splits into two components, and if the direction of observation is perpendicular to the magnetic field, there are three components.

It is related that Zeeman reported his experimental results to H. A. Lorentz, the professor of theoretical physics at the University of Leiden, and that Lorentz went home and derived the theoretical interpretation of the magnetic splittings during the same evening. Lorentz' theory was based on the classical electron theory. He treated the motion of the electron as a three-dimensional harmonic oscillator and calculated the interactions between the magnetic field and the various components of the harmonic oscillator. The theory also predicted the polarizations of the various components, and it was easily verified by Zeeman that the theoretical polarization predictions were consistent with the experimental observations for the *Cd* experiment. Lorentz and Zeeman were joint recipients of the second Nobel prize in physics in 1902.

Unfortunately, in 1897 more precise experiments were performed on the magnetic splittings of the sodium *D* lines, and it was found that the one line splits into four components and the other line splits into six components. This became known as the anomalous Zeeman effect and it defied all theoretical interpretations. Those splittings which followed Lorentz' theoretical predictions were called the normal Zeeman effect. In spite of the various inconsistencies, the normal Zeeman effect in combination with Lorentz' theoretical interpretation constituted a clear proof of the existence of the electron.

The first Nobel prize in physics was awarded in 1901 to Wilhelm Röntgen for the discovery of X-rays. It was found that X-rays are electromagnetic radiation with very short wavelengths, of the order of 1 Å. The emission of an X-ray photon involves the inner electrons of an atom. If one of the inner electrons is ejected by mechanical means, then one of the valence electrons may make a transition to one of the lower orbitals. This downward transition is accompanied by the emission of an X-ray photon. The frequencies of the X-rays that are emitted by an atom were the main source of information on the energies of the inner electrons in the various atoms.

It was found that the X-ray emission spectrum of a particular atom consists of various groups of lines. The group of lines with the highest frequency, corresponding to the energies of the most highly bound electrons, was called the *K*-emission spectrum. The next group of lines is the *L*-emission spectrum. Then comes the *M*-emission spectrum, and so on. The group of electrons that gives rise to the *K* lines is called the *K* shell. The next group is the *L* shell, and so on. It became clear that the electrons in an atom do not all have the same energy. Instead they are divided into shells with quite different energies.

Moseley compared the X-ray emission spectra of different atoms and he discovered a simple relation between the lines of one group, for instance, the *K* spectrum, and the atomic number. It turned out that the spectral lines could be represented as

$$\nu_K = T(Z-p)^2 \tag{10-1}$$

where $Z$ is the atomic number and $T$ and $p$ are constants. There is a similar relation for the $L$ lines, but the constants are different.

The third experimental result is the molecular beam work by Otto Stern and Walther Gerlach in 1921. Stern and Gerlach let a beam of silver atoms pass through an inhomogeneous magnetic field. Each silver atom has a small magnetic moment and according to classical mechanics the magnetic moments should have a random orientation around the magnetic field; this would lead to a broadening of the beam with a Gaussian distribution around the center, due to the magnetic field. In the experiment Stern and Gerlach found that the magnetic field causes the beam to split into two different beams. The experiment presented a clear demonstration of quantization with respect to a magnetic field: about half the atoms had their magnetic moments parallel to the field and the other half antiparallel to the field. From the experimental results Stern and Gerlach could even calculate the magnitude of the magnetic moments. They were equal to

$$\mu_o = \left( \frac{e\hbar}{2mc} \right) \tag{10-2}$$

to within the experimental error. The quantity $\mu_o$ was already known as the Bohr magneton; it is the magnetic moment corresponding to the motion of an orbital electron with an angular momentum $\hbar$.

The Stern-Gerlach results offered clear proof that a silver atom possesses a nonzero angular momentum, but at the time it was not clear whether the angular momentum was associated with the inner core of the atom or with the outer electron. Most scientists were leaning toward the erroneous assumption that the angular momentum was associated with the core, since it was known that the outer electron was in an $s$ state with zero orbital angular momentum.

One aspect of the spectra of alkali atoms was related to this assumption, namely the doublet structure of these spectra. We mentioned already in our discussion of the Zeeman effect that the sodium $D$ line is actually a doublet of two lines, about 6 Å apart. It was believed that these splittings (or doublets) were associated with interactions between the motion of the outer electron and the angular momentum of the core, but it became increasingly difficult to defend this assumption.

It can be seen that in the early twenties there were a number of experimental results that were hard to explain: the anomalous Zeeman effect, the doublet structure of the alkali spectra, the angular momentum of the silver atoms, and so on. Additional questions emerged from the experimental information derived from X-ray spectra. If there are various stationary states available for the electrons in an atom, we would expect all electrons to occupy the state with the lowest energy. This may be a logical assumption, but it was clearly at variance with the experimental results that indicated the existence of the various shells, the $K$ shell, $L$ shell, and so on.

Toward the end of 1922 Wolfgang Pauli became interested in the various problems outlined above. He felt that the questions were related and that it should be possible to derive a comprehensive theory which would supply the answer to the problems of the anomalous Zeeman effect, the doublet structure of the alkali spectra, the shell structure of the atoms, and so on. We discuss Pauli's work and the related work by Bohr and others in the following section.

## 10-2 The Discovery of the Exclusion Principle and the Electron Spin

We mentioned in the previous section that the X-ray spectrum of an atom was the main source of information about the structure of the inner electrons. It could be derived from the X-ray spectrum that the inner electrons could be divided into a $K$ shell, an $L$ shell, an $M$ shell, and so on. According to Bohr the electrons in these shells were characterized by a quantum number $n$. The $K$ shell corresponds to $n=1$, the $L$ shell to $n=2$, and so on. Within each shell the X-ray emission spectrum exhibits a certain fine structure, and Bohr proposed that the electrons within a given shell are characterized by a second quantum number $k$ and eventually by a third quantum number $j$. The conclusion was that each electron in an atom could be characterized by a set of three quantum numbers ($n$, $k$, and $j$).

Meanwhile, Pauli had been studying the various theoretical problems associated with atomic structure. During 1923 and 1924 he considered the Zeeman effect of the alkali metals and the doublet structure of the alkali spectra. It had been assumed earlier that the doublet structure could be ascribed to the magnetic moment of the atomic core, but Pauli showed that this assumption could not be valid and that the angular momentum of the core had to be equal to zero. It followed then that both the Zeeman effect and the doublet structure of the alkali spectrum had to be due to the outer electron. Pauli proposed that this outer electron possessed a certain "two-valued" property. The logical extrapolation of this proposal is the introduction of a new quantum number which can have two possible values, and this is exactly what Pauli suggested in 1925. However, before discussing this we should briefly mention the work by Stoner.

Edmund C. Stoner published a paper on the distribution of electrons among atomic levels where he again characterized each atomic state by three quantum numbers $n$, $k_1$, and $k_2$. He suggested that the number of electrons with a given value of $n$ was double the amount of possible values of $k_1$ and $k_2$. According to this suggestion the $K$ shell could accommodate two electrons, the $L$ shell could have eight, the $M$ shell 18, and so on. In general, the shell corresponding to $n$ could accommodate $2n^2$ electrons.

In 1925 Pauli formulated what became known as the exclusion principle. He assumed that each stationary state is characterized by four quantum numbers, namely $n$, $k_1, k_2$, and a fourth number which can have only two values. Furthermore, he proposed that each stationary state, described by the four quantum numbers, can accommodate only one electron.

Pauli showed that his theoretical proposals could explain the structure of the helium atom and of the alkaline earths. If we place two electrons in two orbitals with identical quantum numbers $(n, k_1, k_2)$, the fourth quantum number $m$ must have different values for the two electrons. This configuration is then nondegenerate. If we place the two electrons in two orbitals with different values $(n, k_1, k_2)$, the fourth quantum numbers may either be the same or different; this gives rise to four possibilities. By means of these arguments Pauli could explain various features of the helium spectrum that had not been understood up to then.

It is somewhat surprising that Pauli did not associate his fourth quantum number with the intrinsic angular momentum of the electron, but for some reason he did not. The idea did occur to R. Kronig, who considered the magnetic moment and the angular momentum of an electron due to its spinning motion around the axis. But when Kronig pursued the idea further, he found that the conclusions were inconsistent with classical electron theory and he rejected the whole hypothesis.

It was left to Goudsmit and Uhlenbeck, who were two graduate students at the University of Leiden, to propose the hypothesis of the spinning electron in a few papers in 1925 and 1926. We will show in the next section that for the orbital motion of the electron the ratio between the magnetic moment $\boldsymbol{\mu}$ and the angular momentum $\mathbf{M}$ is given by

$$\boldsymbol{\mu} = -\frac{e}{2mc}\mathbf{M} \tag{10-3}$$

Goudsmit and Uhlenbeck assumed that for the spinning motion of the electron the ratio between the magnetic moment $\boldsymbol{\mu}_s$ and the angular momentum $\mathbf{M}_s$ is twice as large as for the orbital moment.

We have seen in our discussion of the angular momentum in Section 6-2 that a state with total angular momentum $l$ is $(2l+1)$-fold degenerate. In a magnetic field the state splits into $2l+1$ components. In the experiment by Stern and Gerlach the atomic beam splits into two components, and the spin angular momentum should therefore have a value $\hbar/2$. The value of the corresponding magnetic moment is one Bohr magneton [see Eq. (10-2)], and it can be seen that the ratio between the two is exactly twice the value of Eq. (10-3).

We discuss the classical electromagnetic theory of the spinning electron in the following section. We should add that the mathematical description of the electron spin, based on the analogy with the orbital angular momentum, was proposed by Pauli in 1927. During that year the exclusion principle was also formulated in more general form. We discuss these topics in Sections 10-4 and 10-5.

## 10-3 The Classical Theory of the Spinning Electron

In the present section we derive some expressions for the energy of a spinning electron in the presence of an electromagnetic field. Other expressions will be only stated, because their derivations involve relativity theory which is not discussed in this book.

We first consider the orbital motion of an electron in the presence of a homogeneous magnetic field **H**. We represent the electromagnetic field by means of the potentials defined in Eq. (9-33). It is easily verified that for $\mathbf{E}=\mathbf{0}$ and where **H** is constant we have

$$\mathbf{A}=\tfrac{1}{2}[\mathbf{H}\times\mathbf{r}]$$

$$\phi=0 \tag{10-4}$$

The energy of an electron in the presence of an electromagnetic field is given by Eq. (9-44). If we generalize to a many-electron system we obtain for the Hamiltonian

$$\mathcal{H}=\frac{1}{2m}\sum_j\left(\mathbf{p}_j+\frac{e}{c}\mathbf{A}_j\right)^2-e\phi_j+V \tag{10-5}$$

In the case of a homogeneous magnetic field we substitute Eq. (10-4) and we obtain

$$\mathcal{H}=\frac{1}{2m}\sum_j\mathbf{p}_j^2+V+\frac{e}{2mc}\sum_j\mathbf{p}_j\cdot[\mathbf{H}\times\mathbf{r}_j]+\frac{e^2}{8mc^2}\sum_j|[\mathbf{H}\times\mathbf{r}_j]|^2 \tag{10-6}$$

The term linear in the magnetic field can be written as

$$\mathcal{H}_1=\frac{e}{2mc}\sum_j\mathbf{p}_j\cdot[\mathbf{H}\times\mathbf{r}_j]=\frac{e}{2mc}\mathbf{H}\cdot\sum_j[\mathbf{r}_j\times\mathbf{p}_j] \tag{10-7}$$

The total angular momentum of the orbital motion is defined as

$$\mathbf{M}=\sum_j\mathbf{r}_j\times\mathbf{p}_j \tag{10-8}$$

Also, the energy of a magnetic dipole $\boldsymbol{\mu}$ in the presence of a magnetic field is given by

$$\mathcal{H}_1=-(\boldsymbol{\mu}\cdot\mathbf{H}) \tag{10-9}$$

By comparing Eqs. (10-7), (10-8), and (10-9) we obtain

$$\mathcal{H}_1=\frac{e}{2mc}(\mathbf{H}\cdot\mathbf{M})=-(\mathbf{H}\cdot\boldsymbol{\mu}) \tag{10-10}$$

Obviously, the ratio between $\boldsymbol{\mu}$ and **H** is

$$\boldsymbol{\mu}=-\frac{e}{2mc}\mathbf{M} \tag{10-11}$$

This expression applies to the orbital angular momentum and the corresponding magnetic moment.

We mentioned in the previous section that for a spinning electron the ratio between its magnetic moment $\boldsymbol{\mu}_s$ and its spin angular momentum $\mathbf{S}$ was postulated by Goudsmit and Uhlenbeck as

$$\boldsymbol{\mu}_s = -\frac{e}{mc}\mathbf{S} \tag{10-12}$$

This ratio is twice as large as in the case of Eq. (10-11). The validity of Eq. (10-12) can be derived by means of relativistic arguments. We give some references at the end of this chapter, but we do not present any derivations of Eq. (10-12).

Let us now consider the energy of a spinning electron in an atom in the presence of an electromagnetic field described by the field vectors $\mathbf{E}$ and $\mathbf{H}$. Clearly, if there is a magnetic field $\mathbf{H}$ only, then the energy $H_s$ of the spinning electron is given by

$$H_s = -(\boldsymbol{\mu}_s \cdot \mathbf{H}) = \frac{e}{mc}(\mathbf{H} \cdot \mathbf{S}) \tag{10-13}$$

Next we consider the effect of a nuclear charge $Ze$ on a spinning electron that moves with a velocity $\mathbf{v}$. In the coordinate system where the electron is at rest it appears that the nucleus moves with a velocity $-\mathbf{v}$, and this motion gives rise to a magnetic field

$$\mathbf{H}' = -\frac{Ze}{c}\frac{\mathbf{v}\times\mathbf{r}}{r^3} \tag{10-14}$$

at the position of the electron. According to this argument the spinning electron experiences the field $\mathbf{H}'$ of Eq. (10-14) due to the electric field of the nucleus. The corresponding interaction energy $H_s'$ is derived by substitution of Eq. (10-14) into Eq. (10-13):

$$H_s' = -\frac{e}{mc}\frac{Ze}{mc}\frac{(\mathbf{p}\times\mathbf{r})\cdot\mathbf{S}}{r^3} \tag{10-15}$$

The above expression was used in the early calculations on the interaction between spin and orbital motion, but the expression is incorrect and it leads to discrepancies of a factor 2 between experimental and theoretical results. The reason that Eq. (10-15) is incorrect is that it was derived with respect to a coordinate system where the electron is at rest; it should have been derived with respect to a coordinate system where the nucleus is at rest. According to relativity theory, to an observer at the position of the nucleus it seems that the coordinate axes in which the electron is instantaneously at rest rotate with an angular frequency:

$$\omega = -\frac{1}{2c^2}(\mathbf{v}\times\mathbf{a}) \tag{10-16}$$

The problem was first considered by L. H. Thomas in 1926, and the angular frequency of Eq. (10-16) is known as the Thomas frequency. The Thomas precession has the net effect of reducing the magnetic field $\mathbf{H}'$ of Eq. (10-14) that the electron experiences by a factor 2. The correct expression for $H_s'$ is therefore

$$H_s' = -\frac{Ze^2}{2m^2c^2}\frac{(\mathbf{p}\times\mathbf{r})\cdot\mathbf{S}}{r^3} \tag{10-17}$$

The factor $\frac{1}{2}$ in this expression, which is due to the relativistic precession of the spinning electron, is known as the Thomas factor.

The total energy of one spinning electron in the presence of a nuclear charge $Ze$ and of an electromagnetic field described by $\mathbf{E}$ and $\mathbf{H}$ is given by

$$H_s = \frac{e}{mc}(\mathbf{S}\cdot\mathbf{B}) \tag{10-18}$$

where the vector $\mathbf{B}$ is defined as

$$\mathbf{B} = \mathbf{H} + \frac{1}{2mc}(\mathbf{E}\times\mathbf{p}) + \frac{Ze}{2mc}\frac{(\mathbf{r}\times\mathbf{p})}{r^3} \tag{10-19}$$

Again, the factor $\frac{1}{2}$ in the second and third term of Eq. (10-19) is the Thomas factor. The inclusion of the Thomas factor in the expression for the spin energy brought the theoretical predictions on the structure of atomic spectra into agreement with the experimental results.

Let us now consider the total spin energy of a set of $N$ electrons in an atom with atomic charge $Ze$. This energy is the sum of two terms. The first term is known as the spin-orbit energy and is the sum of the energies of each spin due to the electromagnetic fields it experiences. The second term is the spin-spin energy and is the sum of the interactions between the magnetic dipole moments of the spins.

The spin-orbit energy is derived by generalizing Eqs. (10-18) and (10-19). We write it as

$$H_{so} = \frac{e}{mc}\sum_i (\mathbf{S}_i\cdot\mathbf{B}_i) \tag{10-20}$$

The vector $\mathbf{B}_i$ is a sum of a term similar to Eq. (10-19) and of a term due to the electric and magnetic fields of the other electrons. We write it as

$$\mathbf{B}_i = \mathbf{H}_i + \frac{1}{2mc}(\mathbf{E}_i\times\mathbf{p}_i) + \frac{Ze}{2mc}\frac{\mathbf{r}_i\times\mathbf{p}_i}{r_i^3} + \frac{e}{mc}\sum_{j\neq i}\mathbf{B}_{i,j} \tag{10-21}$$

The vectors $\mathbf{B}_{i,j}$ are given by

$$\mathbf{B}_{i,j} = \frac{(\mathbf{p}_j - \frac{1}{2}\mathbf{p}_i)\times(\mathbf{r}_j - \mathbf{r}_i)}{r_{i,j}^3} \tag{10-22}$$

According to classical electromagnetic theory the energy of interaction between two dipoles $\boldsymbol{\mu}_1$ and $\boldsymbol{\mu}_2$, separated by a vector $\mathbf{r}_{12}$, is given by

$$E_{\text{dip}} = \frac{\boldsymbol{\mu}_1 \cdot \boldsymbol{\mu}_2}{r_{12}^3} - \frac{3(\boldsymbol{\mu}_1 \cdot \mathbf{r}_{12})(\boldsymbol{\mu}_2 \cdot \mathbf{r}_{12})}{r_{12}^5} \tag{10-23}$$

Consequently, the spin-spin interaction energy is given by

$$H_{ss} = \frac{e^2}{2mc^2} \sum_i \sum_{j \neq i} \frac{r_{i,j}^2(\mathbf{S}_i \cdot \mathbf{S}_j) - 3(\mathbf{r}_{i,j} \cdot \mathbf{S}_i)(\mathbf{r}_{i,j} \cdot \mathbf{S}_j)}{r_{i,j}^5} \tag{10-24}$$

## 10-4 The Quantum Mechanical Description of the Electron Spin

The most convenient quantum mechanical description of the spin is based on an analogy with the quantum theory of the orbital angular momentum. In Section 6-2 we saw that the eigenvalues and eigenfunctions of the operator $M^2$, which is the square of the angular momentum operator $\mathbf{M}$, are given by

$$M^2 \psi_{l,m} = l(l+1)\hbar^2 \psi_{l,m} \tag{10-25}$$

The functions $\psi_{l,m}$ are also eigenfunctions of the operator $M_z$, since

$$M_z \psi_{l,m} = m\hbar \psi_{l,m} \tag{10-26}$$

We note that in quantum theory there is a difference between the magnitude $M$ of $\mathbf{M}$ and the maximum projection of $\mathbf{M}$ along the $z$ axis. From Eq. (10-25) we predict that

$$M = \hbar\sqrt{l(l+1)} \tag{10-27}$$

whereas it follows from Eq. (10-26) that the maximum value of $M_z$ is equal to $\hbar l$. This difference is due to the Heisenberg uncertainty principle.

In order to derive the quantum mechanical description of the electron spin, we ought to modify the first hypothesis of Goudsmit and Uhlenbeck somewhat. If we introduce the customary notation $\mathbf{S}$ for the spin angular momentum, we should require that the maximum value of the projection of $\mathbf{S}$ in any given direction is $\frac{1}{2}\hbar$ rather than that the value of $\mathbf{S}$ is equal to $\frac{1}{2}\hbar$.

Formally, we can now construct a quantum mechanical description of the electron spin that is in every respect equivalent to the theory of the angular momentum if we take the quantum number $l$ equal to $\frac{1}{2}$. At first this may seem strange, since the essential condition for the eigenvalues of $M^2$ was that the quantum numbers were all integers. We showed that for noninteger $l$ the functions $Y_{l,m}$ become infinite for $\theta = 0$ and $\theta = \pi$. However, as long as we limit ourselves to the matrix representation of the angular momentum, there is really no need to know the eigenfunctions. The matrix description is concerned only with the transformation properties of the eigenfunctions, and the specific analytical form of these functions is actually immaterial. There is no compelling reason why we cannot consider the transformation properties of functions $\psi_{l,m}$ with $l=\frac{1}{2}$.

If we are not concerned with the specific form of the functions $\psi_{l,m}$, we can easily derive the transformation properties for such functions with half-integer quantum numbers. First we note that if $l=\frac{1}{2}$, the other quantum number $m$ can assume the values $m=+\frac{1}{2}$ and $m=-\frac{1}{2}$ when we adhere to the condition $|m|\leq l$. Our basis set of functions consists, therefore, of

$$\alpha=\psi_{1/2,1/2} \qquad \beta=\psi_{1/2,-1/2} \tag{10-28}$$

From Eq. (10-25) it follows that

$$S^2\alpha=\tfrac{3}{4}\hbar^2\alpha \quad S^2\beta=\tfrac{3}{4}\hbar^2\beta \tag{10-29}$$

The effect of the operators $S_x$, $S_y$, and $S_z$ on $\alpha$ and $\beta$ can be derived from the results of Section 6-2. We derive from Eq. (6-71) that

$$\begin{aligned} S_x\alpha&=\tfrac{1}{2}\hbar\beta \qquad & S_y\alpha&=\tfrac{1}{2}i\hbar\beta \\ S_x\beta&=\tfrac{1}{2}\hbar\alpha \qquad & S_y\beta&=-\tfrac{1}{2}i\hbar\alpha \end{aligned} \tag{10-30}$$

and from Eq. (10-26) that

$$S_z\alpha=\tfrac{1}{2}\hbar\alpha \qquad S_z\beta=-\tfrac{1}{2}\hbar\beta \tag{10-31}$$

In this way we obtain a formal quantum mechanical description of the electron spin, at the basis of which are the two spin functions $\alpha$ and $\beta$. In this approach the operators $S_x$, $S_y$, and $S_z$ can be represented as the matrices

$$S_x=\frac{\hbar}{2}\begin{bmatrix}0 & 1\\ 1 & 0\end{bmatrix} \qquad S_y=\frac{\hbar}{2}\begin{bmatrix}0 & -i\\ i & 0\end{bmatrix} \qquad S_z=\frac{\hbar}{2}\begin{bmatrix}1 & 0\\ 0 & -1\end{bmatrix} \tag{10-32}$$

and the spin functions $\alpha$ and $\beta$ as the column vectors

$$\alpha=\begin{bmatrix}1\\ 0\end{bmatrix} \qquad \beta=\begin{bmatrix}0\\ 1\end{bmatrix} \tag{10-33}$$

In this formal matrix representation of the electron spin, the form of the basis functions has become completely immaterial, and we do not have to worry about it further. Moreover, the conclusions that can be drawn from it are in complete agreement with the experiments. Let us consider, for example, a spinning electron in the presence of a homogeneous magnetic field **H** directed along the $Z$ axis. According to Eq. (10-13) the Hamiltonian for this system is

$$H_s = \frac{e}{mc} H S_z \tag{10-34}$$

It follows immediately from Eq. (10-31) that the eigenvalues and eigenfunctions of this operator are given by

$$H_s \alpha = \frac{e\hbar H}{2mc} \alpha \qquad H_s \beta = -\frac{e\hbar H}{2mc} \beta \tag{10-35}$$

The separation $\Delta E$ between the two energy levels is

$$\Delta E = \frac{e\hbar H}{mc} = 2\mu_o H \tag{10-36}$$

The quantity

$$\mu_o = \frac{e\hbar}{2mc} \tag{10-37}$$

is the same as the Bohr magneton that we introduced in Eq. (10-2). We find that the result (10-36) is in agreement with the Stern-Gerlach experiment and with other experimental results.

We mentioned already in Section 10-2 that the above representation of the spin was first presented by Pauli, and the matrices of Eq. (10-32) are known as Pauli matrices.

Let us now investigate how the wave function of a bound particle ought to be represented if we include the spin in our considerations. In general, we ought to expand the wave function in terms of all possible spin states. In the case of an electron we are fortunate that there are only two spin states, characterized by the functions $\alpha$ and $\beta$, so that the general wave function $\psi(x, y, z; s)$ can always be written as

$$\psi(x, y, z; s) = \psi_+(x, y, z)\alpha + \psi_-(x, y, z)\beta \tag{10-38}$$

The function $\psi(x, y, z; s)$ is an eigenfunction of the complete Hamiltonian $H$ of the system, which is a sum of a large number of contributions $H_o$, $H_1$, $H_2, \ldots,$ and so on, if we wish to consider all relativistic contributions and the various spin energies that we described in the previous section. In general, these various terms are much smaller than the conventional spinless Hamiltonian and to a first approximation they may be neglected. In that case $\psi_+$ and $\psi_-$ should be one and the same eigenfunction of the Hamiltonian $H_o$. We consider the possible effect of the spin-spin and spin-orbit interactions on the eigenvalues and eigenfunctions at a later stage in this chapter and also in Chapter 11.

## 10-5 The Exclusion Principle

The first formulation of the exclusion principle by Pauli was limited to atomic systems only, and it was based on the identification of the various electrons by hydrogenlike quantum numbers. In the present section we discuss the more general formulation of the same exclusion principle which applies to many-electron or many-particle systems in general. This formulation is related to the behavior of the eigenfunction when subjected to permutations of the electrons or the other particles. We discuss the mathematical preliminaries of this formulation in detail, but it may be helpful if we anticipate these discussions and give a summary of the exclusion principle in its most general form; it requires that the electronic eigenfunction of a many-electron system be anti-symmetric with respect to permutations of the electron coordinates. However, before formulating the exclusion principle we should present the quantum mechanical description of many-electron systems.

First, it may be helpful to recall how we construct the Schrödinger equation for one electron moving in a potential field $V(\mathbf{r})$. From the Hamiltonian function

$$H=T+V=\frac{1}{2m}\left(p_x^2+p_y^2+p_z^2\right)+V(\mathbf{r}) \tag{10-39}$$

we construct the Hamiltonian operator

$$H=-\frac{\hbar^2}{2m}\left(\frac{\partial^2}{\partial x^2}+\frac{\partial^2}{\partial y^2}+\frac{\partial^2}{\partial z^2}\right)+V(\mathbf{r}) \tag{10-40}$$

by replacing $\mathbf{p}$ by $(-i\hbar\nabla)$ everywhere. The time-independent Schrödinger equation is then obtained as the eigenvalue problem,

$$H\psi=\varepsilon\psi \tag{10-41}$$

and its solutions are $\varepsilon_n$ and $\psi_n$,

$$H\psi_n=\varepsilon_n\psi_n \tag{10-42}$$

We have not discussed the time-dependent Schrödinger equation

$$H\phi=i\hbar\frac{\partial\phi}{\partial t} \tag{10-43}$$

at any great length, but it is easily verified that its general solution is

$$\phi=\sum_n a_n\psi_n\exp\left(\frac{\varepsilon_n t}{i\hbar}\right) \tag{10-44}$$

The Schrödinger equation for a many-particle system is constructed by following the same procedure exactly. For example, the Hamiltonian function of a helium atom is

$$H=\frac{1}{2m}(p_1^2+p_2^2)-\frac{2e^2}{r_1}-\frac{2e^2}{r_2}+\frac{e^2}{r_{12}} \tag{10-45}$$

after we eliminate the motion of the center of gravity. We replace $\mathbf{p}_1$ by $(-i\hbar\nabla_1)$ and $\mathbf{p}_2$ by $(-i\hbar\nabla_2)$, and we obtain the Hamiltonian operator

$$H=-\frac{\hbar^2}{2m}\left(\frac{\partial^2}{\partial x_1^2}+\frac{\partial^2}{\partial y_1^2}+\frac{\partial^2}{\partial z_1^2}+\frac{\partial^2}{\partial x_2^2}+\frac{\partial^2}{\partial y_2^2}+\frac{\partial^2}{\partial z_2^2}\right)-\frac{2e^2}{r_1}-\frac{2e^2}{r_2}+\frac{e^2}{r_{12}} \tag{10-46}$$

If we use atomic units of length and energy, this operator can be simplified to

$$H=-\tfrac{1}{2}(\Delta_1+\Delta_2)-\frac{2}{r_1}-\frac{2}{r_2}+\frac{1}{r_{12}} \tag{10-47}$$

The Schrödinger equation for the helium atom is therefore

$$-\tfrac{1}{2}(\Delta_1+\Delta_2)\psi-\left(\frac{2}{r_1}+\frac{2}{r_2}-\frac{1}{r_{12}}\right)\psi=E\psi \tag{10-48}$$

The Hamiltonian for an atom or ion with nuclear charge $Ze$ and with $N$ electrons is obtained as

$$H=-\tfrac{1}{2}\sum_{j=1}^{N}\Delta_j-Z\sum_{j=1}^{N}\frac{1}{r_j}+\sum_{j>k}\frac{1}{r_{jk}} \tag{10-49}$$

A molecule containing $s$ nuclei, with charges $Z_\alpha e$ and masses $M_\alpha$, and $N$ electrons is represented by a Hamiltonian

$$H=-\sum_{\alpha=1}^{s}\frac{\hbar^2}{2M_\alpha}\left(\frac{\partial^2}{\partial X_\alpha^2}+\frac{\partial^2}{\partial Y_\alpha^2}+\frac{\partial^2}{\partial Z_\alpha^2}\right)-\sum_{j=1}^{N}\frac{\hbar^2}{2m}\left(\frac{\partial^2}{\partial x_j^2}+\frac{\partial^2}{\partial y_j^2}+\frac{\partial^2}{\partial z_j^2}\right)$$

$$-\sum_{j=1}^{N}\sum_{\alpha=1}^{s}\frac{Z_\alpha e^2}{r_{j,\alpha}}+\sum_{j>k}\frac{e^2}{r_{j,k}}+\sum_{\alpha>\beta}\frac{Z_\alpha Z_\beta e^2}{R_{\alpha,\beta}} \tag{10-50}$$

Here $\mathbf{r}_j=(x_j, y_j, z_j)$ and $\mathbf{R}_\alpha=(X_\alpha, Y_\alpha, Z_\alpha)$ are the coordinates of electron $j$ and nucleus $\alpha$, respectively, and the other quantities are defined as

$$r_{j,\alpha}=|\mathbf{r}_j-\mathbf{R}_\alpha|$$
$$r_{j,k}=|\mathbf{r}_k-\mathbf{r}_j| \tag{10-51}$$
$$R_{\alpha,\beta}=|\mathbf{R}_\beta-\mathbf{R}_\alpha|$$

Let us now consider a system that contains $N$ electrons only, for example, the system that is described by the Hamiltonian (10-49). Since the electrons all play exactly the same role, the Hamiltonian remains unchanged if we permute any of the electrons. We can write this as

$$[H,P]=0 \tag{10-52}$$

The various definitions and properties of the permutation operator were discussed in Section 2-2.

Let $\varepsilon_n$ be one of the eigenvalues of $H$ and $\psi_n$ a corresponding eigenfunction, so that

$$H\psi_n=\varepsilon_n\psi_n \tag{10-53}$$

It follows that we also have

$$P(H\psi_n)=P(\varepsilon_n\psi_n) \tag{10-54}$$

where the operator $P$ represents an arbitrary permutation of the $N$ electrons. It now follows from Eq. (10-52) that we can also write this as

$$H(P\psi_n)=\varepsilon_n(P\psi_n) \tag{10-55}$$

since $H$ and $P$ commute. It follows that not only $\psi_n$ but also the functions $(P\psi_n)$ are eigenfunctions of $H$ belonging to $\varepsilon_n$. If all these functions are different, the eigenvalue $\varepsilon_n$ is $(N!)$-fold degenerate.

The experiments clearly indicate that degeneracies of so large an order do not usually occur in $N$-electron systems, and the above conclusion is therefore incorrect. It is not possible to resolve this discrepancy by making use of the quantum theory which we have discussed so far, and we have to introduce an additional basic assumption in quantum mechanics if we want our predictions for $N$-electron systems to be compatible with the experimental observation. This basic assumption can be derived from the Pauli exclusion principle, which states that two electrons in a central field can never be in states that have the same four quantum numbers. Expressing this in more general form, we require that the wave function $\psi_n$ be antisymmetric with respect to permutations of the electron coordinates. Hence the only solutions of the Schrödinger equation that give physically acceptable eigenfunctions are those which satisfy the condition

$$P\psi_n=\delta_P\psi_n \tag{10-56}$$

The result of this restriction is that only one particular linear combination of the $N!$ functions $(P\psi_n)$ is an acceptable eigenfunction. If $\psi_n$ is already antisymmetric, then all functions $(P\psi_n)$ are the same eigenfunctions. If $\psi_n$ is not antisymmetric, we can use it as a basis for the construction of an antisymmetric function $\psi_n$ if we take

$$\psi_n = (N!)^{-1/2} \sum_P P\delta_P \psi_n \tag{10-57}$$

Here the summation is to be taken over all possible permutations, including the identity permutation. If $\psi_n$ is symmetric with respect to any of the electron permutations, it is impossible to construct an antisymmetric eigenfunction from it because the procedure of Eq. (10-57) would lead to zero for $\psi_n$. We must conclude, therefore, that such symmetric functions $\psi_n$ are inadmissible at all times. It can be seen that this conclusion is identical to the original formulation of the Pauli exclusion principle, since any atomic state in which two electrons have the same set of quantum numbers would be represented by a symmetric wave function, and such states cannot exist.

In applying the exclusion principle, we should be aware that the permutations refer not only to the space coordinates of the electrons but also to the spin coordinates. The four quantum numbers that we mentioned before refer to the eigenvalues of the operators $H, M^2, M_z$, and $S_z$ for each individual electron. We will see that the description of atomic states by means of quantum numbers for the individual electrons is not rigorous, and the original formulation of the Pauli principle has only limited validity.

The general exclusion principle is not restricted to electrons alone; it is valid for any system that contains one or more groups of identical particles. It requires that the wave function of such a system be either symmetric or antisymmetric with respect to any permutation of the coordinates of a group of identical particles. The function should be symmetric if the particles in question have integer spin, and antisymmetric if the particles have half-integer spin.

It may be helpful if we make a few comments on this general formulation of the exclusion principle. We have discussed only the quantum theory of the electron spin in which we took the quantum number $l$ equal to $\frac{1}{2}$ and then made use of the matrix representation of the angular momentum operators. A number of other particles, namely the neutron, the proton, the nuclei $H^3$, $He^3$, $C^{13}$, $N^{15}$, and so on, can be treated in the same manner, and we say that all these particles have spin $\frac{1}{2}$. However, there are some nuclei, such as $H^2$, $Li^6$, and $N^{14}$, in which we obtain a satisfactory description of the spin properties only if we take the quantum number $l$ equal to 1. These particles are said to have spin 1. Certain nuclei, for example, $C^{12}$, have no spin whatsoever, and we say that they have spin 0. If we look through a table of nuclear properties of the elements, we see that many different spins occur in nature; some notable cases are $O^{17}$ and $Al^{27}$, which have spin 5/2, $V^{50}$ with spin 6, and $Ge^{73}$ with spin 9/2. The permutation properties of the wave function depend only on whether the spins of the particles are integer or half-integer and not on the magnitudes of the spin.

As an example of the exclusion principle, let us compare the situation for the three molecules $H_2$, HD, and $D_2$. If we denote the space and spin coordinates for the two nuclei symbolically by $a$ and $b$ and for the two electrons by 1 and 2, we can write any wave function of the molecules as $\psi(1,2;a,b)$. Since the electrons have spin $\frac{1}{2}$, the symmetry condition for all three molecules is that

$$\psi(2,1;a,b) = -\psi(1,2;a,b) \tag{10-58}$$

A proton also has spin $\frac{1}{2}$, so that for the $H_2$ molecule the wave function must satisfy also the condition

$$\psi(1,2;b,a) = -\psi(1,2;a,b) \tag{10-59}$$

A deuteron has spin 1, and therefore we have, for the $D_2$ molecule, the condition

$$\psi(1,2;b,a) = \psi(1,2;a,b) \tag{10-60}$$

In the HD molecule the two nuclei are different, and the two functions $\psi(1,2;a,b)$ and $\psi(1,2;b,a)$ are not related in any way.

The permutation properties of the wave functions are also decisive in determining the type of statistics that the particles obey. Particles with integer spin are distributed according to the Bose-Einstein expression, and particles with half-integer spin follow Fermi-Dirac statistics. This is discussed in detail in any book on statistical mechanics.

## 10-6 Two-Electron Systems

It is not easy to give a general and accurate description of the properties of the wave functions of $N$-electron systems, and we will restrict ourselves first to a two-electron system. To a first approximation we can write the Hamiltonian of such a system as

$$H_o = -\frac{\hbar^2}{2m}(\Delta_1 + \Delta_2) + V(\mathbf{r}_1, \mathbf{r}_2) \tag{10-61}$$

An acceptable eigenfunction of a two-electron system must satisfy two conditions: (1) it must be an eigenfunction, and (2) it must be antisymmetric with respect to a permutation of the two sets of electronic coordinates.

The wave function should also contain the spin variables of the two electrons. It is convenient to study first the dependence on the spin variables before going into the dependence on the space coordinates $\mathbf{r}_1$ and $\mathbf{r}_2$ of the wave function. We have seen that for a one-electron system there are two possible states for the spin, namely $\alpha$ and $\beta$. For a two-electron system there are four possible spin states, namely $\alpha_1\alpha_2, \alpha_1\beta_2, \beta_1\alpha_2$, and $\beta_1\beta_2$. The notation is self-explanatory; for example, the symbol $\alpha_1\beta_2$ refers to the situation in which electron 1 is in spin state $\alpha$ and electron 2 is in spin state $\beta$. Any wave function $\Psi_n$ of the system can now be represented by the general expression

$$\Psi_n = \psi_{n,++}(\mathbf{r}_1,\mathbf{r}_2)\alpha_1\alpha_2 + \psi_{n,+-}(\mathbf{r}_1,\mathbf{r}_2)\alpha_1\beta_2 + \psi_{n,-+}(\mathbf{r}_1,\mathbf{r}_2)\beta_1\alpha_2 + \psi_{n,--}(\mathbf{r}_1,\mathbf{r}_2)\beta_1\beta_2 \quad (10\text{-}62)$$

It will prove to be convenient to rearrange this expression slightly and to write it as

$$\Psi_n = \psi_{n,1}(\mathbf{r}_1,\mathbf{r}_2)\alpha_1\alpha_2 + \psi_{n,2}(\mathbf{r}_1,\mathbf{r}_2)\beta_1\beta_2 + \psi_{n,3}(\mathbf{r}_1,\mathbf{r}_2)[\alpha_1\beta_2 + \beta_1\alpha_2] + \psi_{n,0}(\mathbf{r}_1,\mathbf{r}_2)[\alpha_1\beta_2 - \beta_1\alpha_2] \quad (10\text{-}63)$$

Let us now make use of the exclusion principle and impose the condition

$$P\Psi_n = -\Psi_n \quad (10\text{-}64)$$

By combining Eqs. (10-63) and (10-64) we obtain the following conditions:

$$\begin{aligned} \psi_{n,1}(\mathbf{r}_1,\mathbf{r}_2) &= -\psi_{n,1}(\mathbf{r}_2,\mathbf{r}_1) \\ \psi_{n,2}(\mathbf{r}_1,\mathbf{r}_2) &= -\psi_{n,2}(\mathbf{r}_2,\mathbf{r}_1) \\ \psi_{n,3}(\mathbf{r}_1,\mathbf{r}_2) &= -\psi_{n,3}(\mathbf{r}_2,\mathbf{r}_1) \\ \psi_{n,0}(\mathbf{r}_1,\mathbf{r}_2) &= \psi_{n,0}(\mathbf{r}_2,\mathbf{r}_1) \end{aligned} \quad (10\text{-}65)$$

It follows that the functions $\psi_{n,1}$, $\psi_{n,2}$, and $\psi_{n,3}$ are antisymmetric and that $\psi_{n,0}$ is symmetric with respect to permutations. In the function $\Psi_n$ of Eq. (10-63) the first three terms are each a product of an antisymmetric orbital function and a symmetric spin function, and the last term is a product of a symmetric orbital function and an antisymmetric spin function. We introduce a new notation for the spin functions:

$$\begin{aligned} {}^3\zeta_1(1,2) &= \alpha_1\alpha_2 \\ {}^3\zeta_0(1,2) &= 2^{-1/2}(\alpha_1\beta_2 + \beta_1\alpha_2) \\ {}^3\zeta_{-1}(1,2) &= \beta_1\beta_2 \\ {}^1\zeta(1,2) &= 2^{-1/2}(\alpha_1\beta_2 - \beta_1\alpha_2) \end{aligned} \quad (10\text{-}66)$$

and we write the function $\Psi_n$ in the form

$$\Psi_n = {}^3\Psi_n + {}^1\Psi_n \tag{10-67}$$

where

$$\begin{aligned} {}^3\Psi_n &= \psi_{n,1}{}^3\zeta_1(1,2) + \psi_{n,3}{}^3\zeta_0(1,2) + \psi_{n,2}{}^3\zeta_{-1}(1,2) \\ {}^1\Psi_n &= \psi_{n,0}{}^1\zeta(1,2) \end{aligned} \tag{10-68}$$

Let us first study the properties of the spin functions. If we introduce the new operators

$$\begin{aligned} \mathbf{S} &= \mathbf{S}_1 + \mathbf{S}_2 \\ S^2 &= (\mathbf{S}_1 + \mathbf{S}_2)^2 = S_1^2 + S_2^2 + 2\mathbf{S}_1 \cdot \mathbf{S}_2 \end{aligned} \tag{10-69}$$

it is easily verified that

$$\begin{aligned} S^2\, {}^3\zeta_m(1,2) &= 2\hbar^2\, {}^3\zeta_m(1,2) \\ S_z{}^3\zeta_m(1,2) &= m\hbar{}^3\zeta_m(1,2) \qquad m = 1, 0, -1 \end{aligned} \tag{10-70}$$

and

$$\begin{aligned} S^2{}^1\zeta(1,2) &= 0 \\ S_z{}^1\zeta(1,2) &= 0 \end{aligned} \tag{10-71}$$

All four spin functions of Eq. (10-66) are therefore eigenfunctions of both operators $S^2$ and $S_z$. The three functions ${}^3\zeta_m$ have the same transformation properties with respect to $\mathbf{S}$ as the functions $\psi_{l,m}$ of Section 6-3 have with respect to $\mathbf{M}$, if we take the quantum number $l$ equal to unity.

The behavior of the spin functions can be understood on the basis of a simple physical picture. If we take the direction of the first spin as the quantization axis for the second spin, there are two possibilities: the second spin can point in either the same or the opposite direction of the first spin. In the first case the total spin of the system is equal to unity; it can have three possible orientations in any given direction, so that we have a threefold degeneracy. We speak, therefore, of a triplet state. In the second case the total spin is zero. Since there is no degeneracy, we call this a singlet state.

Let us now proceed to the properties of the orbital functions $\psi_{n,j}$. If we impose the conditions that $\Psi_n$ is an eigenfunction of the operator (10-61),

$$H_o\Psi_n = \lambda_n\Psi_n \tag{10-72}$$

and we substitute Eqs. (10-67) and (10-68), we have

$$H_o\psi_{n,j}=\lambda_n\psi_{n,j} \qquad j=0,1,2,3 \tag{10-73}$$

In general the eigenvalue $\lambda_n$ is nondegenerate so that there is only one eigenfunction. This eigenfunction is either symmetric or antisymmetric with respect to permutation of the orbital electron coordinates $\mathbf{r}_1$ and $\mathbf{r}_2$. If the eigenfunction is symmetric, we denote it by $s_n(\mathbf{r}_1,\mathbf{r}_2)$, and the total wave function, including spin, is

$$^1\Psi(1,2)=s_n(\mathbf{r}_1,\mathbf{r}_2)\,^1\zeta(1,2) \tag{10-74}$$

If the eigenfunction is antisymmetric, we denote it by $a_n(\mathbf{r}_1,\mathbf{r}_2)$. Since there is only one linearly independent orbital function, the total function is now

$$\begin{aligned}&^3\Psi(1,2)=a_n(\mathbf{r}_1,\mathbf{r}_2)\,^3\zeta(1,2)\\&^3\zeta(1,2)=c_1\,^3\zeta_1(1,2)+c_0\,^3\zeta_0(1,2)+c_{-1}\,^3\zeta_{-1}(1,2)\end{aligned} \tag{10-75}$$

The eigenvalues and eigenfunctions of $H_o$ can be separated into two groups: (1) the eigenvalues $^1\lambda_n$ which have the symmetric eigenfunctions $s_n$ and which combine with a singlet spin function, and (2) the eigenvalues $^3\lambda_n$ which have the antisymmetric eigenfunctions $a_n$ and which combine with a triplet spin function. It can be seen that the spin multiplicity is denoted by a superscript on the left side of the wave function. The symbol is 1 for a singlet state, since there is no degeneracy, and it is 3 for a triplet state, because this state is threefold degenerate.

It should be noted that we have neglected the effect of spin-orbit and spin-spin interactions in the above discussion. In Section 10-7 we discuss the quantum mechanical description of the orbital wave functions of the helium atom. In Section 10-8 we consider the effect of the spin interactions.

## 10-7 The Helium Atom

According to Section 10-6 the eigenfunctions of the helium atom are

$$\begin{aligned}&^1\Psi_n=s_n(\mathbf{r}_1,\mathbf{r}_2)\,^1\zeta(1,2)\\&^3\Psi_{k,m}=a_k(\mathbf{r}_1,\mathbf{r}_2)\,^3\zeta_m(1,2)\end{aligned} \tag{10-76}$$

where the orbital functions $s_n$ and $a_k$ are eigenfunctions of the Hamiltonian

$$H_o=-\frac{1}{2}(\Delta_1+\Delta_2)-\frac{2}{r_1}-\frac{2}{r_2}+\frac{1}{r_{12}} \tag{10-77}$$

Here we have introduced atomic units of length and energy. No exact solutions of the corresponding Schrödinger equation have ever been found, and we must limit ourselves to the derivation of approximate eigenfunctions. The customary approach uses the variation theorem, starting with a trial function that contains a number of parameters. Obviously, the results will become more accurate if we use more complex trial functions with an increasing number of parameters. We first investigate a few simple functions and then indicate how to modify the functions in order to obtain more accurate results.

In general, it can be expected that the functions $s_n$ and $a_k$ depend on the coordinates of both electrons in some complicated way, but as a first approximation we choose them separable in the electron coordinates. This means that $s_n$ can be written as

$$s_n(\mathbf{r}_1,\mathbf{r}_2)=f_n(\mathbf{r}_1)f_n(\mathbf{r}_2) \tag{10-78}$$

or

$$s_n(\mathbf{r}_1,\mathbf{r}_2)=f_n(\mathbf{r}_1)g_n(\mathbf{r}_2)+g_n(\mathbf{r}_1)f_n(\mathbf{r}_2) \tag{10-79}$$

and that $a_k(\mathbf{r}_1,\mathbf{r}_2)$ is approximated as

$$a_k(\mathbf{r}_1,\mathbf{r}_2)=f_k(\mathbf{r}_1)g_k(\mathbf{r}_2)-g_k(\mathbf{r}_1)f_k(\mathbf{r}_2) \tag{10-80}$$

These approximations are the basis of the Hartree-Fock equations, which we discuss in Chapter 11.

Let us now make an educated guess as to the approximate form of the functions $f_k$ and $g_k$. Let us imagine for a moment that the term $r_{12}^{-1}$ in the Hamiltonian (10-77) is absent. In this case the Schrödinger equation would be separable into the two sets of electron coordinates, and the functions $f$ and $g$ would be hydrogenlike wave functions with a nuclear charge $Z=2$. The lowest eigenstate of the helium atom would be obtained by placing two electrons with opposite spins in a $(1s)$ state, which we might denote as the $(1s)^2$ state. In the same way we would obtain a singlet $(1s)(2s)$ state, a triplet $(1s)(2s)$ state, and so on. In this approximation each helium state can be constructed from two hydrogenlike states. We know that this approximation is not quite correct, but at least it gives us a starting point for our attempts to construct approximate wave functions. Also, it indicates clearly that the ground state of the helium atom ought to be a singlet state and that we should approximate its wave function as

$$s_o(\mathbf{r}_1,\mathbf{r}_2)=f_o(\mathbf{r}_1)f_o(\mathbf{r}_2) \tag{10-81}$$

Of course, it is not permissible to neglect the term $r_{12}^{-1}$, but at last we can think of a crude way to estimate how its presence might affect the hydrogenlike wave function. If we think of one of the two electrons as moving in the potential field of the nucleus and of the other electron, and if we assume that the charge cloud of the second electron is spherically symmetric, we can use classical electrostatic theory to get a rough idea of what this potential field looks like. Let $\delta(r)$ be the fraction of the probability density of the second electron that is contained in a sphere of radius $r$ around the nucleus. The first electron then experiences a potential

$$V(r) = -\frac{2-\delta(r)}{r} \tag{10-82}$$

at a distance $r$ from the nucleus. It is clear that $\delta(r)$ iz zero for $r=0$ and $\delta(r)$ is unity when $r$ tends to infinity, so that $V(r)$ varies between $(-2/r)$ close to the nucleus and $(-1/r)$ at great distances. On the average the first electron experiences, therefore, a potential

$$V(r) = -\frac{Z}{r} \qquad 1<Z<2 \tag{10-83}$$

Accordingly, we approximate $f_o(r)$ as

$$f_o(r) \simeq e^{-Zr} \tag{10-84}$$

where $Z$ is a number between 1 and 2 which we will try to determine by means of the variation theorem.

Our normalized trial function is

$$s_o(1,2) = \frac{Z^3}{\pi} e^{-Zr_1} \cdot e^{-Zr_2} \tag{10-85}$$

and we wish to evaluate the expectation value

$$E(Z) = \langle s_o(1,2) | H_o | s_o(1,2) \rangle \tag{10-86}$$

as a function of the parameter $Z$. We observe first that

$$\left( -\frac{\Delta_1}{2} - \frac{Z}{r_1} \right) e^{-Zr_1} = -\frac{Z^2}{2} e^{-Zr_1} \tag{10-87}$$

so that we may simplify $E(Z)$ to

$$E(Z) = -Z^2 + (2Z-4)\langle s_o | r_1^{-1} | s_o \rangle + \langle s_o | r_{12}^{-1} | s_o \rangle \tag{10-88}$$

It is easily found that

$$\langle s_o | r_1^{-1} | s_o \rangle = Z \tag{10-89}$$

The other integral can be transformed to

$$\langle s_o | r_{12}^{-1} | s_o \rangle = \frac{Z}{32\pi^2} \int\int \frac{e^{-r_1} e^{-r_2}}{r_{12}} d\mathbf{r}_1 d\mathbf{r}_2 \tag{10-90}$$

The evaluation of this integral is discussed in Appendix *B*; it involves the expansion of the $(1/r_{12})$ term in terms of spherical harmonics. Here we just quote the result:

$$\langle s_o | r_{12}^{-1} | s_o \rangle = \left(\frac{5}{8}\right) Z \tag{10-91}$$

Substitution of Eqs. (10-89) and (10-91) into the expression (10-88) for $E(Z)$ gives

$$E(Z) = Z^2 - \left(\frac{27}{8}\right) Z = \left[Z - \left(\frac{27}{16}\right)\right]^2 - \frac{729}{256} \tag{10-92}$$

It follows that the energy has a minimum for $Z = 1.6875$ and that the minimum is $E(Z_o) = -2.8477$ au. The experimental value is $E_{\text{exp}} = -2.90372$ au, so that the agreement is not too bad.

Let us now investigate how to obtain more accurate energies for the ground state of the helium atom. Our first thought is to use more elaborate trial functions $f_o(r)$, for example,

$$f_o(r) \simeq (1 + ar) e^{-Zr} \tag{10-93}$$

but this approach has its limitations. It has been shown that as long as we stay within the approximation (10-81), the best possible energy value that can be obtained is $E_o = -2.8617$ au. This result is obtained by means of the Hartree-Fock method which we will discuss in Chapter 11. This is not a significant improvement over the result that was obtained from the simple variation function (10-84). If we wish to get better results, we must therefore abandon the approximation (10-81).

Since the ground state of the helium atom has zero angular momentum, it can be shown that its wave function $s_o(\mathbf{r}_1, \mathbf{r}_2)$ is a function only of $r_1$, $r_2$, and $r_{12}$:

$$s_o(\mathbf{r}_1, \mathbf{r}_2) = \psi(r_1, r_2, r_{12}) \tag{10-94}$$

It is obvious that the wave function must contain the variable $r_{12}$, since the repulsion term $r_{12}^{-1}$ in the Hamiltonian (10-77) has the effect of keeping the electrons away from each other. Somehow this behavior should be reflected in the wave functions.

There are two different ways to include the electron repulsion in the variational trial functions. The first one is straightforward; it consists of the expansion of $\psi(r_1, r_2, r_{12})$ as a power series in $r_1, r_2, r_{12}$. This method was applied as early as 1929 by Hylleraas, and it led to satisfactory results. In all these expansions it is customary to introduce the new set of variables

$$s = r_1 + r_2 \qquad t = r_1 - r_2 \qquad u = r_{12} \tag{10-95}$$

Hylleraas' calculation was based on the trial function

$$\psi = e^{-Zs} \sum_k \sum_l \sum_m a_{k,l,m} s^k t^{2l} u^m \tag{10-96}$$

There are practical limitations as to how far this expansion can be extended. Hylleraas restricted his calculation to a six-term trial function, and he obtained an excellent result, $E = -2.90324$.

More recently Kinoshita observed that the Hylleraas expansion (10-96) can never lead to the correct eigenfunction for the helium atom. As a more suitable trial function he proposed, therefore,

$$\psi = e^{-Zs} \sum_{l=0}^{\infty} \sum_{m=0}^{\infty} \sum_{n=0}^{\infty} c_{l,m,n} s^{l-m} u^{m-2n} t^{2n} \tag{10-97}$$

From a 38-term function of this type Kinoshita predicted an energy value of $E = -2.9037225$ au.

The reader might wonder what the practical use of such elaborate calculations is. In this way the nonrelativistic ground-state energy of the helium atom can be determined to within an accuracy of a fraction of a $\text{cm}^{-1}$. The experimental ground-state energy, which can be measured with similar accuracy, is the sum of the nonrelativistic and the relativistic energies. From a comparison between Kinoshita's result and the experimental value we can derive the relativistic contribution to the ground-state energies with an accuracy of about 0.2%. In this way several aspects of relativity theory and quantum field theory could be verified.

At the present time calculations of this type are performed with the aid of high-speed computers. Recently they have been extended to trial functions that contain more than 1000 adjustable parameters. In this way the wave functions of the helium atom can be derived to any degree of accuracy that may be desired.

The second method for including electron repulsion in the wave function is based on what is called configuration interaction. In the case of the helium atom it is considerably less effective than the other approach, but it has the advantage that it can be more conveniently extended to other atoms and also to molecules. We illustrate it by considering the trial function

$$s_o(\mathbf{r}_1, \mathbf{r}_2) = Ae^{-Z(r_1+r_2)} + B(e^{-Z_1 r_1} e^{-Z_2 r_2} + e^{-Z_2 r_1} e^{-Z_1 r_2}) \tag{10-98}$$

This function does not contain $r_{12}$ explicitly, but by choosing the parameters

$Z_1$ and $Z_2$ we can see to it that the second term of Eq. (10-98) represents a situation in which the two electrons are kept apart. For example, if we take $Z_1=2$ and $Z_2=1$, then one of the two electrons is much closer to the nucleus than the other, and on the average the two electrons will be much further apart than in the first term. We can extend this expansion by including terms of the type $(1s)(2s)$, $(2s)^2$, $(2p)^2$, and so on, but the energies that are obtained in this way are usually much less accurate than Kinoshita's value. On the other hand, this approach of expanding the wave function as a linear combination of a number of configurations can supply us with a useful tool for improving wave functions of complex atoms.

## 10-8 Spin-Orbit and Spin-Spin Coupling

In the previous section we considered the eigenvalues of the spinless Hamiltonian (10-77). This means that we neglected the effect of the spin-orbit interaction represented by the Hamiltonian $H_{so}$ of Eq. (10-20) and the effect of the spin-spin interaction represented by the Hamiltonian $H_{ss}$ of Eq. (10-24). We wish to consider the effect of these spin-orbit and spin-spin interactions on the theoretical description of the helium atom, but first we consider the orthogonality properties of the spin functions.

We define the various integrals involving the spin functions $\alpha$ and $\beta$ by using the vector representation (10-33). We have

$$\langle \alpha | \alpha \rangle = [1 \quad 0] \begin{bmatrix} 1 \\ 0 \end{bmatrix} = 1 \tag{10-99}$$

and

$$\langle \alpha | \beta \rangle = [1 \quad 0] \begin{bmatrix} 0 \\ 1 \end{bmatrix} = 0 \tag{10-100}$$

Consequently,

$$\begin{aligned} \langle \alpha | \alpha \rangle &= \langle \beta | \beta \rangle = 1 \\ \langle \alpha | \beta \rangle &= \langle \beta | \alpha \rangle = 0 \end{aligned} \tag{10-101}$$

The orthogonality relations between the two-electron spin functions ${}^1\zeta(1,2)$ and ${}^3\zeta_i(1,2)$ can now be derived by combining Eqs. (10-66) and (10-101):

$$\begin{aligned} \langle {}^1\zeta(1,2) | {}^3\zeta_i(1,2) \rangle &= 0 \qquad & i &= -1,0,1 \\ \langle {}^3\zeta_i(1,2) | {}^3\zeta_j(1,2) \rangle &= \delta_{i,j} \qquad & i,\, j &= -1,0,1 \end{aligned} \tag{10-102}$$

Let us now proceed to derive the eigenvalues and eigenfunctions of the helium atom Hamiltonian, including the spin. The total Hamiltonian is given by

$$H = H_o + H_{so} + H_{ss} \tag{10-103}$$

where $H_o$ is the spinless Hamiltonian (10-77), $H_{so}$ is the spin-orbit energy, and $H_{ss}$ is the spin-spin energy. We have seen in the previous section that the eigenvalue spectrum of $H_o$ consists of a set of singlet states with eigenvalues ${}^1\lambda_k$ and eigenfunctions

$${}^1\Psi_k(1,2) = s_k(\mathbf{r}_1, \mathbf{r}_2)\,{}^1\zeta(1,2) \tag{10-104}$$

and of a set of triplet states with eigenvalues ${}^3\lambda_n$ and eigenfunctions

$${}^3\Psi_n(1,2) = a_n(\mathbf{r}_1, \mathbf{r}_2)\,{}^3\zeta(1,2) \tag{10-105}$$

The various orbital and spin functions are described by Eqs. (10-74) and (10-75).

The eigenvalues and eigenfunctions of the operator $H$ are most conveniently derived by means of the perturbation expansions of Sections 7-4 and 7-5. In the case of the helium atom the perturbations are very small and we need only consider the first-order perturbations. We first consider the perturbation of a singlet state since the singlet states are usually nondegenerate.

We denote the perturbed eigenfunction of the operator $H$ by ${}^1\Phi_k$. The relation between the perturbed function ${}^1\Phi_k$ and the unperturbed function ${}^1\Psi_k$ is then given by

$$\begin{aligned} {}^1\Phi_k = {}^1\Psi_k &- \sum_{l \neq k} \frac{\langle {}^1\Psi_l | H_{so} + H_{ss} | {}^1\Psi_k \rangle}{\lambda_l - \lambda_k} \, {}^1\Psi_l \\ &- \sum_n \frac{\langle {}^3\Psi_n | H_{so} + H_{ss} | {}^1\Psi_k \rangle}{\lambda_n - \lambda_k} \, {}^3\Psi_n \end{aligned} \tag{10-106}$$

We will not discuss the evaluation of these terms in detail, but we wish to note that the matrix elements of $H_{so}$ between singlet and triplet states are, in principle, nonzero. The result is that the eigenfunction of the atomic Hamiltonian $H$ is no longer a pure singlet function. Due to the perturbation of the spin-orbit coupling a small amount of triplet character is mixed with the singlet function.

The perturbation of the triplet states is more complicated because each triplet level is at least threefold degenerate. However, one feature of the perturbation is similar to the singlet states. Due to the spin-orbit interactions small amounts of singlet functions are mixed with the unperturbed triplet functions. Here also, the true eigenfunction is no longer a pure triplet function, although it has small amounts of singlet functions mixed in.

The most important application of these perturbation results is in the theoretical prediction of transition probabilities between singlet and triplet states. First we consider the transition moment between a singlet state 0 and a triplet state $n$, based on the unperturbed eigenfunctions (10-104) and (10-105). The dipole moment operator for a two-electron system is defined as

$$\boldsymbol{\mu}=e\mathbf{r}_1+e\mathbf{r}_2 \tag{10-107}$$

and it is independent of the spin operators. It follows that

$$\begin{aligned}\boldsymbol{\mu}_{0,n}&=\langle {}^1\Psi_0|\boldsymbol{\mu}|{}^3\Psi_n\rangle\\&=\langle s_0(\mathbf{r}_1,\mathbf{r}_2)|\boldsymbol{\mu}|a_n(\mathbf{r}_1,\mathbf{r}_2)\rangle\langle {}^1\zeta(1,2)|{}^3\zeta(1,2)\rangle=0\end{aligned} \tag{10-108}$$

To a first approximation, neglecting spin interaction energies, the transition moment between a singlet and a triplet state is equal to zero.

If we calculate the same transition moment from the perturbed singlet function ${}^1\Phi_k$ of Eq. (10-106) and from the perturbed triplet function ${}^3\Phi_n$, the result is

$$\boldsymbol{\mu}_{0,n}=\langle {}^1\Phi_0|\boldsymbol{\mu}|{}^3\Phi_n\rangle \tag{10-109}$$

Now the transition moment is different from zero because the triplet functions that are mixed with the function ${}^1\Psi_0$, due to spin-orbit coupling, give rise to finite contributions to the transition moment, in the same way as the singlet terms that are mixed with ${}^3\Psi_n$ give finite contributions. It follows that in higher approximation the transition moment between a singlet and a triplet state becomes different from zero as a result of the spin-orbit interaction perturbations.

A spectroscopic transition between two states of different spin multiplicity is called spin-forbidden. The observed transition probabilities are due to spin-orbit coupling. In the case of the helium atom the spin-orbit coupling is very small, and initially no singlet-triplet transitions were observed. It was believed initially that there were two different species of helium, ortho and para helium, since no transitions between the singlet and the triplet levels could be observed experimentally.

Finally, we consider the effect of spin-orbit coupling on a triplet level. As an example we take the $(1s)(2p)$ triplet configuration of the helium atom. This configuration is denoted by the symbol $2\ {}^3P$.

The spin-orbit Hamiltonian is described by Eqs. (10-20), (10-21), and (10-22), and it is generally assumed that the major part of $H_{so}$ can be represented as

$$H'_{so} = f(r_1, r_2, r_{12})(\mathbf{L} \cdot \mathbf{S}) \tag{10-110}$$

where **L** is the orbital angular momentum operator and **S** is the spin angular momentum operator. The important feature of Eq. (10-110) is that the magnitude of the spin-orbit interaction depends on the relative orientation of the vector **L** with respect to the vector **S**.

Let us now interpret the fine structure of the $^3P$ configuration. The $2p$ state is threefold degenerate and the triplet spin state is also threefold degenerate, so that the $^3P$ configuration is ninefold degenerate. The orbital angular momentum is one unit of $\hbar$ and the spin angular momentum is also one unit of $\hbar$. We assume now that one of the two vectors, for instance, **L**, can be taken as the axis of quantization for the second vector **S**. We have sketched the situation in Fig. 10-1, and it can be seen that the vector **S** may be parallel, perpendicular, or antiparallel to **L**. The total angular momentum $J$ can therefore assume the values 2, 1, or 0 in units $\hbar$. In Fig. 10-2 we have sketched the experimental fine structure of the 2 $^3P$ state of the helium atom. In the absence of exterior fields the level is split into three levels, corresponding to $J=0$, $J=1$, and $J=2$. The separation between the levels $J=0$ and $J=2$ is almost 1 $\text{cm}^{-1}$. If we apply a homogeneous magnetic field, the $J=2$ level splits into five levels, the $J=1$ state splits into three components, and the $J=0$ level remains single. The magnitude of the splittings can be calculated from the operators $H_{so}$ and $H_{ss}$, and the theoretical predictions agree reasonably well with experiment.

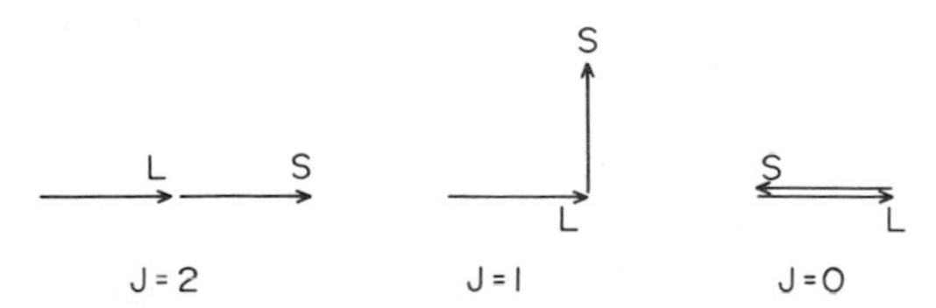


**Fig. 10-1** Coupling between an orbital angular momentum vector $L=1$ and a spin angular momentum vector $S=1$ due to spin-orbit coupling.

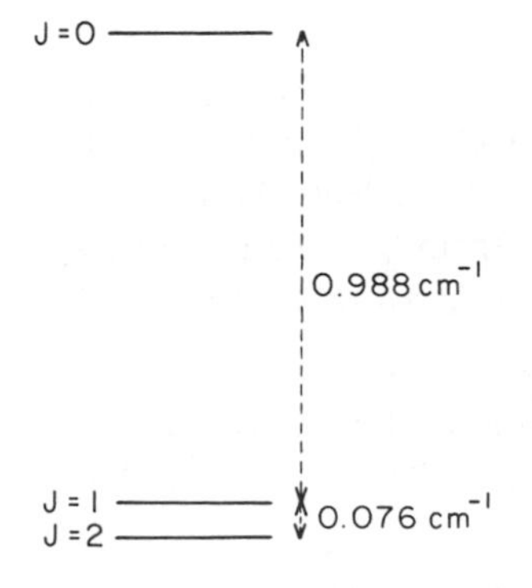


**Fig. 10-2** The experimental fine structure of the 2 $^3P$ state of the helium atom.

## Problems

**1** Determine the energy eigenvalues and eigenfunctions of a triplet state in a magnetic field **H** in an arbitrary direction. The Hamiltonian of this system is

$$H_s = \frac{e\hbar}{mc}(\mathbf{S}\cdot\mathbf{H})$$

where $\mathbf{S} = \mathbf{S}_1 + \mathbf{S}_2$.

**2** The zero-field splitting of the components of a triplet spin state is described by a Hamiltonian

$$H_{zf} = D\left(S_z^2 - \frac{S^2}{3}\right) - E\left(S_x^2 - S_y^2\right)$$

Derive the eigenvalues and eigenfunctions of the operator $H_{zf}$ with respect to the three triplet spin functions.

**3** Electron spin resonance is performed with radiation of 3.15 cm wavelength. Determine the magnitude of the magnetic field for which the separation of the two energy levels of a free electron corresponds to radiation of the above wavelength.

**4** The $(2s, 2p)$ state of the hydrogen atoms is eightfold degenerate if we neglect spin-orbit coupling. Calculate the matrix elements of the spin-orbit operator $H_{so}$ of the hydrogen atom with respect to the eight eigenfunctions of the $n=2$ state, and calculate the energy splittings of the $n=2$ state due to the spin-orbit perturbation.

**5** Evaluate the integral

$$I = \int\int r_{12}^{-1} \exp(-r_1 - r_2)\, d\mathbf{r}_1\, d\mathbf{r}_2$$

by means of conventional integration methods. The integral can be calculated by first integrating over the coordinates of electron 2 by introducing polar coordinates with the vector $\mathbf{r}_1$ as reference axis.

**6** The eigenfunction of the singlet $(1s)(2p_z)$ configuration can be constructed by assuming that the $1s$ electron experiences an effective nuclear charge $Z=2$ and the $2p$ electron experiences an effective nuclear charge $Z=1$. Calculate the transition moment for the transition from thc singlet $(1s)(2p_z)$ state to the $(1s)^2$ ground state and derive the lifetime of the singlet $(1s)(2p_z)$ state.

**7** Give a qualitative analysis of the fine structure of the triplet $(1s)(3d)$ configuration of the helium atom. Predict the number of levels and the degeneracy of each level.

## Recommended Reading

The properties of the electron spin and the exclusion principle are discussed in most quantum mechanics textbooks. An interesting approach is presented by Kramers, based on analogies between spinors and rotations in space. There are also many textbooks on the subject of statistical mechanics. We mention the book by Tolman because it gives an excellent description of the basic principles of the relation between quantum statistics and classical statistical mechanics. The book by Bethe and Salpeter presents a detailed quantitative discussion of the various spin interactions that we have outlined only qualitatively. Finally, we mention the book by Keesom because of its historical interest.

H. A. Kramers, *Quantum Mechanics*, North-Holland Publishing Co., Amsterdam, 1958.

R. C. Tolman, *The Principles of Statistical Mechanics*, Oxford University Press, London, 1938.

H. A. Bethe and E. E. Salpeter, *Quantum Mechanics of One- and Two-Electron Atoms*, Springer-Verlag, Berlin, 1957.

W. H. Keesom, *Helium*, Elsevier, Amsterdam, 1942.

CHAPTER ELEVEN

# Atomic Structure

## 11-1 Introduction

In our historical introduction to the previous chapter, Section 10-1, we described some of the early developments in the interpretation of atomic spectra. An important aspect of the interpretation is the identification of each electron by a set of quantum numbers. It can be seen that these quantum numbers bear some resemblance to the hydrogen atom quantum numbers that were discussed in Section 6-5 and that were used to identify the stationary states of the hydrogen atom. It may be recalled that the energy levels of a single electron, moving in a spherically symmetric Coulomb potential field, are identified by a quantum number $n$ which can assume the values $n=1,2,3,4,\ldots,$ and so on. The angular momentum of each stationary state is described by a quantum number $l$, which has the possible values $l=0,1,2,\ldots,n-1$. Finally, the direction of the angular momentum is denoted by a quantum number $m$, which has the possible values $m=0,\pm1,\pm2,\ldots,\pm l$.

If the electron moves in a potential field that is not purely Coulombic, its energy levels will depend on both quantum numbers $n$ and $l$. In the interpretation of atomic spectra, $n$ was called the principle quantum number and $l$ was called the azimuthal quantum number. Again, each energy level described by $(n,l)$ is $(2l+1)$-fold degenerate because the quantum number $l$ can assume $2l+1$ different values. It is customary to denote the energy levels by a number and a letter; the number describes the value of $n$ and the letter describes the value of $l$. The notation for $l$ is as follows: $l=0$ is denoted by $s$, $l=1$ by $p$, $l=2$ by $d$, $l=3$ by $f$, $l=4$ by $g$, and so on. Some energy levels are described in Table 6-1.

A stationary state of an atom can be described by giving the distribution of the individual electrons over the available one-electron states. This is called thc configuration of the stationary state. In the ground-state configuration all electrons are in the lowest possible one-electron states. The ground-state configuration is also known as the normal state of the atom. In the previous chapter we saw that the ground-state configuration of the helium atom is $(1s)^2$. We also encountered various excited-state configurations which were denoted by $(1s)(2s),(1s)(2p),(1s)(3d),\ldots,$ etc.

**Table 11-1** Ground-State Configurations of Selected Elements

| | | | |
|---|---|---|---|
| 1 | H | $^2S_{1/2}$ | $(1s)$ |
| 2 | He | $^1S_0$ | $(1s)^2$ |
| 3 | Li | $^2S_{1/2}$ | $(1s)^2(2s)$ |
| 4 | Be | $^1S_0$ | $(1s)^2(2s)^2$ |
| 5 | B | $^2P_{1/2}$ | $(1s)^2(2s)^2(2p)$ |
| 6 | C | $^3P_0$ | $(1s)^2(2s)^2(2p)^2$ |
| 7 | N | $^4S_{3/2}$ | $(1s)^2(2s)^2(2p)^3$ |
| 8 | O | $^3P_2$ | $(1s)^2(2s)^2(2p)^4$ |
| 9 | F | $^2P_{3/2}$ | $(1s)^2(2s)^2(2p)^5$ |
| 10 | Ne | $^1S_0$ | $(1s)^2(2s)^2(2p)^6$ |
| 11 | Na | $^2S_{1/2}$ | $\cdots(2s)^2(2p)^6(3s)$ |
| 12 | Mg | $^1S_0$ | $\cdots(2s)^2(2p)^6(3s)^2$ |
| 13 | Al | $^2P_{1/2}$ | $\cdots(2s)^2(2p)^6(3s)^2(3p)$ |
| 14 | Si | $^3P_0$ | $\cdots(2s)^2(2p)^6(3s)^2(3p)^2$ |
| 15 | P | $^4S_{3/2}$ | $\cdots(2s)^2(2p)^6(3s)^2(3p)^3$ |
| 16 | S | $^3P_2$ | $\cdots(2s)^2(2p)^6(3s)^2(3p)^4$ |
| 17 | Cl | $^2P_{3/2}$ | $\cdots(2s)^2(2p)^6(3s)^2(3p)^5$ |
| 18 | A | $^1S_0$ | $\cdots(2s)^2(2p)^6(3s)^2(3p)^6$ |
| 19 | K | $^2S_{1/2}$ | $\cdots(3s)^2(3p)^6(4s)$ |
| 20 | Ca | $^1S_0$ | $\cdots(3s)^2(3p)^6(4s)^2$ |
| 21 | Sc | $^2D_{3/2}$ | $\cdots(3s)^2(3p)^6(3d)(4s)^2$ |
| 22 | Ti | $^3F_2$ | $\cdots(3s)^2(3p)^6(3d)^2(4s)^2$ |
| 23 | V | $^4F_{3/2}$ | $\cdots(3s)^2(3p)^6(3d)^3(4s)^2$ |
| 24 | Cr | $^7S_3$ | $\cdots(3s)^2(3p)^6(3d)^5(4s)$ |
| 25 | Mn | $^6S_{5/2}$ | $\cdots(3s)^2(3p)^6(3d)^5(4s)^2$ |
| 26 | Fe | $^5D_4$ | $\cdots(3s)^2(3p)^6(3d)^6(4s)^2$ |
| 27 | Co | $^4F_{9/2}$ | $\cdots(3s)^2(3p)^6(3d)^7(4s)^2$ |
| 28 | Ni | $^3F_4$ | $\cdots(3s)^2(3p)^6(3d)^8(4s)^2$ |
| 29 | Cu | $^2S_{1/2}$ | $\cdots(3s)^2(3p)^6(3d)^{10}(4s)$ |
| 30 | Zn | $^1S_0$ | $\cdots(3s)^2(3p)^6(3d)^{10}(4s)^2$ |
| 31 | Ga | $^2P_{1/2}$ | $\cdots(3s)^2(3p)^6(3d)^{10}(4s)^2(4p)$ |
| 32 | Ge | $^3P_0$ | $\cdots(3s)^2(3p)^6(3d)^{10}(4s)^2(4p)^2$ |
| 33 | As | $^4S_{3/2}$ | $\cdots(3s)^2(3p)^6(3d)^{10}(4s)^2(4p)^3$ |
| 34 | Se | $^3P_2$ | $\cdots(3s)^2(3p)^6(3d)^{10}(4s)^2(4s)^4$ |
| 35 | Br | $^2P_{3/2}$ | $\cdots(3s)^2(3p)^6(3d)^{10}(4s)^2(4p)^5$ |
| 36 | Kr | $^1S_0$ | $\cdots(3s)^2(3p)^6(3d)^{10}(4s)^2(4p)^6$ |
| 37 | Rb | $^2S_{1/2}$ | $\cdots(4s)^2(4p)^6(5s)$ |
| 38 | Sr | $^1S_0$ | $\cdots(4s)^2(4p)^6(5s)^2$ |
| 39 | Y | $^2D_{3/2}$ | $\cdots(4s)^2(4p)^6(4d)(5s)^2$ |
| 40 | Zr | $^3F_2$ | $\cdots(4s)^2(4p)^6(4d)^2(5s)^2$ |
| 41 | Cb | $^6D_{1/2}$ | $\cdots(4s)^2(4p)^6(4d)^4(5s)$ |
| 42 | Mo | $^7S_3$ | $\cdots(4s)^2(4p)^6(4d)^5(5s)$ |
| 43 | Tc | $^6S_{7/2}$ | $\cdots(4s)^2(4p)^6(4d)^5(5s)^2$ |
| 44 | Ru | $^5F_5$ | $\cdots(4s)^2(4p)^6(4d)^7(5s)$ |

**Table 11-1** (*Continued*)

| | | | |
|---|---|---|---|
| 45 | Rh | $^4F_{9/2}$ | $\cdots(4s)^2(4p)^6(4d)^8(5s)$ |
| 46 | Pd | $^1S_0$ | $\cdots(4s)^2(4p)^6(4d)^{10}$ |
| 47 | Ag | $^2S_{1/2}$ | $\cdots(4s)^2(4p)^6(4d)^{10}(5s)$ |
| 48 | Cd | $^1S_0$ | $\cdots(4s)^2(4p)^6(4d)^{10}(5s)^2$ |
| 49 | In | $^2P_{1/2}$ | $\cdots(4s)^2(4p)^6(4d)^{10}(5s)^2(5p)$ |
| 50 | Sn | $^3P_0$ | $\cdots(4s)^2(4p)^6(4d)^{10}(5s)^2(5p)^2$ |
| 51 | Sb | $^4S_{3/2}$ | $\cdots(4s)^2(4p)^6(4d)^{10}(5s)^2(5p)^3$ |
| 52 | Te | $^3P_2$ | $\cdots(4s)^2(4p)^6(4d)^{10}(5s)^2(5p)^4$ |
| 53 | I | $^2P_{3/2}$ | $\cdots(4s)^2(4p)^6(4d)^{10}(5s)^2(5p)^5$ |
| 54 | Xe | $^1S_0$ | $\cdots(4s)^2(4p)^6(4d)^{10}(5s)^2(5p)^6$ |

In deriving the ground-state configurations of the elements we must consider the Pauli exclusion principle which allows us to place two electrons only in each orbital. In Table 11-1 we list the normal states of about half the elements, up to xenon. We can order the one-electron states according to their energies, and it follows from Table 11-1 that this gives the following scheme:

$$\varepsilon(1s)<\varepsilon(2s)<\varepsilon(2p)<\varepsilon(3s)<\varepsilon(3p)<\varepsilon(4s)<\varepsilon(3d)<\varepsilon(4p)$$
$$<\varepsilon(5s)<\varepsilon(4d)<\varepsilon(5p)<\cdots \text{ etc.} \tag{11-1}$$

We shall see that the total atomic energy is not exactly equal to the sum of the individual one-electron energies. Nevertheless, Eq. (11-1) offers us a guideline for the assignment of the electrons in an atom or for the prediction of the ground-state configuration. Starting with hydrogen, the ground-state configuration of each successive atom can be derived by adding one electron to the configuration of the preceding atom in an orbit defined by the quantum numbers having the lowest energy that is compatible with the exclusion principle. This procedure is known as the building-up principle (Aufbauprinzip).

In Table 11-1 we have also listed the symbols that describe the values of the angular momentum vectors in the ground states of the various atoms. The capital letters $S$, $P$, $D$, $F$, and so on describe the value of the orbital angular momentum $\mathbf{L}$, with $S$ denoting $L=0$, $P$ denoting $L=1$, $D$ denoting $L=2$, $F$ denoting $L=3$, and so on. The superscript on the left refers to the spin angular momentum; the superscript is equal to the spin multiplicity $2S+1$, where $S$ is the magnitude of the spin angular momentum. The value of the total angular momentum is described by the subscript on the right; its value depends on the relative orientation of $\mathbf{L}$ and $\mathbf{S}$.

The wave function of an atom can in first approximation be represented as an antisymmetrized product of one-electron functions. By definition, a one-electron function is called an orbital. If the one-electron function is part of an atomic wave function, it is called an atomic orbital.

In the previous discussion we identified each electron by a set of quantum numbers $(n, l, m)$. From now on we identify each electron by an atomic orbital. Naturally, the atomic orbital should be compatible with the quantum numbers. The quantum numbers $l$ and $m$ describe the angular momentum of the electron and they determine the angular dependence of the orbital. Consequently, for any given atomic configuration we should be able to predict the angular dependence of the individual atomic orbitals. This leaves only the radial dependence of each orbital to be determined.

Let us now describe how the atomic wave functions are constructed from the individual atomic orbitals. We denote the individual atomic orbitals by $\chi_1, \chi_2, \chi_3, \ldots,$ and so on and the corresponding energy parameters by $\varepsilon_1, \varepsilon_2, \varepsilon_3, \ldots,$ and so on. We assume that

$$\varepsilon_1 \leq \varepsilon_2 \leq \varepsilon_3 \leq \cdots \text{ and so on} \tag{11-2}$$

The simplest situation is a closed-shell ground state, where we have an even number $2N$ of electrons and where we place a pair of electrons in each of the orbitals $\chi_1, \chi_2, \chi_3, \ldots, \chi_N$. The properly antisymmetrized atomic wave function, including spin, is then given by

$$\Psi_o = \left[\frac{1}{(2N)!}\right]^{1/2} \sum_P P\delta_P[\chi_1(\mathbf{r}_1)\alpha(1)\chi_1(\mathbf{r}_2)\beta(2)\chi_2(\mathbf{r}_3)\alpha(3)$$
$$\times \chi_2(\mathbf{r}_4)\beta(4)\cdots\chi_N(\mathbf{r}_{2N-1})\alpha(2N-1)\chi_N(\mathbf{r}_{2N})\beta(2N)] \tag{11-3}$$

or, in abbreviated form, by

$$\Psi_o = [(2N)!]^{-1/2} \sum_P P\delta_P\left[\prod_{i=1}^{N} \chi_i(\mathbf{r}_{2i-1})\chi_i(\mathbf{r}_{2i})\alpha(2i-1)\beta(2i)\right] \tag{11-4}$$

It was first observed by Slater that this expression can also be written as a determinant, namely as

$$\begin{vmatrix} \chi_1(1)\alpha(1) & \chi_1(1)\beta(1) & \cdots & \cdots & \cdots & \cdots & \chi_N(1)\alpha(1) & \chi_N(1)\beta(1) \\ \chi_1(2)\alpha(2) & \chi_1(2)\beta(2) & \cdots & \cdots & \cdots & \cdots & \chi_N(2)\alpha(2) & \chi_N(2)\beta(2) \\ \chi_1(3)\alpha(3) & \chi_1(3)\beta(3) & \cdots & \cdots & \cdots & \cdots & \chi_N(3)\alpha(3) & \chi_N(3)\beta(3) \\ \cdots & \cdots & \cdots & \cdots & \cdots & \cdots & \cdots & \cdots \\ \cdots & \cdots & \cdots & \cdots & \cdots & \cdots & \cdots & \cdots \\ \cdots & \cdots & \cdots & \cdots & \cdots & \cdots & \cdots & \cdots \\ \chi_1(2N)\alpha(2N) & \chi_1(2N)\beta(2N) & \cdots & \cdots & \cdots & \cdots & \chi_N(2N)\alpha(2N) & \chi_N(2N)\beta(2N) \end{vmatrix} \tag{11-5}$$

Determinants of this type are known as Slater determinants.

If the configuration of the electrons in the atom is different from a closed-shell ground-state configuration, the atomic wave function takes a more complex form. For instance, if we excite one of the electrons from a lower orbital $\chi_j$ to a higher orbital $\chi_n$, the configuration is $(\chi_1)^2(\chi_2)^2\cdots(\chi_{j-1})^2(\chi_{j+1})^2\cdots(\chi_N)^2(\chi_j)(\chi_n)$. This configuration is either a triplet or a singlet spin state. The singlet function ${}^1\Psi(j\rightarrow n)$ is given by

$$
{}^1\Psi(j\rightarrow n)=[2\cdot(2N)!]^{-1/2}\sum_P P\delta_P\left\{\left[\prod_{i\neq j}^{N-1}\chi_i(2i-1)\chi_i(2i)\alpha(2i-1)\beta(2i)\right]\right.
$$

$$
\left.\times[\chi_j(2N-1)\chi_n(2N)+\chi_n(2N-1)\chi_j(2N)]\alpha(2N-1)\beta(2N)\right\} \tag{11-6}
$$

and the triplet function ${}^3\Psi(j\rightarrow n)$ is given by

$$
{}^3\Psi(j\rightarrow n)=[(2N)!]^{-1/2}\sum_P P\delta_P\left\{\left[\prod_{i\neq j}^{N-1}\chi_i(2i-1)\chi_i(2i)\alpha(2i-1)\beta(2i)\right]\right.
$$

$$
\left.\times\chi_j(2N-1)\chi_n(2N)\,{}^3\zeta(2N-1,2N)\right\} \tag{11-7}
$$

Here it is assumed that the triplet spin function is normalized to unity.

If the atomic configuration is degenerate in first approximation, the atomic wave function can no longer be represented by a single Slater determinant. For instance, the ground-state configuration of the carbon atom is $(1s)^2(2s)^2(2p)^2$, and this configuration corresponds to a number of different atomic eigenstates, determined by the values of the orbital and spin angular momentum values $L$ and $S$ and of the value of the total angular momentum $J$. We discuss the problems relating to these degenerate atomic configurations in Section 11-4. In Sections 11-2 and 11-3 we consider the problem of deriving the detailed form of the atomic orbitals and we confine ourselves to closed-shell atomic configurations only.

Once we have introduced the approximation of representing the atomic wave function as an antisymmetrized product of one-electron functions, the atomic orbitals, we can determine the exact form of the atomic orbitals from the variational principle without making use of additional approximations. We discuss this procedure in Section 11-3. It is known as the Hartree-Fock method, the Self-Consistent Field method, or the SCF method. The atomic orbitals that are derived by means of this procedure are known as SCF orbitals; they are the most accurate atomic orbitals within the approximation of Eq. (11-4). More approximate atomic orbitals can be obtained by means of arguments similar to those given in Section 10-7, where the potential fields acting upon each electron are approximated by Coulomb-type potentials. We discuss this in Section 11-2.

Finally, we wish to make some observations on the orthogonality properties of the atomic orbitals. For this purpose it is helpful if we consider the atomic wave function written in the form of a Slater determinant, as in Eq. (11-5). We discussed the properties of determinants in Section 2-4, and we showed in Eq. (2-63) that the value of a determinant remains unchanged if we add an arbitrary linear combination of a number of columns to another column. By making use of this property we can transform to a different set of atomic orbitals $\chi_i$ without changing the atomic eigenfunction $\Psi_o$. It follows that the choice of atomic orbitals is not unique. If there are no degeneracies, we can select a unique set of atomic orbitals by imposing the condition that they form an orthonormal set:

$$\langle \chi_i | \chi_j \rangle = \delta_{i,j} \tag{11-8}$$

As we shall see in Section 11-3, this choice simplifies the formulation of the SCF method, and we will assume from now on that the atomic orbitals we consider always form orthonormal sets.

## 11-2 Slater Orbitals

In Section 11-3 we derive the "exact" form of the SCF orbitals, but it is often possible to predict the approximate form of the atomic orbitals by means of arguments that are similar to our discussion of the helium atom in Section 10-7. Here we considered the $(1s)^2$ ground-state configuration and we argued that the $1s$ atomic orbital could be approximated by the function

$$\chi(1s) = \left( \frac{Z^3}{\pi} \right)^{1/2} \exp(-Zr) \tag{11-9}$$

By making use of the variation theorem, we found that the lowest energy is obtained for $Z = 1.6875$. From similar considerations we can also derive approximate one-electron functions for more complex atoms.

We may recall that the physical reasoning that we used in constructing the function (11-9) was that each electron moves in the potential field of the nucleus and of the second electron. We argued that the effect of the repulsion by the second electron can be roughly represented as a shielding of the nuclear charge 2 by an amount $\sigma$ due to the fraction of the charge cloud of the second electron that is situated between the nucleus and the first electron. In the case of the helium atom it was found that $\sigma = 2 - Z = 0.3125$.

We can use the same arguments for the construction of approximate atomic orbitals for more complex atoms. For example, in the case of neon, which has the configuration $(1s)^2(2s)^2(2p)^6$, we can use the approximation

$$\chi(1s)=\left(\frac{Z_1^3}{\pi}\right)^{1/2}\exp(-Z_1 r)$$

$$\chi(2s)=\left(\frac{Z_2^3}{32\pi}\right)^{1/2}(Z_2 r-2)\exp\left(\frac{-Z_2 r}{2}\right)$$

$$\chi(2p_\alpha)=\left(\frac{Z_3^5}{32\pi}\right)^{1/2} r_\alpha \exp\left(\frac{-Z_3 r}{2}\right) \qquad \alpha=x,y,z \tag{11-10}$$

The values of the effective nuclear charges $Z_1$, $Z_2$, and $Z_3$ can then be determined by means of the variation theorem. A calculation of this type was performed for the carbon atom. In this case $Z_1$ is very close to the nuclear charge 6, whereas $Z_2$ and $Z_3$ are slightly larger than 3. Shortly thereafter, during 1930, Slater proposed some simple algebraic rules for estimating the values of the effective nuclear charges that are obtained in this way. First, however, it was pointed out by Slater that instead of the exact hydrogenlike wave functions, we can just as well take the simplified functions

$$\chi_{n,l,m}(r,\theta,\phi)=r^{n-1}\exp\left(\frac{-Z'r}{n}\right)\psi_{l,m}(\theta,\phi) \tag{11-11}$$

as the basis of our approximate description. In the case of the hydrogen atom the exact eigenfunctions $\psi_{n,l,m}(r,\theta,\phi)$ have the advantage of being orthogonal to one another, but this advantage disappears if we substitute different nuclear charges in the different states. The functions $\chi_{n,l,m}(r,\theta,\phi)$, which are now generally known as Slater orbitals, form just as satisfactory a basis set for describing atomic states as the exact hydrogen eigenfunctions $\psi_{n,l,m}(r,\theta,\phi)$, and since the Slater orbitals are easier to deal with than the $\psi_{n,l,m}$, we might as well use them

In order to estimate the effective nuclear charges $Z'$ we write them as

$$Z'=Z-\sigma \tag{11-12}$$

where $Z$ is the exact nuclear charge and $\sigma$ represents the shielding of the nucleus by the other electrons. The value of $\sigma$ depends on the state of the electron that we are concerned with and on the states of the other electrons that are present in the atom. According to Slater's rules, $\sigma$ is obtained as a sum of the shielding contributions of these other electrons. These contributions are

1. Nothing from any electron that has a principal quantum number $n$ that is higher than the one that we consider.
2. An amount 0.35 from each electron that has the same principal quantum number as the electron that we consider, except that when we consider a $(1s)$ electron, the contribution from the other $(1s)$ electron is 0.30.

3 An amount 0.85 from each electron that has a principal quantum number $n$ that is 1 less than the quantum number of the electron that we consider if the latter is an $s$ or a $p$ electron, and an amount 1.00 from each electron whose principal quantum number is less by 1 than the electron that we consider if the latter is in a $d, f$, or $g$ state.

4 An amount 1.00 from each electron with a principal quantum number that is less by 2 or more than the quantum number of the electron considered.

We illustrate these rules for the sulfur atom. According to Table 11-1 the atomic configuration here is $(1s)^2(2s)^2(2p)^6(3s)^2(3p)^4$, and the nuclear charge is 16. The effective nuclear charges $Z'$ for the various electrons are

$$\begin{aligned} Z'(1s) &= 16-0.30=15.70 \\ Z'(2s) &= 16-7\times0.35-2\times0.85=11.85 \\ Z'(2p) &= Z'(2s)=11.85 \\ Z'(3s) &= 16-2\times1-8\times0.85-5\times0.35=5.45 \\ Z'(3p) &= Z'(3s)=5.45 \end{aligned} \tag{11-13}$$

For the helium atom Slater's rules predict an effective nuclear charge $Z'=1.70$, which is reasonably close to the value $Z'=1.6875$ that we obtained in Chapter 10 from the variation theorem.

These Slater orbitals have found wide applicability in atomic and molecular problems. They are obviously very convenient for performing crude order-of-magnitude calculations of atomic properties. They are also useful in more precise theoretical work. We will see that in some methods for deriving more accurate atomic wave functions we make use of iterative procedures in which we start with approximate orbitals that are transformed progressively into more and more accurate orbitals. Here the Slater orbitals supply us with a convenient starting point for the iterative process.

## 11-3 The Hartree-Fock Method

In Section 11-1 we wrote the atomic wave function as an antisymmetrized product of one-electron orbitals. The specific form of the atomic wave functions for the case of a closed-shell configuration was presented in Eqs. (11-3) and (11-4). The goal of the Hartree-Fock method, or the Self-Consistent Field Method, is the derivation of the analytical form of the atomic orbitals from the variational principle without introducing additional approximations. This goal is achieved by varying one atomic orbital at a time until the expectation value of the total Hamiltonian with respect to all atomic orbitals is minimized.

In the case of atomic structure the computational effort required in the SCF procedure is of manageable proportion because the angular dependence of the atomic orbitals is known. For instance, the ground-state configuration of the argon atom is $(1s)^2(2s)^2(2p)^6(3s)^2(3p)^6$. This atom has 18 electrons, but in the SCF method we have to determine only five unknown functions of one variable each, namely the $(1s)$, $(2s)$, and $(3s)$ atomic orbitals and the radial parts of the $(2p)$ and $(3p)$ orbitals. Naturally, we should have some idea of the analytical form of these orbitals, but we might approximate them initially by making use of the Slater orbitals which we introduced in the previous section. From these approximate orbitals we construct an effective potential field acting on the electrons, and by making use of the variational principle we can then derive a set of differential equations for the atomic orbitals. By substituting the solutions of the Hartree-Fock equations back into the expressions for the effective potential we obtain a set of more precise differential equations which must again be solved. The final result is obtained when the solutions of the equations become identical with the orbitals that were used in determining the effective potential. At that point the solutions are consistent with themselves, hence the name "Self-Consistent Field method."

The mathematics in deriving the Hartree-Fock equations is fairly complicated. First we must derive the expectation value of the Hamiltonian, expressed in terms of atomic orbitals, and then we must transform these various expressions in order to obtain convenient mathematical equations. We limit ourselves to a discussion of a closed-shell ground state; other configurations lead to similar sets of equations, but they are somewhat more complicated.

We have shown in Eq. (11-4) that a closed-shell atomic eigenfunction can be written in the form

$$\Psi_o = [(2N)!]^{-1/2} \sum_P P\delta_P \left[ \prod_{k=1}^{N} \phi_k(2k-1)\alpha(2k-1)\phi_k(2k)\beta(2k) \right] \quad (11\text{-}14)$$

if we approximate the atomic eigenfunction as an antisymmetrized product of one-electron orbitals. According to the considerations of Section 11-1 we assume that the atomic orbitals form an orthonormal set

$$\langle \phi_j | \phi_k \rangle = \delta_{j,k} \quad (11\text{-}15)$$

This assumption leads to unique solutions and it also makes the following mathematical derivations much more tractable.

The first step in our derivation of the Hartree-Fock equations is to express the expectation value of the energy:

$$E = \langle \Psi_0 | H_o | \Psi_0 \rangle \quad (11\text{-}16)$$

and the overlap integral:

$$S = \langle \Psi_0 | \Psi_0 \rangle \quad (11\text{-}17)$$

in terms of the one-electron orbitals $\phi_k$. Here $H_o$ is the atomic Hamiltonian that we presented in Eq. (10-50). We derive the Hartree-Fock equations for a more general situation where the Hamiltonian is given by

$$H_o = -\frac{\hbar^2}{2m}\sum_j \Delta_j + \sum_j V(\mathbf{r}_j) + \sum_{i>j} \frac{e^2}{r_{ij}} \tag{11-18}$$

Here the first term represents the kinetic energy of the electrons, the second term is the sum of the potential energies of the electrons, and the last term is the sum of the electronic repulsion terms. It is convenient to write $H_o$ as a sum of one-electron terms and as a sum of two-electron terms:

$$H_o = \sum_j G(j) + \sum_{i>j} \Omega(i, j) \tag{11-19}$$

The specific form of the one-electron operators $G(j)$ and of the two-electron operators $\Omega(i, j)$ follows from a comparison of Eqs. (11-18) and (11-19).

First we consider the overlap integral $S$ as defined by Eq. (11-17). By substituting Eq. (11-14) we obtain

$$S = \{(2N)!\}^{-1} \left\langle \sum_P P\delta_P \left[ \prod_{k=1}^{N} \phi_k(2k-1)\alpha(2k-1)\phi_k(2k)\beta(2k) \right] \right| \left. \sum_P P\delta_P \left[ \prod_{k=1}^{N} \phi_k(2k-1)\alpha(2k-1)\phi_k(2k)\beta(2k) \right] \right\rangle \tag{11-20}$$

which we can also write as

$$S = \left\langle \sum_P P\delta_P \left[ \prod_{k=1}^{N} \phi_k(2k-1)\alpha(2k-1)\phi_k(2k)\beta(2k) \right] \right| \left. \prod_{k=1}^{N} \phi_k(2k-1)\alpha(2k-1)\phi_k(2k)\beta(2k) \right\rangle \tag{11-21}$$

We consider first the contribution from the identity permutation, which is

$$\left\langle \prod_{k=1}^{N} \phi_k(2k-1)\alpha(2k-1)\phi_k(2k)\beta(2k) \right| \left. \prod_{k=1}^{N} \phi_k(2k-1)\alpha(2k-1)\phi_k(2k)\beta(2k) \right\rangle = 1 \tag{11-22}$$

Here each set of electron coordinates occurs in the same functions on the left and on the right, and the result is unity because of Eq. (11-15). For any other permutation, different from the identity permutation, there are at least two "misfits," where for one set of electron coordinates the functions on the left and on the right differ either in the orbital part or in the spin part. Because of the orthogonality of the orbitals $\phi_k$ and of the spin functions $\alpha$ and $\beta$, the contribution from any one of these permutations is zero. Hence $S$ is equal to unity.

We use the same technique to evaluate $E$. First we write $E$ as a sum of two terms:

$$E = E_1 + E_2 \tag{11-23}$$

which are defined as

$$E_1 = \left\langle \Psi_o \middle| \sum_j G(j) \middle| \Psi_o \right\rangle \tag{11-24}$$

and

$$E_2 = \left\langle \Psi_o \middle| \sum_{i>j} \Omega(i, j) \middle| \Psi_o \right\rangle \tag{11-25}$$

By drawing an analogy with Eq. (11-21) we can write the term $E_1$ as

$$E_1 = \left\langle \sum_P P\delta_P \left[ \prod_{k=1}^{N} \phi_k(2k-1)\alpha(2k-1)\phi_k(2k)\beta(2k) \right] \middle| \sum_j G(j) \middle| \prod_{k=1}^{N} \phi_k(2k-1)\alpha(2k-1)\phi_k(2k)\beta(2k) \right\rangle \tag{11-26}$$

Again we consider first the diagonal term of the Slater determinant, which gives

$$\begin{aligned} E_1' &= \left\langle \prod_{k=1}^{N} \phi_k(2k-1)\alpha(2k-1)\phi_k(2k)\beta(2k) \middle| \sum_j G(j) \middle| \prod_{k=1}^{N} \phi_k(2k-1)\alpha(2k-1)\phi_k(2k)\beta(2k) \right\rangle \\ &= \sum_{k=1}^{N} \{ \langle \phi_k\alpha | G | \phi_k\alpha \rangle + \langle \phi_k\beta | G | \phi_k\beta \rangle \} \\ &= 2\sum_{k=1}^{N} \langle \phi_k | G | \phi_k \rangle = 2\sum_{k=1}^{N} G_k \end{aligned}$$

$$G_k = \langle \phi_k(i) | G(i) | \phi_k(i) \rangle \tag{11-27}$$

Each of the other permutations creates at least two "misfits"; that is, there are at least two sets of electronic coordinates that occur in different functions on the left and right sides of the operators in Eq. (11-27). Since each term of the operator depends only on one set of electronic coordinates, the contributions of all these other permutations are zero. Our final result is therefore

$$E_1 = 2\sum_{k=1}^{N} G_k \tag{11-28}$$

The other energy term $E_2$ can be written as

$$E_2 = \left\langle \sum_P P\delta_P \left[ \prod_{k=1}^{N} \phi_k(2k-1)\alpha(2k-1)\phi_k(2k)\beta(2k) \right] \middle| \sum_{i>j} \Omega(i,j) \middle| \prod_{k=1}^{N} \phi_k(2k-1)\alpha(2k-1)\phi_k(2k)\beta(2k) \right\rangle \tag{11-29}$$

Here the contribution from the diagonal term is

$$E_2' = \left\langle \prod_{k=1}^{N} \phi_k(2k-1)\alpha(2k-1)\phi_k(2k)\beta(2k) \middle| \sum_{i>j} \Omega(i,j) \middle| \prod_{k=1}^{N} \phi_k(2k-1)\alpha(2k-1)\phi_k(2k)\phi(2k) \right\rangle \tag{11-30}$$

If we define the integrals $J_{k,l}$ as

$$J_{k,l} = \langle \phi_k(i)\phi_l(j) | \Omega(i,j) | \phi_k(i)\phi_l(j) \rangle \tag{11-31}$$

it is easily verified that $E_2'$ can be expressed as

$$E_2' = \sum_{k=1}^{N} \left( J_{k,k} + 2\sum_{l\neq k} J_{k,l} \right) \tag{11-32}$$

Let us now determine the contributions to $E_2$ from the other permutations. First we consider the permutations that correspond to a pairwise interchange of two sets of electron coordinates. It can be seen that these permutations give rise to two "misfits" in the expression for $E_2$. If the permutation involves a pair of electrons in which one electron has an even subscript and the other an odd subscript, the discrepancy is located in the spin functions, and the result of the integration is zero. On the other hand, if the permutation involves a pair of electrons whose subscripts are both even or both odd, the discrepancy is located in the orbital functions. In the present case the operator is a sum of two-electron terms, so that for each pair of "misfits" that is restricted to the orbital functions only, there exists one term in the Hamiltonian that gives a nonzero contribution. For example, if the permutation consists of an interchange of electrons $2k$ and $2l$, its contribution to $E_2$ is

$$E_2(k,l) = -\langle \phi_l(2k)\phi_k(2l)|\Omega(2k,2l)|\phi_k(2k)\phi_l(2l)\rangle \qquad (11\text{-}33)$$

It can be deduced that the sum of all these contributions to $E_2$ can be written in the form

$$E_2'' = -\sum_{k=1}^{N} \sum_{l \neq k} K_{k,l} \qquad (11\text{-}34)$$

if we define

$$K_{k,l} = \langle \phi_l(i)\phi_k(j)|\Omega(i,j)|\phi_k(i)\phi_l(j)\rangle \qquad (11\text{-}35)$$

It is easily verified that

$$K_{k,l} = K_{l,k} \qquad (11\text{-}36)$$

since the operator is symmetric in $i$ and $j$.

The contributions to $E_2$ of all other permutations, which go beyond a simple pairwise interchange of two electrons, are all zero. These permutations give rise to at least three "misfits," and since each term in the Hamiltonian depends only on two sets of electron coordinates, we find that the integrations all give zero. The total result for $E_2$ is obtained by adding Eqs. (11-32) and (11-34):

$$E_2 = \sum_{k=1}^{N} J_{k,k} + \sum_{l \neq k} (2J_{k,l} - K_{k,l}) \qquad (11\text{-}37)$$

We can write this in a slightly different form. Since

$$J_{k,k} = K_{k,k} \qquad (11\text{-}38)$$

We can add a term $J_{k,k}$ to Eq. (11-37) if at the same time we also subtract a term $K_{k,k}$. The results then becomes

$$E_2 = \sum_{k=1}^{N} \sum_{l=1}^{N} (2J_{k,l} - K_{k,l}) \qquad (11\text{-}39)$$

The total expectation value $E$ is now obtained as the sum of $E_1$ and $E_2$:

$$E = 2\sum_{k=1}^{N} \left[ G_k + \sum_{l=1}^{N} (J_{k,l} - \tfrac{1}{2}K_{k,l}) \right] \qquad (11\text{-}40)$$

We now impose the condition that $E$ remains invariant under arbitrary small variations $\delta\phi_k$ in any of the orbitals $\phi_k$, with the auxiliary condition that the change in $\phi_k$ does not affect the orthonormality of the one-electron orbitals. Let us first determine the change $\delta E_k$ in $E$ due to the variation $\delta\phi_k^*$. According to Eq. (11-40) we have

$$\delta E_k = 2\left(\delta G_k + \delta J_{k,k} + 2\sum_{l\neq k}\delta J_{k,l} - \tfrac{1}{2}\delta K_{k,k} - \sum_{l\neq k}\delta K_{k,l}\right) \tag{11-41}$$

Here, the variations in the Coulomb integrals are

$$\begin{aligned}\delta J_{k,k} &= 2\langle\delta\phi_k(i)\phi_k(j)|\Omega(i,j)|\phi_k(i)\phi_k(j)\rangle\\ \delta J_{k,l} &= \langle\delta\phi_k(i)\phi_l(j)|\Omega(i,j)|\phi_k(i)\phi_l(j)\rangle\end{aligned} \tag{11-42}$$

the variations in the exchange integrals are

$$\begin{aligned}\delta K_{k,k} &= 2\langle\delta\phi_k(i)\phi_k(j)|\Omega(i,j)|\phi_k(i)\phi_k(j)\rangle\\ \delta K_{k,l} &= \langle\delta\phi_k(i)\phi_l(j)|\Omega(i,k)|\phi_l(i)\phi_k(j)\rangle\end{aligned} \tag{11-43}$$

and $\delta G_k$ is

$$\delta G_k = \langle\delta\phi_k(i)|G(i)|\phi_k(i)\rangle \tag{11-44}$$

We now introduce the Coulomb operator $J_n(i)$ and the exchange operator $K_n(i)$. The Coulomb operator is defined as

$$J_n(i) = \int \phi_n^*(j)\Omega(i,j)\phi_n(j)\,d\mathbf{r}_j \tag{11-45}$$

and the exchange operator as

$$K_n(i)\psi(i) = \left[\int \phi_n^*(j)\Omega(i,j)\psi(j)\,d\mathbf{r}_j\right]\phi_n(i) \tag{11-46}$$

Substitution of Eq. (11-45) into Eq. (11-42) gives

$$\begin{aligned}\delta J_{k,k} &= 2\langle\delta\phi_k(i)|J_k(i)|\phi_k(i)\rangle\\ \delta J_{k,l} &= \langle\delta\phi_k(i)|J_l(i)|\phi_k(i)\rangle\end{aligned} \tag{11-47}$$

and substitution of Eqs. (11-46) into Eq. (11-43) gives

$$\begin{aligned}\delta K_{k,k} &= 2\langle\delta\phi_k(i)|K_k(i)|\phi_k(i)\rangle\\ \delta K_{k,l} &= \langle\delta\phi_k(i)|K_l(i)|\phi_k(i)\rangle\end{aligned} \tag{11-48}$$

If follows from Eqs. (11-44), (11-47), and (11-48) that Eq. (11-41) can be transformed to

$$\delta E_k = 2\left\langle \delta\phi_k(i)|G(i) + \sum_l [2J_l(i) - K_l(i)]|\phi_k(i)\right\rangle \tag{11-49}$$

It is customary to introduce the Hartree-Fock Hamiltonian operator $F$, which is defined as

$$F = G + \sum_l (2J_l - K_l) \tag{11-50}$$

This enables us to rewrite Eq. (11-49) as

$$\delta E_k = 2\langle \delta\phi_k|F|\phi_k\rangle \tag{11-51}$$

We should remember that the one-electron operator $F$ contains the one-electron orbitals $\phi_k$ that we seek to determine. In the derivation above we have required that the orthonormality of the orbitals not be affected by the changes in $\phi_k$, so that we imposed the conditions

$$\langle \delta\phi_k|\phi_l\rangle = 0 \qquad l = 1, 2, \ldots, N \tag{11-52}$$

It follows from our discussion in Section 7-7 that the variational problem $\delta E_k = 0$ with the auxiliary restraints (11-52) leads to

$$\delta E_k - 2\sum_l \lambda_{k,l}\delta_{k,l} = 0 \tag{11-53}$$

where $\lambda_{k,l}$ are a set of $N$ Lagrangian multipliers. Substitution of Eqs. (11-51) and (11-52) into Eq. (11-53) gives

$$2\left\langle \delta\phi_k \middle| F - \sum_l \lambda_{k,l} \middle| \phi_l \right\rangle = 0 \tag{11-54}$$

The electronic orbitals $\phi_k$ must therefore satisfy the equations

$$F\phi_k = \sum_l \lambda_{k,l}\phi_l \tag{11-55}$$

These are the Hartree-Fock equations for the orbitals $\phi_k$.

The solution of these equations is not easy, since the operators depend on the solutions and since Eq. (11-55) represents a set of $N$ coupled differential equations. Usually we can transform the set of functions $\phi_k$ by means of a unitary transformation, so that the matrix $\lambda_{k,l}$ becomes diagonal, and in this case the equations reduce to

$$F\phi_k = \lambda_k\phi_k \tag{11-56}$$

The customary approach to the solution of these equations is by iterative procedures. We start with a set of approximate orbitals and use them to construct the operator $F$. We then solve the equations, and in this way we obtain a new, improved set of orbitals. We now construct a new set of operators from this solution and solve the equations again. We keep repeating this procedure until the solutions of the equations are identical to the orbitals that we have used for the construction of the operators. In this case we say that the orbitals are self-consistent. Hence the name self-consistent field method is used for the Hartree-Fock method.

In the case of an atomic wave function the angular parts of the various atomic orbitals are usually known, and the only unknown functions are the radial components of the various orbitals. The Hartree-Fock equations reduce then to a one-dimensional problem. A differential equation in one variable can always be solved numerically, no matter what the form of the potential is. The various atomic SCF orbitals are all known in numerical form. Most of them were derived by Hartree during the period between 1925 and 1940.

It may be helpful to point out certain features of the Hartree-Fock method that are different from what might be expected at first glance. First it should be noted that the total energy of the system is not equal to the sum of the Hartree-Fock parameters $\lambda_k$ which are derived from Eq. (11-56). The values of these parameters, the orbital energies, are easily derived from Eqs. (11-50) and (11-56):

$$\begin{aligned}\lambda_k &= \langle \phi_k | F | \phi_k \rangle \\ &= \langle \phi_k | G | \phi_k \rangle + \sum_l \langle \phi_k | 2J_l - K_l | \phi_k \rangle \\ &= G_k + \sum_l (2J_{k,l} - K_{k,l}) \end{aligned} \tag{11-57}$$

The total energy of the system is given by Eq. (11-40). It can be written as

$$E = \sum_{k=1}^{N} (G_k + \lambda_k) \tag{11-58}$$

Clearly, this is different from the sum of the orbital energies over the occupied orbitals.

The Hartree-Fock energies $\lambda_k$ do have a physical significance; the parameter $\lambda_k$ represents the ionization energy which is required to remove an electron in the orbital $\phi_k$ from the closed-shell configuration. This property was first proved by Tjalling C. Koopmans in 1930, and it is now known as Koopmans' theorem. In order to prove its validity we derive the expectation value of the energy for the configuration $(\phi_1)^2(\phi_2)^2 \cdots (\phi_{N-1})^2(\phi_N)$. By analogy with Eq.

(11-40) we find that

$$E_N = 2\sum_{k=1}^{N-1}\left[G_k + \sum_{l=1}^{N-1}\left(J_{k,l} - \tfrac{1}{2}K_{k,l}\right)\right]$$

$$+ G_N + 2\sum_{l=1}^{N-1}\left(J_{N,l} - \tfrac{1}{2}K_{N,l}\right)$$

$$= 2\sum_{k=1}^{N-1}\left[G_k + \sum_{l=1}^{N}\left(J_{k,l} - \tfrac{1}{2}K_{k,l}\right)\right] + G_N$$

$$= \sum_{k=1}^{N-1}\left(G_k + \lambda_k\right) + G_N$$

$$= E - \lambda_N \tag{11-59}$$

The difference in energy between the configurations $(\phi_1)^2(\phi_2)^2\cdots(\phi_{N-1})^2(\phi_N)$ and $(\phi_1)^2(\phi_2)^2\cdots(\phi_N)^2$ is exactly equal to the Hartree-Fock orbital energy $\lambda_N$.

The Hartree-Fock theory of excitation energies is somewhat more complicated. We consider the situation where one electron is excited from an orbital $\phi_j$ to an orbital $\phi_n$ as compared to the closed-shell ground state. These excited state configurations can have either a singlet or a triplet spin. The corresponding antisymmetrized atomic wave functions were described in Eqs. (11-6) and (11-7).

First we derive the expression for the expectation value of the Hamiltonian (11-18) or (11-19) with respect to the wave functions (11-6) or (11-7). We use the same arguments that we used for deriving Eq. (11-40) and we only list the results. The singlet energy ${}^1E(j\rightarrow n)$ is given by

$${}^1E(j\rightarrow n) = 2\sum_{i\neq j}^{N}\left[G_i + \sum_{k\neq j}^{N}\left(J_{i,k} - \tfrac{1}{2}K_{i,k}\right)\right] + G_j + G_n$$

$$+ \sum_{i\neq j}^{N}\left(2J_{i,j} - K_{i,j}\right) + \sum_{i\neq j}^{N}\left(2J_{i,n} - K_{i,n}\right) + J_{j,n} + K_{j,n} \tag{11-60}$$

and the triplet energy ${}^3E(j\rightarrow n)$ is given by

$${}^3E(j\rightarrow n) = 2\sum_{k\neq j}^{N}\left[G_i + \sum_{k\neq j}^{N}\left(J_{i,k} - \tfrac{1}{2}K_{i,k}\right)\right] + G_j + G_n$$

$$+ \sum_{i\neq j}^{N}\left(2J_{i,j} - K_{i,j}\right) + \sum_{i\neq j}^{N}\left(2J_{i,n} - K_{i,n}\right) + J_{j,n} - K_{j,n} \tag{11-61}$$

We should realize that the exact Hartree-Fock energies ${}^1E(j\rightarrow n)$ and ${}^3E(j\rightarrow n)$ are derived by varying the expressions (11-60) or (11-61) with respect to all one-electron orbitals $\phi_i$. These new atomic orbitals are different from the old set that we derived for the ground-state configuration because the Hartree-Fock operators are slightly different in the two situations.

Naturally, such an exact Hartree-Fock calculation of the ground state and of the lower excited states of a given system is very laborious because it involves separate SCF calculations for each atomic configuration. In many cases use is made of an approximate procedure which is less precise but a lot more convenient. Let us consider again the Hartree-Fock equation (11-56) corresponding to the ground-state configuration. The Hartree-Fock operator $F$ in this equation has a set of eigenvalues $\lambda_k$ and corresponding eigenfunctions $\phi_k$, which correspond to the orbitals of the ground-state configurations. The operator $F$ has additional eigenvalues and eigenfunctions which can be derived without too much effort; these solutions are referred to as virtual orbitals. It is possible to use these virtual orbitals, which we denote by $\phi_n$ or $\phi_m$ for the construction of excited states. The excitation energies can then be derived from Eqs. (11-60) and (11-61) by substituting these virtual orbitals. The results are

$$
\begin{aligned}
{}^1E(j\rightarrow n)-E&=\lambda_n-\lambda_j-J_{j,n}+2K_{j,n}\\
{}^3E(j\rightarrow n)-E&=\lambda_n-\lambda_j-J_{j,n}
\end{aligned}
\tag{11-62}
$$

It may seem at first sight that a more precise result for the various ionization energies can also be derived by performing separate SCF calculations for the corresponding configurations. However, the simpler approach of Eq. (11-59), where the ionization energies are taken equal to the Hartree-Fock orbital energies $\lambda_k$, leads to better agreement with the experimental results than the more complex treatments. In fact, it has been shown that the simple Koopman's theorem is more accurate than the more involved treatments, but we will not attempt to describe these arguments.

In deriving the Hartree-Fock procedures we have limited ourselves to the simplest possible configuration: a closed-shell ground state where each orbital accommodates two electrons. It should be clear that the Hartree-Fock equations are quite different for other types of configurations. The corresponding derivations are similar to the case that we have considered, and many different situations have been reported in the literature. It would lead us too far to look into these different cases.

The situation becomes even more complicated if the atomic configuration involves degenerate atomic orbitals. For instance, if we consider the $(1s)^2(2s)^2(2p)^2$ configuration of the carbon atom, we should recognize that the $2p$ orbital is threefold degenerate and that there are many different atomic configurations that have the same energy in first approximation. The theoretical methods for analyzing these degenerate configurations are known as multiplet theory. We discuss some of them in the following section.

## 11-4 Multiplet Theory

In the previous section we were mainly concerned with nondegenerate atomic configurations, but we should realize that many atomic configurations are highly degenerate in first approximation. For instance, an atomic $p$ orbital is threefold degenerate and an atomic $d$ orbital is fivefold degenerate. If we include the electron spin, we find that a $(p)(d)$ configuration is $4\times3\times5=60$-fold degenerate to a first approximation. Similarly, a $(2p)(3p)$ configuration is 36-fold degenerate, and so on.

The atomic eigenstates belonging to a given configuration are derived from the Slater determinants belonging to each separate state by diagonalizing the Hamiltonian operator with respect to these eigenstates. This would involve the diagonalization of a $36\times36$ matrix in the case of a $(p)(p')$ configuration; clearly that is a laborious task. Fortunately, there are easier procedures for deriving the atomic eigenstates. These procedures are based on theorems that are derived from symmetry considerations. We call them "multiplet theory" and we present them in this section.

The atomic Hamiltonian $H$ is the sum of the orbital Hamiltonian $H_o$, which we defined in Eq. (11-18), and of the spin-orbit Hamiltonian $H_{so}$ and spin-spin Hamiltonian $H_{ss}$,

$$H=H_o+H_{so}+H_{ss} \tag{11-63}$$

Usually the matrix elements of $H_o$ are much larger than the matrix elements of $H_{so}$; the matrix elements of $H_{ss}$ are even smaller. It is therefore reasonable to diagonalize the Slater determinants first with respect to the orbital Hamiltonian $H_o$ only and to consider the spin-orbit coupling later as a small perturbation. This approximation is known as Russell-Saunders coupling and it is valid for all but the very heavy atoms.

The use of multiplet theory is based on a theorem that can be derived either from group theory or from the properties of operators that we discussed in Section 4-3. We consider the total angular momentum $\mathbf{L}$ of all electrons in an atom, which is defined by

$$\mathbf{L}=\sum_i \mathbf{L}_i=\sum_i (\mathbf{r}_i\times\mathbf{p}_i) \tag{11-64}$$

It is easily verified that the components of $\mathbf{L}$ and $L^2$ all commute with the Hamiltonian $H_o$:

$$[L_x,H_o]=[L_y,H_o]=[L_z,H_o]=0$$

$$[L^2,H_o]=0 \tag{11-65}$$

According to our considerations in Section 4-3, Eqs. (4-80) to (4-90), the operators $H_o$, $L^2$, and $L_z$ must all have the same eigenfunctions. We expect therefore that the eigenfunctions of $H_o$ can be obtained by deriving the eigenfunctions of the operators $L^2$ and $L_z$. These conclusions are consistent with a theorem that is derived from group theory. Here it can be shown that the matrix elements of $H_o$ between states with different angular momentum eigenvalues are all zero. Again, it follows that the eigenstates of the operator $H_o$ are obtained by deriving the eigenfunctions and eigenvalues of the operators $L^2$ and $L_z$ from the set of Slater determinants belonging to the same configuration. The purpose of multiplet theory is the derivation of the eigenfunctions of $L^2$ and $L_z$. We shall see that this can be achieved by making use of the properties of the angular momentum operators which we discussed in Section 6-2.

We should point out that matrix elements of $H_o$ between states with different spin multiplicity are orthogonal to each other. It follows that the eigenfunctions of $H_o$ must be eigenfunctions also of the operators $S^2$ and $S_z$. The latter are defined by

$$\mathbf{S} = \sum_i \mathbf{S}_i \tag{11-66}$$

Our goal is thus the selection of the eigenfunctions of the operators $L^2$, $L_z$, $S^2$, and $S_z$ from the set of Slater determinants belonging to one particular configuration. Before discussing this problem we first show how the eigenvalues of the operators $L^2$ and $S^2$ can be derived by means of the old vector model of the atom.

We make use of the Russell-Saunders coupling scheme, where it is assumed that the coupling between the orbital angular momentum vectors of the different electrons is much stronger than the coupling between the orbital and spin angular momenta of the electrons. This assumption is valid if the spin-orbit interaction energy is smaller than the matrix elements of $H_o$. This condition is true for all except the heaviest atoms.

In the vector model we find the possible values of the angular momentum of two electrons by taking the one angular momentum as the axis of quantization for the second electron. For example, in the configuration $(p)(p')$ the possible values of the total angular momentum are $L=0$, $L=1$, and $L=2$. The possible values of the spin angular momentum are $S=0$ and $S=1$. Following the notation of Section 11-1 we find that the possible eigenstates of the configuration $(p)(p')$ are denoted by $^1S$, $^1P$, $^1D$ and $^3S$, $^3P$ and $^3D$. If we add a third electron, the possible values of the total angular momentum are obtained by taking the angular momentum of the other two electrons as the axis of quantization of the third-electron angular momentum. According to the Russell-Saunders coupling scheme we couple the various orbital angular momentum vectors in order to obtain the value of the total angular momentum

**Table 11-2** Eigenstates of Configurations Involving Nonequivalent Electrons

| Configuration | Eigenstates | | | | | | | | |
|---|---|---|---|---|---|---|---|---|---|
| $(s)(s')$ | $^1S$ | $^3S$ | | | | | | | |
| $(s)(p)$ | $^1P$ | $^3P$ | | | | | | | |
| $(s)(d)$ | $^1D$ | $^3D$ | | | | | | | |
| $(p)(p')$ | $^1S$ | $^1P$ | $^1D$ | $^3S$ | $^3P$ | $^3D$ | | | |
| $(p)(d)$ | $^1P$ | $^1D$ | $^1F$ | $^3P$ | $^3D$ | $^3F$ | | | |
| $(d)(d')$ | $^1S$ | $^1P$ | $^1D$ | $^1F$ | $^1G$ | $^3S$ | $^3P$ | $^3D$ | $^3F$ | $^3G$ |

**Table 11-3** Eigenstates of Configurations Involving Equivalent Electrons

| Configuration | Eigenstates | | | | | | |
|---|---|---|---|---|---|---|---|
| $(s^2)$ | $^1S$ | | | | | | |
| $(p^2)$ | $^1S$ | $^1D$ | $^3P$ | | | | |
| $(p^3)$ | $^2P$ | $^2D$ | $^4S$ | | | | |
| $(p^4)$ | $^1S$ | $^1D$ | $^3P$ | | | | |
| $(p^5)$ | $^2P$ | | | | | | |
| $(p^6$ | $^1S$ | | | | | | |
| $(d^2)$ | $^1S$ | $^1D$ | $^1G$ | $^3P$ | $^3F$ | | |
| $(d^3)$ | $^2P$ | $^2D$ | $^2F$ | $^2G$ | $^2H$ | $^4P$ | $^4F$ |

**L** of the atom, and we also couple the electron spins in order to obtain the value of the total spin angular momentum **S**. Ultimately we must also consider the coupling between **L** and **S**, and we discuss this in Section 11-5.

We have listed the possible eigenstates of various configurations in Table 11-2. Here we consider configurations where the electrons are in nonequivalent orbitals, for instance $(2p)(3p)$. If we consider configurations involving equivalent orbitals, we must take the exclusion principle into account, for example, in the configuration $(2p)^2$ the states $^1P$, $^3S$, and $^3D$ must be excluded. We have listed various possible eigenstates of configurations involving equivalent electrons in Table 11-3.

Let us now proceed to the derivation of the angular momentum eigenfunctions. We mentioned already that they can be obtained by using various properties of the angular momentum operators and their eigenfunctions which we discussed in Section 6-2. As an example, we consider the two configurations $(p)(p')$ and $(p^2)$; the latter case can be derived from the former by taking the exclusion principle into account.

First we define the various wave functions belonging to the configuration $(p)(p')$. We assume that the system has a set of $N$ closed-shell doubly occupied orbitals $\chi_1, \chi_2, \ldots, \chi_N$ and two additional electrons in the orbitals $p_i$

and $p_j'$. The corresponding Slater determinant wave functions are

$$\Phi(i\alpha;\, j\alpha)=[(2N+2)!]^{-1/2}\sum_P P\delta_P[\chi_1(1)\alpha(1)\chi_1(2)\beta(2)\chi_2(3)\alpha(3)\chi_2(4)\beta(4)$$
$$\cdots\chi_N(2N-1)\alpha(2N-1)\chi_N(2N)\beta(2N)$$
$$\times p_i(2N+1)\alpha(2N+1)p_j'(2N+2)\alpha(2N+2)]$$
$$\Phi(i\alpha;\, j\beta)=[(2N+2)!]^{-1/2}\sum_P P\delta_P[\chi_1(1)\alpha(1)\chi_1(2)\beta(2)\chi_2(3)\alpha(3)\chi_2(4)\beta(4)$$
$$\cdots\chi_N(2N-1)\alpha(2N-1)\chi_N(2N)\beta(2N)$$
$$\times p_i(2N+1)\alpha(2N+1)p_j'(2N+2)\beta(2N+2)]$$
$$\Phi(i\beta;\, j\alpha)=\cdots \text{ and so on} \qquad (11\text{-}67)$$

Here $i$ and $j$ can assume the values 1, 0, and $-1$, and we have a total of $3\times3\times4=36$ different functions $\Phi$.

In Section 6-2 we introduced the operators $M_1=M_x+iM_y$ and $M_{-1}=M_x-iM_y$ and we found that the effect of these operators on the eigenfunctions $\psi_{l,m}$ is to increase or decrease the quantum number $m$ by unity. The relations are described by Eqs. (6-63), (6-64), and (6-70) for positive values of $m$. We report the results here again:

$$M_1\psi_{l,m}=h[l(l+1)\quad m(m+1)]^{1/2}\psi_{l,m+1}$$
$$M_{-1}\psi_{l,m}=\hbar[l(l+1)-m(m-1)]^{1/2}\psi_{l,m-1} \qquad (11\text{-}68)$$

These various relations were derived from the commutator relations of the angular momentum operators, and it is easily shown that the same relations are valid for the operators **L** and **S**. In particular,

$$L_{-1}\Psi(L,M_L)=\hbar[L(L+1)-M_L(M_L-1)]^{1/2}\Psi(L,M_L-1)$$
$$S_{-1}\Psi(S,M_S)=\hbar[S(S+1)-M_S(M_S-1)]^{1/2}\Psi(S,M_S-1) \qquad (11\text{-}69)$$

where

$$L_{-1}=L_x-iL_y \qquad S_{-1}=S_x-iS_y$$

and $M_L$ and $M_S$ are the quantum numbers denoting the eigenvalues of $L_z$ and $S_z$, respectively.

Let us first consider the function $\Phi(1\alpha;\, 1\alpha)$. This function is an eigenfunction of $L_z$ with eigenvalue 2 and it is an eigenfunction of $S_z$ with eigenvalue 1.

We denote the eigenfunctions of the four operators $L^2$, $L_z$, $S^2$, and $S_z$ by $\Psi(L, M_L; S, M_S)$, where the four quantities $L$, $S$, $M_L$, and $M_S$ denote the eigenvalues. It is clear that $\Phi(1\alpha;\ 1\alpha)$ must be an eigenfunction of $L^2$ with eigenvalue $L=2$ since $M_L=2$. Also, $\Phi(1\alpha;\ 1\alpha)$ must be an eigenfunction of $S^2$ with eigenvalue $S=1$ since $M_S=1$. Therefore, we conclude that

$$\Psi(2,2;\ 1,1)=\Phi(1\alpha;\ 1\alpha) \tag{11-70}$$

We use Eq. (11-69) for the derivation of some additional eigenfunctions. We omit the constant $\hbar$ and we find

$$L_{-1}\Psi(2,2;\ 1,1)=2\Psi(2,1;\ 1,1)$$

$$S_{-1}\Psi(2,2;\ 1,1)=\sqrt{2}\,\Psi(2,2;\ 1,0) \tag{11-71}$$

It can also be derived from Eq. (11-68) that

$$L_{-1}\Phi(1\alpha;\ 1\alpha)=\sqrt{2}\,[\Phi(0\alpha;\ 1\alpha)+\Phi(1\alpha;\ 0\alpha)] \tag{11-72}$$

and

$$S_{-1}\Phi(1\alpha;\ 1\alpha)=[\Phi(1\alpha;\ 1\beta)+\Phi(1\beta;\ 1\alpha)] \tag{11-73}$$

By comparing Eqs. (11-71), (11-72), and (11-73) we find that

$$\Psi(2,1;\ 1,1)=2^{-1/2}[\Phi(0\alpha;\ 1\alpha)+\Phi(1\alpha;\ 0\alpha)]$$

$$\Psi(2,2;\ 1,0)=2^{-1/2}[\Phi(1\alpha;\ 1\beta)+\Phi(1\beta;\ 1\alpha)] \tag{11-74}$$

The other wave functions belonging to the $^3D$ state (with $L=1$; $S=1$) can be derived by repeated applications of the above procedure. As an example we show the derivation of the function $\Psi(2,0;\ 1,1)$. From Eq. (11-69) we find that

$$L_{-1}\Psi(2,1;\ 1,1)=6^{-1/2}\Psi(2,0;\ 1,1) \tag{11-75}$$

and from Eq. (11-74) we obtain

$$\begin{aligned} L_{-1}\Psi(2,1;\ 1,1) &= 2^{-1/2}\{L_{-1}\Phi(0\alpha;\ 1\alpha)+L_{-1}\Phi(1\alpha;\ 0\alpha)\} \\ &= 2^{-1/2}[2^{1/2}\Phi(-1\alpha;\ 1\alpha)+2^{1/2}\Phi(0\alpha;\ 0\alpha) \\ &\quad +2^{1/2}\Phi(0\alpha;\ 0\alpha)+2^{1/2}\Phi(1\alpha;\ -1\alpha)] \\ &= \Phi(1\alpha;\ -1\alpha)+2\Phi(0\alpha;\ 0\alpha)+\Phi(-1\alpha;\ 1\alpha) \end{aligned} \tag{11-76}$$

By comparing Eqs. (11-75) and (11-76) we derive that

$$\Psi(2,0;\ 1,1)=6^{-1/2}[\Phi(1\alpha;\ -1\alpha)+2\Phi(0\alpha;\ 0\alpha)+\Phi(-1\alpha;\ 1\alpha)] \tag{11-77}$$

We have listed the various results in Table 11-4.

Let us now attempt to determine some of the other eigenfunctions. There are two functions with the quantum numbers $M_L=1$ and $M_S=1$, namely $\Phi(1\alpha;\ 0\alpha)$ and $\Phi(0\alpha;\ 1\alpha)$. These functions must belong to the state $S=1$ and to either $L=2$ or $L=1$. Consequently, the functions $\Psi(2,1;\ 1,1)$ and $\Psi(1,1;\ 1,1)$ must be linear combinations of the two functions:

$$\Psi(2,1;\ 1,1)=A\Phi(1\alpha;\ 0\alpha)+B\Phi(0\alpha;\ 1\alpha)$$

$$\Psi(1,1;\ 1,1)=C\Phi(1\alpha;\ 0\alpha)+D\Phi(0\alpha;\ 1\alpha) \tag{11-78}$$

The function $\Psi(2,1;\ 1,1)$ is known, and since the other function must be orthogonal, it is easily found that

$$\Psi(1,1;\ 1,1)=2^{-1/2}[\Phi(1\alpha;\ 0\alpha)-\Phi(0\alpha;\ 1\alpha)] \tag{11-79}$$

This same result can also be derived by an alternative procedure, which is based on the use of projection operators. This procedure has been extensively used by Per-Olov Löwdin. We note that

$$L^2\Psi(2,1;\ 1,1)=2(2+1)\Psi(2,1;\ 1,1)$$

$$L^2\Psi(1,1;\ 1,1)=1(1+1)\Psi(1,1;\ 1,1) \tag{11-80}$$

or

$$(L^2-6)\Psi(2,1;\ 1,1)=0$$

$$(L^2-2)\Psi(1,1;\ 1,1)=0 \tag{11-81}$$

Obviously, if we have a function that is a linear combination of $\Psi(2,1;\ S,M_S)$ and $\Psi(1,1;\ S,M_S)$, for example, the function $\Phi(1\alpha;\ 0\alpha)$, we find that

$$(L^2-6)\Phi(1\alpha;\ 0\alpha)=\lambda\Psi(1,1;\ 1,1)$$

$$(L^2-2)\Phi(1\alpha;\ 0\alpha)=\mu\Psi(2,1;\ 1,1) \tag{11-82}$$

We evaluate Eq. (11-82) by writing the operator $L^2$ as

$$L^2=L_1L_{-1}+L_z^2-L_z \tag{11-83}$$

**Table 11-4** Eigenfunctions for the Configuration $(p)(p')$

| Symmetry | Eigenfunctions |
|---|---|
| $^3D$ | $\Psi(2,2;\ 1,1)=\Phi(1\alpha;\ 1\alpha)$ |
| | $\Psi(2,2;\ 1,0)=2^{-1/2}[\Phi(1\alpha;\ 1\beta)+\Phi(1\beta;\ 1\alpha)]$ |
| | $\Psi(2,2;\ 1,-1)=\Phi(1\beta;\ 1\beta)$ |
| | $\Psi(2,1;\ 1,1)=2^{-1/2}[\Phi(1\alpha;\ 0\alpha)+\Phi(0\alpha;\ 1\alpha)]$ |
| | $\Psi(2,1;\ 1,0)=\frac{1}{2}[\Phi(1\alpha;\ 0\beta)+\Phi(1\beta;\ 0\alpha)+\Phi(0\alpha;\ 1\beta)+\Phi(0\beta;\ 1\alpha)]$ |
| | $\Psi(2,1;\ 1,-1)=2^{-1/2}[\Phi(1\beta;\ 0\beta)+\Phi(0\beta;\ 1\beta)]$ |
| | $\Psi(2,0;\ 1,1)=6^{-1/2}[\Phi(1\alpha;\ -1\alpha)+2\Phi(0\alpha;\ 0\alpha)+\Phi(-1\alpha;\ 1\alpha)]$ |
| | $\Psi(2,0;\ 1,0)=(12)^{-1/2}[\Phi(1\alpha;\ -1\beta)+\Phi(1\beta;\ -1\alpha)+2\Phi(0\alpha;\ 0\beta)+2\Phi(0\beta;\ 0\alpha)$ $+\Phi(-1\alpha;\ 1\beta)+\Phi(-1\beta;\ 1\alpha)]$ |
| | $\Psi(2,0;\ 1,-1)=6^{-1/2}[\Phi(1\beta;\ -1\beta)+2\Phi(0\beta;\ 0\beta)+\Phi(-1\beta;\ 1\beta)]$ |
| | $\Psi(2,-1;\ 1,1)=2^{-1/2}[\Phi(0\alpha;\ -1\alpha)+\Phi(-1\alpha;\ 0\alpha)]$ |
| | $\Psi(2,-1;\ 1,0)=\frac{1}{2}[\Phi(0\alpha;\ -1\beta)+\Phi(0\beta;\ -1\alpha)+\Phi(-1\alpha;\ 0\beta)+\Phi(-1\beta;\ 0\alpha)]$ |
| | $\Psi(2,-1;\ 1,-1)=2^{-1/2}[\Phi(0\beta;\ -1\beta)+\Phi(-1\beta;\ 0\beta)]$ |
| | $\Psi(2,-2;\ 1,1)=\Phi(-1\alpha;\ -1\alpha)$ |
| | $\Psi(2,-2;\ 1,0)=2^{-1/2}[\Phi(-1\alpha;\ -1\beta)+\Phi(-1\beta;\ -1\alpha)]$ |
| | $\Psi(2,-2;\ 1,-1)=\Phi(-1\beta;\ -1\beta)$ |
| $^3P$ | $\Psi(1,1;\ 1,1)=2^{-1/2}[\Phi(1\alpha;\ 0\alpha)-\Phi(0\alpha;\ 1\alpha)]$ |
| | $\Psi(1,1;\ 1,0)=\frac{1}{2}[\Phi(1\alpha;\ 0\beta)+\Phi(1\beta;\ 0\alpha)-\Phi(0\alpha;\ 1\beta)-\Phi(0\beta;\ 1\alpha)]$ |
| | $\Psi(1,1;\ 1,-1)=2^{-1/2}[\Phi(1\beta;\ 0\beta)-\Phi(0\beta;\ 1\beta)]$ |
| | $\Psi(1,0;\ 1,1)=2^{-1/2}[\Phi(1\alpha;\ -1\alpha)-\Phi(-1\alpha;\ 1\alpha)]$ |
| | $\Psi(1,0;\ 1,0)=\frac{1}{2}[\Phi(1\alpha;\ -1\beta)+\Phi(1\beta;\ -1\alpha)-\Phi(-1\alpha;\ 1\beta)-\Phi(1\beta;\ 1\alpha)]$ |
| | $\Psi(1,0;\ 1,-1)=2^{-1/2}[\Phi(1\beta;\ -1\beta)-\Phi(-1\beta;\ 1\beta)]$ |
| | $\Psi(1,-1;\ 1,1)=2^{-1/2}[\Phi(0\alpha;\ -1\alpha)-\Phi(-1\alpha;\ 0\alpha)]$ |
| | $\Psi(1,-1;\ 1,0)=\frac{1}{2}[\Phi(0\alpha;\ -1\beta)+\Phi(0\beta;\ -1\alpha)-\Phi(-1\alpha;\ 0\beta)-\Phi(-1\beta;\ 0\alpha)]$ |
| | $\Psi(1,-1;\ 1,1)=2^{-1/2}[\Phi(0\beta;\ -1\beta)-\Phi(-1\beta;\ 0\beta)]$ |
| $^3S$ | $\Psi(0,0;\ 1,1)=3^{-1/2}[\Phi(1\alpha;\ -1\alpha)-\Phi(0\alpha;\ 0\alpha)+\Phi(-1\alpha;\ 1\alpha)]$ |
| | $\Psi(0,0;\ 1,0)=6^{-1/2}[\Phi(1\alpha;\ -1\beta)+\Phi(1\beta;\ -1\alpha)-\Phi(0\alpha;\ 0\beta)-\Phi(0\beta;\ 0\alpha)$ $+\Phi(-1\alpha;\ 1\beta)+\Phi(-1\beta;\ 1\alpha)]$ |
| | $\Psi(0,0;\ 1-1)=3^{-1/2}[\Phi(1\beta;\ -1\beta)-\Phi(0\beta;\ 0\beta)+\Phi(-1\beta;\ 1\beta)]$ |
| $^1D$ | $\Psi(2,2;\ 0,0)=2^{-1/2}[\Phi(1\alpha;\ 1\beta)-\Phi(1\beta;\ 1\alpha)]$ |
| | $\Psi(2,1;\ 0,0)=\frac{1}{2}[\Phi(1\alpha;\ 0\beta)-\Phi(1\beta;\ 0\alpha)+\Phi(0\alpha;\ 1\beta)-\Phi(0\beta;\ 1\alpha)]$ |
| | $\Psi(2,0;\ 0,0)=(12)^{-1/2}[\Phi(1\alpha;\ -1\beta)-\Phi(1\beta;\ -1\alpha)+2\Phi(0\alpha;\ 0\beta)-2\Phi(0\beta;\ 0\alpha)$ $+\Phi(-1\alpha;\ 1\beta)-\Phi(-1\beta;\ 1\alpha)]$ |
| | $\Psi(2,-1;\ 0,0)=\frac{1}{2}[\Phi(-1\alpha;\ 0\beta)-\Phi(-1\beta;\ 0\alpha)+\Phi(0\alpha;\ -1\beta)-\Phi(0\beta;\ -1\alpha)]$ |
| | $\Psi(2,-2;\ 0,0)=2^{-1/2}[\Phi(-1\alpha;\ -1\beta)-\Phi(-1\beta;\ -1\alpha)]$ |

**Table 11-4** (*Continued*)

| Symmetry | Eigenfunctions |
|---|---|
| $^1P$ | $\Psi(1,1;\,0,0)=\frac{1}{2}[\Phi(1\alpha;\,0\beta)-\Phi(1\beta;\,0\alpha)-\Phi(0\alpha;\,1\beta)+\Phi(0\beta;\,1\alpha)]$ |
| | $\Psi(1,0;\,0,0)=\frac{1}{2}[\Phi(1\alpha;\,-1\beta)-\Phi(1\beta;\,-1\alpha)-\Phi(-1\alpha;\,1\beta)+\Phi(-1\beta;\,1\alpha)]$ |
| | $\Psi(1,-1;\,0,0)=\frac{1}{2}[\Phi(0\alpha;\,-1\beta)-\Phi(0\beta;\,-1\alpha)-\Phi(-1\alpha;\,0\beta)+\Phi(-1\beta;\,0\alpha)]$ |
| $^1S$ | $\Psi(0,0;\,0,0)=6^{-1/2}[\Phi(1\alpha;\,-1\beta)-\Phi(1\beta;\,-1\alpha)-\Phi(0\alpha;\,0\beta)+\Phi(0\beta;\,0\alpha)]+$ $\Phi(-1\alpha;\,1\beta)-\Phi(-1\beta;\,1\alpha)\}$ |

following Eq. (6-49). We have

$$L_{-1}\Phi(1\alpha;\,0\alpha)=2^{1/2}[\Phi(0\alpha;\,0\alpha)+\Phi(1\alpha;\,-1\alpha)] \tag{11-84}$$

and

$$L_1L_{-1}\Phi(1\alpha;\,0\alpha)=2\Phi(1\alpha;\,0\alpha)+2\Phi(0\alpha;\,1\alpha)+2\Phi(1\alpha;\,0\alpha)$$

Consequently,

$$L^2\Phi(1\alpha;\,0\alpha)=4\Phi(1\alpha;\,0\alpha)+2\Phi(0\alpha;\,1\alpha)+\Phi(1\alpha;\,0\alpha)-\Phi(1\alpha;\,0\alpha) \tag{11-85}$$

and

$$(L^2-6)\Phi(1\alpha;\,0\alpha)=-2\Phi(1\alpha;\,0\alpha)+2\Phi(0\alpha;\,1\alpha)$$

$$(L^2-2)\Phi(1\alpha;\,0\alpha)=2\Phi(1\alpha;\,0\alpha)+2\Phi(0\alpha;\,1\alpha) \tag{11-86}$$

By comparing this result with Eq. (11-82) we obtain

$$\Psi(1,1;\,1,1)=2^{-1/2}[\Phi(1\alpha;\,0\alpha)-\Phi(0\alpha;\,1\alpha)]$$

$$\Psi(2,1;\,1,1)=2^{-1/2}[\Phi(1\alpha;\,0\alpha)+\Phi(0\alpha;\,1\alpha)] \tag{11-87}$$

The projection operator method may seem somewhat complicated, but it has the advantage that it can also be applied to more complex situations. For instance, it can be used for the derivation of the eigenfunctions $\Psi(2,0;\,1,1)$, $\Psi(1,0;\,1,1)$, and $\Psi(0,0;\,1,1)$ from the function $\Phi(0\alpha;\,0\alpha)$.

By combining the various procedures that we discussed above, we can determine all 36 different eigenfunctions belonging to the configuration $(p)(p')$. We have listed the results in Table 11-4.

The second example that we discuss here is the configuration $(p)^2$, where the two electrons are in the same $2p$ orbital (the carbon ground-state configuration $(1s)^2(2s)^2(2p)^2$ also belongs to this category). The eigenstates and eigenfunctions of the $(p)^2$ configuration can be derived from Table 11-4 for the $(p)(p')$ configuration by taking the exclusion principle into account. For instance, if the two $p$ orbitals are equivalent, the functions $\Phi(1\alpha;\ 1\alpha)$ and $\Phi(1\beta;\ 1\beta)$ are both zero. It can be verified that all eigenfunctions belonging to the $^3D$ configuration are zero. According to the exclusion principle the $^3D$ state is not allowed if the $p$ orbitals are equivalent, and this is confirmed by the result that all corresponding eigenfunctions in Table 11-4 are zero. Similarly, it is found that all eigenfunctions belonging to the $^3S$ and $^1P$ states are zero. We have listed all nonzero eigenfunctions of the $(p)^2$ configuration in Table 11-5.

It can be seen from the two examples that we have discussed how the eigenfunctions of the operators $L^2$ and $S^2$ are derived by making use of the ladder operators $L_1$, $L_{-1}$, $S_1$, and $S_{-1}$ and by making use of projection operators. We cannot present a unique and straightforward procedure for the derivation of the eigenvalues, but by combining the various procedures that we have used it is possible to derive the eigenfunctions of other configurations, such as $(d)(d')$, $(d)^2$, $(p^2)(d)$, and so on. The results of such configurations have been derived and they are available in the literature. We do not attempt to present them here.

Finally, we should consider the effects of spin-orbit coupling on the eigenstates that we have derived. We discuss this problem in Section 11-5.

**Table 11-5** Eigenfunctions for the Configuration $(p)^2$

| Symmetry | Eigenfunctions |
|---|---|
| $^3P$ | $\Psi(1,1;\ 1,1)=\Phi(1\alpha;\ 0\alpha)$ |
| | $\Psi(1,1;\ 1,0)=2^{-1/2}[\Phi(1\alpha;\ 0\beta)+\Phi(1\beta;\ 0\alpha)]$ |
| | $\Psi(1,1,1,-1)=\Phi(1\beta;\ 0\beta)$ |
| | $\Psi(1,0;\ 1,1)=\Phi(1\alpha;\ -1\alpha)$ |
| | $\Psi(1,0;\ 1,0)=2^{-1/2}[\Phi(1\alpha;\ -1\beta)+\Phi(1\beta;\ -1\alpha)]$ |
| | $\Psi(1,0;\ 1,-1)=\Phi(1\beta;\ -1\beta)$ |
| | $\Psi(1,-1;\ 1,1)=\Phi(0\alpha;\ -1\alpha)$ |
| | $\Psi(1,-1;\ 1,0)=2^{-1/2}[\Phi(0\alpha;\ -1\beta)+\Phi(0\beta;\ -1\alpha)]$ |
| | $\Psi(1,-1;\ 1,-1)=\Phi(0\beta;\ -1\beta)$ |
| $^1D$ | $\Psi(2,2;\ 0,0)=\Phi(1\alpha;\ 1\beta)$ |
| | $\Psi(2,1;\ 0,0)=2^{-1/2}[\Phi(1\alpha;\ 0\beta)-\Phi(1\beta;\ 0\alpha)]$ |
| | $\Psi(2,0;\ 0,0)=6^{-1/2}[\Phi(1\alpha;\ -1\beta)+2\Phi(0\alpha;\ 0\beta)-\Phi(1\beta;\ -1\alpha)]$ |
| | $\Psi(2,-1;\ 0,0)=2^{-1/2}[\Phi(0\alpha;\ -1\beta)-\Phi(0\beta;\ -1\alpha)]$ |
| | $\Psi(2,-2;\ 0,0)=\Phi(-1\alpha;\ -1\beta)$ |
| $^1S$ | $\Psi(0,0;\ 0,0)=3^{-1/2}[\Phi(1\alpha;\ -1\beta)-\Phi(1\beta;\ -1\alpha)-\Phi(0\alpha;\ 0\beta)]$ |

## 11-5 Spin-Orbit Coupling and Clebsch-Gordon Coefficients

In the Russell-Saunders coupling scheme the spin-orbit interaction is treated as a small perturbation. Its net effect on the energy levels is the splitting of the degenerate energy levels with a given value of the quantum numbers $L$ and $S$. The major part of the spin-orbit coupling transforms as the scalar product $\mathbf{L}\cdot\mathbf{S}$ of the two angular momentum vectors, and the magnitude of the spin-orbit coupling depends on the relative orientation of the two vectors. The total angular momentum of the atom is denoted by $\mathbf{J}$:

$$\mathbf{J}=\mathbf{L}+\mathbf{S} \tag{11-88}$$

In the vector model the magnitude of $\mathbf{J}$ is again obtained by quantizing the smaller of the two vectors $\mathbf{L}$ or $\mathbf{S}$ with respect to the larger one. The magnitude of $\mathbf{J}$ depends on the relative orientation of $\mathbf{L}$ and $\mathbf{S}$ and we expect that states with different $J$ values have slightly different energies due to spin-orbit coupling. In Table 11-1 we have followed the standard notation, where the $J$ value of a given atomic state is denoted by a subscript on the right side of the letter denoting the $L$ value.

We mentioned in Section 10-1 the Zeeman effect of the sodium $D$ line, and it may be helpful to interpret this situation qualitatively by making use of the vector model. We will present a more rigorous, mathematical description of the coupling between $\mathbf{L}$ and $\mathbf{S}$ later in this section.

It can be seen in Table 11-1 that the ground-state configuration of the sodium atom is $(1s)^2(2s)^2(2p)^6(3s)$, in this configuration $L=0$ and $S=\frac{1}{2}$. The total angular momentum $J$ must have the value $J=\frac{1}{2}$, and the ground state is described by the symbol ${}^2S_{1/2}$ as indicated in Table 11-1. The sodium $D$ line corresponds to a transition from the ${}^2S_{1/2}$ ground state to the states belonging to the configuration $(1s)^2(2s)^2(2p)^6(3p)$. The latter state has the quantum numbers $L=1$ and $S=\frac{1}{2}$. There are two possible values of the quantum number $J$, namely $J=\frac{1}{2}$ and $J=\frac{3}{2}$; this result is easily derived by quantizing $\mathbf{S}$ with respect to $\mathbf{L}$. Consequently, there are two different atomic states ${}^2P_{1/2}$ and ${}^2P_{3/2}$ belonging to the configuration $(1s)^2(2s)^2(2p)^6(3p)$.

The two ${}^2P$ states have slightly different energies due to spin-orbit perturbation, and consequently, there are two lines corresponding to the ${}^2S\rightarrow{}^2P$ transition. The sodium $D_1$ line at 5895.93 Å corresponds to the ${}^2S_{1/2}\rightarrow{}^2P_{1/2}$ transition and the sodium $D_2$ line at 5889.96 Å corresponds to the ${}^2S_{1/2}\rightarrow$ ${}^2P_{3/2}$ transition. The difference in energy between the two ${}^2P$ energy levels, 17.19 $cm^{-1}$ in magnitude, is due to the spin-orbit perturbation.

In Fig. 11-1 we have sketched the splitting of the three energy levels due to a magnetic field and the corresponding spectroscopic transitions. The level ${}^2S_{1/2}$ splits into two levels, corresponding to $M_J=\frac{1}{2}, -\frac{1}{2}$. The level ${}^2P_{1/2}$ also splits into two components, corresponding to $M_J=\frac{1}{2}, -\frac{1}{2}$, and the level ${}^2P_{3/2}$ splits into four components with $M_J=\frac{3}{2}, \frac{1}{2}, -\frac{1}{2}, -\frac{3}{2}$. The selection rules for

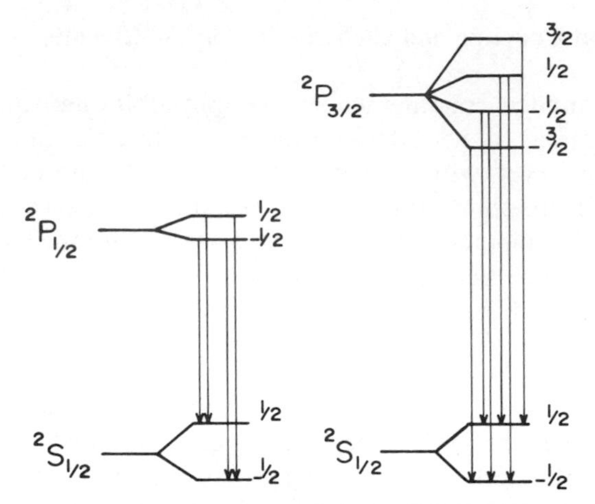


**Fig. 11-1** The anomalous Zeeman effect of the sodium $D$ line. In the absence of a magnetic field the $^2P$ level is split into two states $^2P_{1/2}$ and $^2P_{3/2}$. We show the splittings of all states due to a magnetic field.

the transitions are $\Delta M = 0$ for transistions that are polarized parallel to the magnetic field and $\Delta M = \pm 1$ for transitions that are polarized perpendicular to the magnetic field.

The problem of deriving the eigenfunctions of the operators $J^2$ and $J_z$ from the eigenfunctions of the operators $L^2$ and $L_z$ and of the operators $S^2$ and $S_z$ is similar to the problems that we considered in the previous Section 11-4. The problem is often referred to as the addition of two angular momentum vectors in quantum mechanics. In the previous section we considered the addition of two electron angular momenta to one atomic angular momentum, but we also had to take into account the spin angular momentum and the exclusion principle. The present situation is easier to deal with since it concerns only the addition of the orbital and spin angular momentum vectors without additional difficulties due to the exclusion principle.

First we define our basis set of functions. In the previous section we derived the eigenfunctions of the orbital and spin angular momentum vectors. These functions satisfy the relations

$$
\begin{aligned}
L^2\Psi(L, M_L; S, M_S) &= L(L+1)\Psi(L, M_L; S, M_S) \\
S^2\Psi(L, M_L; S, M_S) &= S(S+1)\Psi(L, M_L; S, M_S) \\
L_z\Psi(L, M_L; S, M_S) &= M_L\Psi(L, M_L; S, M_S) \\
S_z\Psi(L, M_L; S, M_S) &= M_S\Psi(L, M_L; S, M_S)
\end{aligned}
\tag{11-89}
$$

The resultant vector $\mathbf{J}$ is defined by Eq. (11-88). We define its eigenfunctions

$\Psi(J, M)$ by

$$J^2\Psi(J, M)=J(J+1)\Psi(J, M)$$

$$J_z\Psi(J, M)=M\Psi(J, M) \tag{11-90}$$

The problem of expressing the eigenfunctions $\Psi(J, M)$ in terms of the eigenfunctions $\Psi(L, M_L; S, M_S)$ has been solved and the solutions are available. At the end of this chapter we list various books where the solutions are derived. Here (and in Appendix C) we only list the result.

In order to describe the result we note first that

$$\begin{aligned} J_z\Psi(L, M_L; S, M_S) &= (L_z+S_z)\Psi(L, M_L; S, M_S) \\ &= (M_L+M_S)\Psi(L, M_L; S, M_S) \end{aligned} \tag{11-91}$$

In order to describe the eigenfunctions of $J_z$ we introduce the Kronecker symbol

$$\begin{aligned} \delta(M, M_L+M_S) &= 0 \quad \text{if } M_L+M_S \neq M \\ &= 1 \quad \text{if } M_L+M_S = M \end{aligned} \tag{11-92}$$

The eigenfunctions $\Psi(J, M)$, corresponding to given specific values of the quantum numbers $L$ and $S$, can now be represented as

$$\Psi(J, M)=\Sigma_{M_L}\Sigma_{M_S}\delta(M, M_L+M_S)C(L, S, J; M_L, M_S, M)\Psi(L, M_L; S, M_S) \tag{11-93}$$

The coefficients $C(L, S, J; M_L, M_S, M)$ are known as Clebsch-Gordon coefficients. Their values can be derived from different procedures, from group-theoretical considerations, from a combination of projection and ladder operators, or from the properties of spherical harmonics. Various tables of their values have been reported in the books listed at the end of this chapter.

We list the values of some coefficients in Appendix C, namely the case where $S=\frac{1}{2}$ and where $S=1$. Here it is assumed that $L \geq S$ or $L>0$. It should be noted that for $S=\frac{1}{2}$ the quantum number $J$ has two possible values, namely $J=L+\frac{1}{2}$ and $J=L-\frac{1}{2}$, and the quantum number $M_S$ has the values $M_S=\frac{1}{2}$ and $M_S=-\frac{1}{2}$. If $S=1$, the possible values of $J$ are $J=L+1$, $J=L$, and $J=L-1$, and the possible values of $M_S$ are $M_S=1$, $M_S=0$, and $M_S=-1$.

An analytical expression for the values of the Clebsch-Gordon coefficients has been reported in the book by Wigner. We present this expression here:

$$C(L,S,J;M_L,M_S,M)=$$

$$(2J+1)^{1/2}\left[\frac{(J+L-S)!(J-L+S)!(L+S-J)!(J+M)!(J-M)!}{(J+L+S+1)!(L+M_L)!(L-M_L)!(S+M_S)!(S-M_S)!}\right]^{1/2}$$

$$\times\sum_n \frac{(-1)^{n+S+M_S}(S+J+M_L-n)!(L-M_L+n)!}{n!(J-L+S-n)!(J+M-n)!(n+L-S-M)!} \tag{11-94}$$

Here the summation is to be performed over those integer values of the index $n$ for which none of the factorials in the denominator has a negative argument.

## Problems

**1** Derive the antisymmetrized eigenfunctions belonging to the $(1s)(2p)$ configuration of the helium atom, neglecting the spin-orbit interaction. The functions should be eigenfunctions of the operators $L^2$ and $S^2$. After these eigenfunctions are obtained derive the eigenfunctions of the operators $J^2$ and $J_z$ belonging to the $^3P$ state.

**2** Derive the eigenfunctions of the $2p$ state of the hydrogen atom, first without considering spin-orbit interaction and then with spin-orbit interaction. The latter eigenfunctions are assumed to be the eigenfunctions of the operators $J^2$ and $J_z$.

**3** Derive the eigenfunctions belonging to the $^3S$ state in Table 11-4 by making use of projection operators.

**4** Consider the configuration $(s)(s')(s'')$ and derive the eigenfunctions of the operators $S^2$ and $S_z$ that belong to this configuration.

**5** Derive the Slater orbitals corresponding to the ground-state configuration of the neon atom.

**6** Derive the Slater orbitals for the valence electrons in the sodium atom and in the potassium atom and derive the expectation values of the variable $r$ with respect to the valence orbitals for both atoms.

## Recommended Reading

The following list contains three mathematics books that contain a description and a derivation of the Clebsch-Gordon coefficients. They are Wigner's text on group theory, the general book by Margenau and Murphy, and the book by

Rose on the theory of angular momentum operators. Herzberg's text offers a very nice nonmathematical description of atomic theory. The two books by Slater treat the same topic from a much more mathematical and rigorous viewpoint. The book by Hartree gives a fairly complete description of the Hartree-Fock method; it also contains a number of tables of Hartree-Fock orbitals.

E. U. Condon and G. H. Shortley, *The Theory of Atomic Spectra*, Cambridge University Press, London, 1935.

D. R. Hartree, *The Calculation of Atomic Structure*, John Wiley & Sons, Inc., New York, 1957.

G. Herzberg, *Atomic Spectra and Atomic Structure*, Dover Publications, New York, 1944.

H. Margenau and G. M. Murphy, *The Mathematics of Physics and Chemistry*, Vol. Two, D. Van Nostrand Co., Inc., Princeton, NJ, 1964.

M. E. Rose, *Elementary Theory of Angular Momentum*, John Wiley & Sons, Inc., New York, 1957.

John C. Slater, *Quantum Theory of Atomic Structure*, Vol. I and Vol. II, McGraw-Hill Book Co., Inc., New York, 1960.

Eugen Wigner, *Gruppentheorie und ihre Anwendung auf die Quantenmechanik der Atomspektren*, Friedr. Vieweg & Sohn Akt.-Ges., Braunschweig, 1931.

APPENDIX A

# The Transformation of the Laplace Operator into Polar Coordinates

We wish to transform the Laplace operator

$$\Delta=\frac{\partial^2}{\partial x^2}+\frac{\partial^2}{\partial y^2}+\frac{\partial^2}{\partial z^2} \tag{A-1}$$

into polar coordinates $(r, \theta, \phi)$, which are defined as

$$\begin{aligned} x&=r\sin\theta\cos\phi \\ y&=r\sin\theta\sin\phi \\ z&=r\cos\theta \end{aligned} \tag{A-2}$$

We perform this transformation in two steps. The first is given by

$$\begin{aligned} x&=\rho\cos\phi \\ y&=\rho\sin\phi \end{aligned} \tag{A-3}$$

We have

$$\begin{aligned} \frac{\partial f}{\partial \rho}&=\frac{\partial x}{\partial \rho}\frac{\partial f}{\partial x}+\frac{\partial y}{\partial \rho}\frac{\partial f}{\partial y} \\ \frac{\partial f}{\partial \phi}&=\frac{\partial x}{\partial \phi}\frac{\partial f}{\partial x}+\frac{\partial y}{\partial \phi}\frac{\partial f}{\partial y} \end{aligned} \tag{A-4}$$

or

$$\begin{aligned} \frac{\partial f}{\partial \rho}&=\cos\phi\frac{\partial f}{\partial x}+\sin\phi\frac{\partial f}{\partial y} \\ \frac{\partial f}{\partial \phi}&=-\rho\sin\phi\frac{\partial f}{\partial x}+\rho\cos\phi\frac{\partial f}{\partial y} \end{aligned} \tag{A-5}$$

From these equations we derive

$$\frac{\partial f}{\partial x}=\cos\phi\frac{\partial f}{\partial\rho}-\frac{\sin\phi}{\rho}\frac{\partial f}{\partial\phi}$$

$$\frac{\partial f}{\partial y}=\sin\phi\frac{\partial f}{\partial\rho}+\frac{\cos\phi}{\rho}\frac{\partial f}{\partial\phi} \tag{A-6}$$

In the same way, we obtain for the second derivatives

$$\begin{aligned}\frac{\partial^2 f}{\partial x^2}&=\cos^2\phi\frac{\partial^2 f}{\partial\rho^2}-\frac{2\sin\phi\cos\phi}{\rho}\frac{\partial^2 f}{\partial\rho\,\partial\phi}\\&\quad+\frac{\sin^2\phi}{\rho^2}\frac{\partial^2 f}{\partial\phi^2}+\frac{\sin^2\phi}{\rho}\frac{\partial f}{\partial\rho}+\frac{2\sin\phi\cos\phi}{\rho^2}\frac{\partial f}{\partial\phi}\end{aligned} \tag{A-7}$$

$$\begin{aligned}\frac{\partial^2 f}{\partial y^2}&=\sin^2\phi\frac{\partial^2 f}{\partial\rho^2}+\frac{2\sin\phi\cos\phi}{\rho}\frac{\partial^2 f}{\partial\rho\,\partial\phi}\\&\quad+\frac{\cos^2\phi}{\rho^2}\frac{\partial^2 f}{\partial\phi^2}+\frac{\cos^2\phi}{\rho}\frac{\partial f}{\partial\rho}-\frac{2\sin\phi\cos\phi}{\rho^2}\frac{\partial f}{\partial\phi}\end{aligned}$$

Addition of these two equations gives

$$\frac{\partial^2 f}{\partial x^2}+\frac{\partial^2 f}{\partial y^2}=\frac{\partial^2 f}{\partial\rho^2}+\frac{1}{\rho}\frac{\partial f}{\partial\rho}+\frac{1}{\rho^2}\frac{\partial^2 f}{\partial\phi^2} \tag{A-8}$$

The Laplace operator $\Delta$ of Eq. (A-1) can now be written as

$$\Delta=\frac{\partial^2}{\partial\rho^2}+\frac{\partial^2}{\partial z^2}+\frac{1}{\rho}\frac{\partial}{\partial\rho}+\frac{1}{\rho^2}\frac{\partial^2}{\partial\phi^2} \tag{A-9}$$

The second step of our transformation is

$$z=r\cos\theta$$

$$\rho=r\sin\theta \tag{A-10}$$

By analogy to Eq. (A-8) we find

$$\frac{\partial^2}{\partial\rho^2}+\frac{\partial^2}{\partial z^2}=\frac{\partial^2}{\partial r^2}+\frac{1}{r}\frac{\partial}{\partial r}+\frac{\partial}{r^2}\frac{\partial^2}{\partial\theta^2} \tag{A-11}$$

Similarly, we derive from the second Eq. (A-6) that

$$\frac{\partial}{\partial\rho}=\sin\theta\frac{\partial}{\partial r}+\frac{\cos\theta}{r}\frac{\partial}{\partial\theta} \tag{A-12}$$

Consequently, we have

$$\frac{1}{\rho}\frac{\partial}{\partial\rho}=\frac{1}{r}\frac{\partial}{\partial r}+\frac{\cos\theta}{r^2\sin\theta}\frac{\partial}{\partial\theta} \tag{A-13}$$

In addition, we have

$$\frac{1}{\rho^2}\frac{\partial^2}{\partial\phi^2}=\frac{1}{r^2\sin^2\theta}\frac{\partial^2}{\partial\phi^2} \tag{A-14}$$

Substitution of Eqs. (A-1), (A-13), and (A-14) into Eq. (A-9) gives, finally,

$$\Delta=\frac{\partial^2}{\partial r^2}+\frac{2}{r}\frac{\partial}{\partial r}+\frac{1}{r^2}\frac{\partial}{\partial\theta^2}+\frac{\cos\theta}{r^2\sin\theta}\frac{\partial}{\partial\theta}+\frac{1}{r^2\sin^2\theta}\frac{\partial^2}{\partial\phi^2} \tag{A-15}$$

APPENDIX B

# The Calculation of $\int\int e^{-r_1}(1/r_{12})e^{-r_2}\,d\mathbf{r}_1\,d\mathbf{r}_2$

In this appendix we evaluate the integral

$$I=\int\int e^{-r_1}\left(\frac{1}{r_{12}}\right)e^{-r_2}\,d\mathbf{r}_1\,d\mathbf{r}_2 \tag{B-1}$$

We used the result of this integration in our discussion of the helium atom in Section 10-6.

The integral is calculated by introducing the two sets of polar coordinates $(r_1,\theta_1,\phi_1)$ and $(r_2,\theta_2,\phi_2)$ and by expanding $r_{12}$ as a power series in $r_1$ and $r_2$. Such an expansion is possible by making use of the theory of Legendre polynomials which we discussed in Section 5-10.

We show in Fig. B-1 that the two points are described by the vectors $\mathbf{r}_1$ and $\mathbf{r}_2$. We denote the angle between $\mathbf{r}_1$ and $\mathbf{r}_2$ by $\theta_{12}$. According to the cosine rule we can express $r_{12}$ as

$$r_{12}=\left(r_1^2+r_2^2-2r_1r_2\cos\theta_{12}\right)^{1/2} \tag{B-2}$$

Let $R$ and $r$ now be the greater and the smaller of $r_1$ and $r_2$. We then have

$$\frac{1}{r_{12}}=\frac{1}{R}\left(1-2h\cos\theta_{12}+h^2\right)^{-1/2} \qquad h=\frac{r}{R} \tag{B-3}$$

This function is just the generating function (5-185) for the Legendre polynomials, so that we can expand

$$\frac{1}{r_{12}}=\frac{1}{R}\sum_{n=0}^{\infty}\left(\frac{r}{R}\right)^n P_n(\cos\theta_{12}) \tag{B-4}$$

It is possible to express $P_n(\cos\theta_{12})$ in terms of the polar angles $\theta_1,\theta_2,\phi_1$, and $\phi_2$, but we do not need to do this in order to calculate the integral $I$. It is

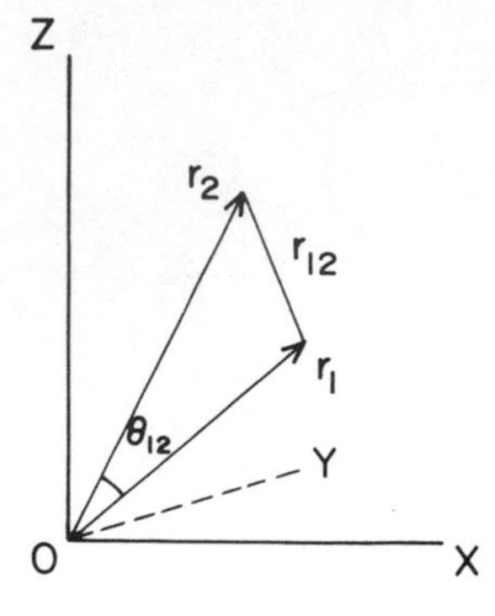


**Fig. B-1** Coordinate systems for the evaluation of the integral $I$ of Eq. (B-1).

permissible to take the vector $\mathbf{r}_1$ as the axis of reference for the definition of the polar coordinates of the second electron. The integral $I$ can then be written as

$$I=\int e^{-r_1}d\mathbf{r}_1\int\int\int\frac{e^{-r_2}}{R}\sum_{n=0}^{\infty}\left(\frac{r}{R}\right)^n P_n(\cos\theta_{12})r_2^2\sin\theta_{12}\,dr_2\,d\theta_{12}\,d\phi_2 \quad \text{(B-5)}$$

Because of the orthogonality of the Legendre polynomials we have

$$\int P_o(\cos\theta_{12})\sin\theta_{12}\,d\theta_{12}=2$$

$$\int P_n(\cos\theta_{12})\sin\theta_{12}\,d\theta_{12}=0 \qquad n\neq 0 \quad \text{(B-6)}$$

and the integral $I$ reduces to

$$I=4\pi\int\int\int e^{-r_1}r_1^2\sin\theta_1\,dr_1\,d\theta_1\,d\phi_1\int e^{-r_2}\left(\frac{r_2^2}{R}\right)dr_2 \quad \text{(B-7)}$$

or

$$I=16\pi^2\int_o^{\infty}r_1^2e^{-r_1}dr_1\int_o^{\infty}\left(\frac{r_2^2}{R}\right)e^{-r_2}dr_2 \quad \text{(B-8)}$$

Here $R$ is the larger of $r_1$ and $r_2$ so that

$$R=r_1 \qquad r_2\leq r_1$$

$$R=r_2 \qquad r_2\geq r_1 \quad \text{(B-9)}$$

In order to evaluate $I$ we write it as

$$I=16\pi^2\int_o^{\infty}r_1e^{-r_1}dr_1\int_o^{r_1}r_2^2e^{-r_2}dr_2+16\pi^2\int_o^{\infty}r_1^2e^{-r_1}dr_1\int_{r_1}^{\infty}r_2e^{-r_2}dr_2 \quad \text{(B-10)}$$

We define the auxiliary functions

$$A_n(a)=\int_a^\infty x^n e^{-x}\,dx \tag{B-11}$$

and we note that

$$A_n(a)=-\int_a^\infty x^n\,d(e^{-x})=a^n e^{-a}+nA_{n-1}(a)$$

$$A_o(a)=\int_a^\infty e^{-x}\,dx=e^{-a} \tag{B-12}$$

From these results we derive that

$$\int_o^{r_1} r_2^2 e^{-r_2}\,dr_2=\int_o^\infty r_2^2 e^{-r_2}\,dr_2-\int_{r_1}^\infty r_2^2 e^{-r_2}\,dr_2=2-A_2(r_1)$$

$$=2-r_1^2 e^{-r_1}-2r_1 e^{-r_1}-2e^{-r_1}$$

$$\int_{r_1}^\infty r_2 e^{-r_2}\,dr_2=A_1(r_1)=r_1 e^{-r_1}+e^{-r_1} \tag{B-13}$$

Substitution into Eq. (B-10) gives

$$I=16\pi^2\Big[2\int_o^\infty r_1 e^{-r_1}\,dr_1-\int_o^\infty (r_1^3+2r_1^2+2r_1)e^{-2r_1}\,dr_1$$

$$+\int_o^\infty (r_1^3+r_1^2)e^{-2r_1}\,dr_1\Big]=16\pi^2\cdot\left(\frac{5}{8}\right)=20\pi^2 \tag{B-14}$$

We obtained Eq. (10-91) by substituting the above result into Eq. (10-90).

APPENDIX C

# Values of Clebsch-Gordon Coefficients

## $C(L, S, J; M_L, M_S, M)$ for the Two Cases $S=\frac{1}{2}$, $L>0$ and $S=1$, $L>0$

**Case 1** $S=\frac{1}{2}$

Possible values for $J$ are $J=L+\frac{1}{2}$, $J=L-\frac{1}{2}$. Possible values for $M_S$ and $M_L$ are

$$M_S=\tfrac{1}{2},\ M_L=M-\tfrac{1}{2} \qquad \text{and} \qquad M_S=-\tfrac{1}{2},\ M_L=M+\tfrac{1}{2}$$

$$C\left(L,\tfrac{1}{2},L+\tfrac{1}{2};M-\tfrac{1}{2},\tfrac{1}{2},M\right)=\left[\frac{L+M+\frac{1}{2}}{2L+1}\right]^{1/2}$$

$$C\left(L,\tfrac{1}{2},L+\tfrac{1}{2};M+\tfrac{1}{2},-\tfrac{1}{2},M\right)=\left[\frac{L-M+\frac{1}{2}}{2L+1}\right]^{1/2}$$

$$C\left(L,\tfrac{1}{2},L-\tfrac{1}{2};M-\tfrac{1}{2},\tfrac{1}{2},M\right)=\left[\frac{L-M+\frac{1}{2}}{2L+1}\right]^{1/2}$$

$$C\left(L,\tfrac{1}{2},L-\tfrac{1}{2};M+\tfrac{1}{2},-\tfrac{1}{2},M\right)=\left[\frac{L+M+\frac{1}{2}}{2L+1}\right]^{1/2}$$

**Case 2** $S=1$

Possible values for $J$ are $J=L+1$, $J=L$, $J=L-1$. Possible values for $M_S$ are $M_S=1$, $M_S=0$, $M_S=-1$.

$$C(L,1,L+1;M-1,1,M)=\left[\frac{(L+M)(L+M+1)}{(2L+1)(2L+2)}\right]^{1/2}$$

$$C(L,1,L+1;M,0,M)=\left[\frac{(L-M+1)(L+M+1)}{(2L+1)(L+1)}\right]^{1/2}$$

$$C(L,1,L+1;M+1,-1,M)=\left[\frac{(L-M)(L-M+1)}{(2L+1)(2L+2)}\right]^{1/2}$$

$$C(L,1,L;M-1,1,M)=-\left[\frac{(L+M)(L-M+1)}{2L(L+1)}\right]^{1/2}$$

$$C(L,1,L;M,0,M)=\left[\frac{M}{L(L+1)}\right]^{1/2}$$

$$C(L,1,L;M+1,-1,M)=\left[\frac{(L-M)(L+M+1)}{2L(L+1)}\right]^{1/2}$$

$$C(L,1,L-1;M-1,1,M)=\left[\frac{(L-M)(L-M+1)}{2L(2L+1)}\right]^{1/2}$$

$$C(L,1,L-1;M,0,M)=-\left[\frac{(L-M)(L+M)}{L(2L+1)}\right]^{1/2}$$

$$C(L,1,L-1;M+1,-1,M)=\left[\frac{(L+M+1)(L+M)}{2L(2L+1)}\right]^{1/2}$$

APPENDIX D

# Fundamental Constants in Terms of esu or cgs Units

| Symbol | Description | Magnitude |
|---|---|---|
| $c$ | Velocity of light | $2.997928 \times 10^{10}$ cm/sec |
| $N$ | Avogadro's number | $6.0248 \times 10^{23}$ g/mole |
| $k$ | Boltzmann constant | $1.38046 \times 10^{-16}$ erg/degree |
| $e$ | Electronic charge | $4.80281 \times 10^{-10}$ esu |
| $h$ | Planck's constant | $6.6251 \times 10^{-27}$ erg sec |
| $\hbar$ | Dirac constant ($h/2\pi$) | $1.05442 \times 10^{-27}$ erg sec |
| $m$ | Electron mass | $9.1085 \times 10^{-28}$ g |
| $M/m$ | Ratio of proton/electron mass | 1836.11 |
| $R_H$ | Hydrogen atom Rydberg constant | 109 677.58 $cm^{-1}$ |
| $R_\infty$ | Rydberg constant for inf. mass | 109 737.3 $cm^{-1}$ |
| $a_o$ | Bohr radius ($\hbar^2/me^2$) | $0.529166 \times 10^{-8}$ cm |
| $\mu_o$ | Bohr magneton ($e\hbar/2mc$) | $0.92729 \times 10^{-20}$ erg $gauss^{-1}$ |

APPENDIX E

# Conversion Factors

$$1\ \text{eV} = 8066.0\ \text{cm}^{-1} = 1.60204 \times 10^{-12}\ \text{erg}$$

$$1\ \text{cm}^{-1} = 1.23977 \times 10^{-4}\ \text{eV} = 1.98616 \times 10^{-16}\ \text{erg}$$

$$1\ \text{erg} = 6.24204 \times 10^{11}\ \text{eV} = 5.03404 \times 10^{15}\ \text{cm}^{-1}$$

# Index